Linear Circuits, Systems and Signal Processing

ELECTRICAL ENGINEERING AND ELECTRONICS

A Series of Reference Books and Textbooks

60. Battery Technology Handbook, *edited by H. A. Kiehne*
61. Network Modeling, Simulation, and Analysis, *edited by Ricardo F. Garzia and Mario R. Garzia*
62. Linear Circuits, Systems and Signal Processing: Advanced Theory and Applications, *edited by Nobuo Nagai*
63. High-Voltage Engineering: Theory and Practice, *edited by M. Khalifa*
64. Large-Scale Systems Control and Decision Making, *edited by Hiroyuki Tamura and Tsuneo Yoshikawa*

Additional Volumes in Preparation

Distributed Computer Control for Industrial Automation, *edited by D. Popovic and Vijay P. Bhatkar*

Industrial Power Distribution and Illuminating Systems, *Kao Chen*

Computer-Aided Analysis of Active Circuits, *Adrian Ioinovici*

Mathematical Modeling and Simulation for Electronic Design, *edited by Ricardo F. Garzia*

Electrical Engineering-Electronics Software

1. Transformer and Inductor Design Software for the IBM PC, *Colonel Wm. T. McLyman*
2. Transformer and Inductor Design Software for the Macintosh, *Colonel Wm. T. McLyman*
3. Digital Filter Design Software for the IBM PC, *Fred J. Taylor and Thanos Stouraitis*

Linear Circuits, Systems and Signal Processing

Advanced Theory and Applications

edited by

Nobuo Nagai

Research Institute of Applied Electricity
Hokkaido University
Sapporo, Japan

CRC Press
Taylor & Francis Group
Boca Raton London New York

CRC Press is an imprint of the
Taylor & Francis Group, an **informa** business

First published 1990 by Marcel Dekker, INC.

Published 2019 by CRC Press
Taylor & Francis Group
6000 Broken Sound Parkway NW, Suite 300
Boca Raton, FL 33487-2742

© 1990 by Taylor & Francis Group, LLC
CRC Press is an imprint of Taylor & Francis Group, an Informa business

First issued in paperback 2019

No claim to original U.S. Government works

ISBN 13: 978-0-367-45090-8 (pbk)
ISBN 13: 978-0-8247-8185-9 (hbk)

Visit the Taylor & Francis Web site at
http://www.taylorandfrancis.com

and the CRC Press Web site at
http://www.crcpress.com

Library of Congress Cataloging in Publication Data

Linear circuits, systems and signal processing : advanced theory and
    applications / [edited by] Nobuo Nagai.
        p.   cm.   (Electrical engineering and electronics ; 62)
    Includes bibliographical references.
    ISBN 0-8247-8185-6 (alk. paper)
    1. Electrical circuits, Linear.   2. Electric network synthesis.
3. Signal processing.   I. Nagai, Nobuo.   II. Series.
TK454.L55   1989
621.319'2–dc20   89-23740
                                                              CIP

Preface

This book contains a wide range of subjects concerning linear circuits and systems, including their applications to digital filters. The aim of this book is to document the significant progress in studies of these areas in Japan. Each contribution is written by an expert who has played a major role in the particular topic covered. Since this is not intended to be a textbook, basic materials are kept to a minimum.

Part I comprises two chapters about system-theoretical considerations of network functions. Input-output characterizations of systems such as passivity are viewed from the modern state-space approach. The increased mutual influence between network theory and linear dynamical system theory is revealed. Chapter 1 broadly reviews a variety of state-space characterizations of linear passive networks. Chapter 2 focuses on an elegant system-theoretical synthesis method of RC active networks with a prescribed transfer function or matrix. These circuit structures and design procedures are closely connected to system-theoretical representations of RC networks.

In Part II, rational approximations in circuit and system theory are considered. These topics are treated and reviewed as mathematical tools for characterizing linear systems. The reader will learn how these methods are useful in applications. Chapter 3 considers the classical interpolation problems in terms of many renewed aspects that have been receiving more and more attention in the last several years. These techniques, known as Nevanlinna-Pick and Carathéodory theorems, are relevant to interpolation for some classes of bounded or passive functions. In Chapter 4, multipoint Padé approximation is discussed. Padé-type approximations are one of the powerful methods for rational approximations important in system modeling, circuit synthesis, and filter design.

Chapters 5 and 7 consider some systems' order-reduction methods relating to a stability-reserving property. In connection with this problem, recently developed algebraic stability criteria are also discussed. Chapters 5 and 7 consider scalar and matrix cases, respectively. Chapter 6 is concerned with the generalization of the rational bounded real function, which plays an important role in discrete-time system analysis and design. Canonical factorization problems for the nonrational matrix-valued bounded type of function are also considered.

Part III deals with the digital lattice filters used in digital signal processing (DSP). The study of DSP is becoming important in many fields. The autoregressive (AR) stochastic model has played an important role in DSP since the linear prediction theory was first developed. Furthermore, approaches based on a more complex stochastic model, e.g., an ARMA model, have been proposed recently. The algorithms derived with these approaches are employed in speech analysis and recognition, system identification and communication. However, the characteristics of this stochastic model and properties of the proposed algorithms have not yet been explored. In order to ascertain these characteristics, we need to fully understand linear prediction theory, circuit theory, and system theory. For example, if we wish to verify the stability of an estimated stochastic model, we apply the stability check given in either circuit theory or system theory to a linear prediction problem.

The lattice structure used in DSP is based on linear prediction theory. This structure is closely related to the stability test used in system theory. On the other hand, it has been shown recently that the design of an AR lattice structure is equivalent to the Richards theorem presented in circuit theory. Thus, this structure has in common the properties discussed in circuit theory, system theory, and prediction theory.

An ARMA model can represent energy concentration and energy dispersion in the spectrum domain. Thus, this model approximates the properties of the given stochastic data as accurately as possible. However, if we identify the reference model yielding the given data by using the ARMA model, we have to solve a nonlinear problem. The algorithm needed to solve it therefore becomes fairly complex. The notations used in its prediction and estimation are complicated.

Chapter 8 presents the circular lattice method for multichannel DSP with applications to RLS (recursive least-squares) filtering and ARMA modeling. In Chapter 9, the nonlinear problem in ARMA identification is solved by using an adaptive method. A new ARMA lattice structure is presented in Chapter 10. This chapter indicates that the ARMA lattice structure based on prediction theory is quite useful for DSP.

Chapter 11 shows the reader how to use the wave digital filter presented in circuit theory for DSP. Chapter 12 considers complex reflection coefficients, a complex scattering matrix, etc., for a complex transmission-line circuit, in order to synthesize positive or bounded digital filters with complex transmission-line circuit elements. In Chapter 13, one application of the lattice structure is shown. This algorithm is most suitable for speech analysis and reconstruction.

We aimed for consistency of style and terminology in all chapters, but several key differences do exist, such as

Identity matrix
    1_n (Chapters 11 and 12)
    I (other chapters)
Analytic region for complex variable z or z^{-1}
    analytic in $|z| \leq 1$ (Chapter 3)
    analytic in $|z| \geq 1$ (other chapters)

The editor is very grateful for this opportunity to present Japan's contribution to this field, as otherwise our work might remain unnoticed owing to the language barrier.

NOBUO NAGAI

Contents

Contributors

NOZOMU HAMADA Department of Electrical Engineering, Faculty of Science and Technology, Keio University, Yokohama, Japan

KAZUMI HORIGUCHI Department of Electronics Engineering, Faculty of Science and Technology, Kinki University, Kowakae, Higashiosaka, Japan

YUJIRO INOUYE Department of Control Engineering, Faculty of Engineering Science, Osaka University, Toyonaka, Osaka, Japan

HIDENORI KIMURA Department of Mechanical Engineering for Computer-Controlled Machinery, Faculty of Engineering, Osaka University, Suita, Osaka, Japan

NAOKI MATSUMOTO Department of Electronics and Communication, Faculty of Engineering, Meiji University, Kawasaki, Japan

NOBUHIRO MIKI Research Institute of Applied Electricity, Hokkaido University, Sapporo, Japan

YOSHIKAZU MIYANAGA Department of Electronics Engineering, Faculty of Engineering, Hokkaido University, Sapporo, Japan

NOBUO NAGAI Research Institute of Applied Electricity, Hokkaido University, Sapporo, Japan

HIROSHI NAGAOKA* Department of Information Technology, Faculty of Engineering, Tokyo Engineering University, Hachioji, Tokyo, Japan

HIDEAKI SAKAI Department of Applied Mathematics and Physics, Faculty of Engineering, Kyoto University, Kyoto, Japan

---

*Present affiliation: Department of Information Engineering, Faculty of Engineering, Hokkaido University, Sapporo, Japan.

MASAKIYO SUZUKI Research Institute of Applied Electricity, Hokkaido University, Sapporo, Japan

SHIN-ICHI TAKAHASHI Department of Electrical Engineering, Faculty of Science and Technology, Keio University, Yokohama, Japan

TOHRU TAKAHASHI Department of Electronics, Faculty of Engineering, Fukuoka Institute of Technology, Fukuoka, Japan

Part I

System-Theoretical Representation of Network Functions

1

System-Theoretical Representations of Linear Passive Networks

NAOKI MATSUMOTO

SHIN-ICHI TAKAHASHI

Meiji University, Kawasaki, Japan   Keio University, Yokohama, Japan

1-1 INTRODUCTION

In modern network theory, several function matrices and associated lemmas
are important and worthy not only for network theory but for other engineering theories as well. These are represented by positive real matrices and
positive real lemmas and by bounded real matrices and bounded real lemmas.
The positive and bounded real matrices both arise from a study of immittance and scattering matrices of linear time-invariant lumped passive networks. Various properties of these matrices have been derived from the
theory of complex variable functions. By using these properties, network
theorists have developed a broad literature on approximation theory and
synthesis theory of passive networks [1,2].

By the 1950s, classical network theory based on the frequency-
domain method was almost complete. In the 1960s, however, a new wave
came to network theory, with the advent of the state space method [3,4].
This method takes note of the internal states of the network and describes
the network with state vector differential equations [5-7] (called simply
"state equations"). The state equations of the network provide both a time-
domain representation of the network character and a system-theoretical
representation of the network. Positive and bounded real lemmas are the
system-theoretical representations of positive and bounded real matrices,
respectively. These lemmas show the necessary and sufficient conditions
that the state equations of positive and bounded real matrices must satisfy.

To date, many system theoretical synthesis network methods have
been derived from these lemmas [8-13,17], but positive real lemmas are
more famous as a guarantee of the asymptotic stability of adaptive control
systems or certain classes of nonlinear feedback control systems [19-21].
As for bounded real lemmas, recent studies show that the discrete version
of lossless bounded real lemmas is useful for the design of low-sensitivity
digital filters [32-34,41]. These facts tell us that the system-theoretical
representations of passive network functions are important not only for network theory but also for adjacent engineering applications. The purpose of
this chapter is to gather and introduce the system-theoretical representations of various network functions, including positive real matrices and
bounded real matrices. First we introduce the case of continuous time and
later, the case of discrete time. The systems considered are all one-
dimensional real coefficient systems. Each system-theoretical representation is given in the form of a theorem. Due to size restrictions, the
proofs are omitted; instead, interesting properties of the theorem are
stated. In the final section, future problems, in particular those related to
system-theoretical representations of multidimensional and complex coefficient systems, are discussed.

1-2 PRELIMINARIES

In this chapter $Z(s)$ denotes the $p \times p$ matrix with elements of real rational functions of complex variable s, and

$$\frac{dx}{dt} = Ax + Bu$$

$$y = Cx + Du \tag{1-1}$$

denotes a minimal realization of $Z(s)$ with $Z(\infty) < \infty$, where $Z \in R^{n \times n}$, $A \in R^{n \times n}$, $B \in R^{n \times p}$, $C \in R^{p \times n}$, and $D \in R^{p \times p}$; $u \in R^p$ is the input, $y \in R^p$ is the output, and $x \in R^n$ is the state variable vector. The relation between Eq. (1-1) and $Z(s)$ is given by

$$Z(s) = D + C[sI - A]^{-1} B \tag{1-2}$$

where I is an identity matrix of order n. Equation (1-1) is often abbreviated as $\{A \ B \ C \ D\}$, and we call it a minimal realization of $Z(s)$. We often use $Z(s)$ as the impedance matrix of a p-port network. $Z(s)^T$ means the transpose of $Z(s)$ and $Z^*(s)$ means the complex conjugate of $Z(s)$. For a square matrix P, $P > 0$ ($P \geq 0$) means that P is positive (semipositive or nonnegative) definite symmetric and $P < 0$ ($P \leq 0$) means that P is negative (seminegative or nonpositive) definite symmetric. Furthermore, $Z^*(s)^T + Z(s) > 0$ $[Z^*(s)^T + Z(s) \geq 0]$ means that $Z^*(s)^T + Z(s)$ is positive (semipositive) definite Hermitian [15].

1-3 SYMMETRIC TRANSFER FUNCTION MATRICES

Let $Z(s)$ be an impedance matrix of a p-port network N which consists of resistors, capacitors, inductors, and ideal transformers. Then the network N is reciprocal and $Z(s)$ becomes a symmetric matrix. A system-theoretical representation of a symmetric transfer function matrix follows.

Theorem 1-1 [8]. Let $Z(s)$ be a $p \times p$ matrix of real rational functions of complex variable s with $Z(\infty) < \infty$, and $\{A \ B \ C \ D\}$ be a minimal realization of $Z(s)$. Then $Z(s)$ is symmetric if and only if there exists a nonsingular symmetric matrix P such that

$$A^T P = PA \tag{1-3a}$$

$$PB = \pm C^T \tag{1-3b}$$

$$D^T = D \tag{1-3c}$$

∎

This theorem first appeared in Youla and Tissi [8] and applied to the synthesis of the passive reciprocal networks. When Eqs. (1-3) with the plus sign hold, we obtain

$$PB = C^T$$

$$PAB = A^T PB = A^T C^T \qquad (1-4)$$

$$PA^2 B = (A^T)^2 C^T$$

$$\vdots$$

Hence P is given by

$$P = (V_O V_C^T)(V_C V_C^T)^{-1} \qquad (1-5)$$

where

$$V_C = [B \ AB \ \cdots \ A^{n-1}B] \qquad (1-6)$$

$$V_O = [C^T \ A^T C^T \ \cdots \ (A^T)^{n-1} C^T] \qquad (1-7)$$

and n is the McMillan degree of $Z(s)$. V_C is a well-known controllability matrix and V_O is an observability matrix. On the other hand, it follows from (1-3) with the plus sign that

$$P^{-1} A^T P = A$$

$$P^{-1} C^T = B$$

$$B^T P = C \qquad (1-8)$$

$$D^T = D$$

Equations (1-8) imply that a minimal realization $\{A \ B \ C \ D\}$ is similar to a minimal realization $\{A^T \ C^T \ B^T \ D^T\}$, which means that $Z(s)^T = Z(s)$. For the complete proof of the theorem, see [8].

1-4 LOSSLESS MATRICES

Definition 1-1. When the transfer function matrix $Z(s)$ satisfies the relation $Z(-s)^T + Z(s) = 0$, we call $Z(s)$ a lossless matrix.   ∎

Theorem 1-2. Let $\{A \ B \ C \ D\}$ be a minimal realization of $Z(s)$. Then $Z(s)$ is lossless if and only if there exists a nonsingular symmetric matrix P such that

$$A^T P + PA = 0$$

$$PB = C^T \tag{1-9}$$

$$D^T + D = 0$$

∎

The proof of Theorem 1-2 is quite similar to that of Theorem 1-1. If Eqs. (1-9) hold, it follows that

$$P^{-1}(-A^T)P = A$$

$$P^{-1}C^T = B$$

$$B^T P = C \tag{1-10}$$

$$-D^T = D$$

Equations (1-10) imply that a minimal realization of $-Z(-s)^T$ and a minimal realization of $Z(s)$ are similar. Hence $Z(-s)^T + Z(s) = 0$.

1-5 POSITIVE REAL MATRICES AND LOSSLESS POSITIVE REAL MATRICES

1-5-1 Definitions

The impedance matrix or admittance matrix of the network, which consists of resistors, inductors, capacitors, ideal transformers, and ideal gyrators, is always a positive real matrix. Conversely, any positive real matrix is realizable as an impedance matrix or admittance matrix of a network with passive elements as stated above. Moreover, the impedance matrix or admittance matrix of the network, which consists of lossless passive elements—inductors, capacitors, ideal transformers, and ideal gyrators— becomes a lossless positive real matrix. Definitions of positive real and lossless positive real matrices follow.

Definition 1-2 [16, 17]. A $p \times p$ matrix of real rational functions $Z(s)$ is positive real if and only if

(i) All elements of $Z(s)$ are analytic in $Re[s] > 0$.

(ii) $Z^*(s)^T + Z(s) \geq 0$ in $Re[s] \geq 0$.   ∎

Definition 1-3 [16, 17]. A $p \times p$ matrix of real rational functions $Z(s)$ is lossless positive real if and only if

(i) All elements of $Z(s)$ are analytic in $Re[s] > 0$.

(ii) $Z^*(s)^T + Z(s) \geq 0$.

(iii) $Z^*(j\omega)^T + Z(j\omega) = 0$ for all $j\omega$ except the poles of $Z(s)$.   ∎

It should be noted that a lossless positive real matrix $Z(s)$ can be expanded in the following form [17]:

$$Z(s) = sL + J + \sum_{i=0}^{q} \frac{A_i s + B_i}{s^2 + \omega_i^2} + \frac{C}{s} \tag{1-11}$$

where L, A_i, and C are real and nonnegative definite symmetric, and J and B_i are real skew-symmetric matrices.

1-5-2 Theorems

System-theoretical representations of positive real matrices are as follows.

Theorem 1-3 [16, 17] (Positive Real Lemma). Let $\{A \ B \ C \ D\}$ be a minimal realization of $Z(s)$. Then $Z(s)$ is positive real if and only if there exist real matrices L and W_0 and a positive definite symmetric matrix P such that

$$A^T P + PA = -L^T L$$

$$PB = C^T - L^T W_0 \tag{1-12}$$

$$D^T + D = W_0^T W_0$$

                                                                        ∎

In Theorem 1-3, the first equation of (1-12) with $P > 0$ guarantees that all eigenvalues of A have nonpositive real parts. If we define $W(s)$ as

$$W(s) = W_0 + L[sI - A]^{-1}B \tag{1-13}$$

it follows from (1-12) with $P = P^T$ that

$$Z*(s)^T + Z(s) = W*(s)^T W(s) + 2 \, Re[s] \, B^T [s*I - A^T]^{-1} P[sI - A]^{-1} B \qquad (1\text{-}14)$$

and

$$Z(-s)^T + Z(s) = W(-s)^T W(s) \qquad (1\text{-}15)$$

Therefore, if $P > 0$, $Z*(s)^T + Z(s) \geq 0$ in $Re[s] \geq 0$.

System-theoretical representations of lossless positive real matrices are given as follows.

Theorem 1-4 [16, 17] (Lossless Positive Real Lemma). Let $\{A \ B \ C \ D\}$ be a minimal realization of $Z(s)$. $Z(s)$ is lossless positive real if and only if there exists a positive definite symmetric matrix P such that

$$A^T P + PA = 0$$

$$PB = C^T \qquad (1\text{-}16)$$

$$D^T + D = 0$$

∎

We should note that Theorem 1-4 resembles Theorem 1-2. The only difference between the two theorems lies in whether or not the nonsingular symmetric matrix P is positive definite.

1-6 STRICTLY POSITIVE REAL MATRICES

Definition 1-4 [14, 18]. A $p \times p$ matrix of real rational functions $Z(s)$ is called strictly positive real if and only if the following conditions hold:

(i) $Z(s - \sigma)$ is positive real for some $\sigma > 0$, or equivalently,
(ii) All the elements of $Z(s)$ are analytic in $Re[s] \geq 0$.
(iii) $Z*(s)^T + Z(s) > 0$ in $Re[s] \geq 0$. ∎

Now let $\{A \ B \ C \ D\}$ be a minimal realization of $Z(s)$. Then a minimal realization of $Z(s - \sigma)$ is given by $\{\hat{A} \ B \ C \ D\}$, where

$$\hat{A} = A + \sigma I \qquad (1\text{-}17)$$

If $Z(s - \sigma)$ is positive real, it follows from Theorem 1-3 that there exist matrices L and W_0 and a positive definite symmetric matrix P such that

$$\hat{A}^T P + P\hat{A} = -L^T L \qquad (1\text{-}18a)$$

$$PB = C^T - L^T W_0 \tag{1-18b}$$

$$D^T + D = W_0^T W_0 \tag{1-18c}$$

Equation (1-18a) can be rewritten as

$$A^T P + PA = -L^T L - 2\sigma P \tag{1-19}$$

Therefore, a system-theoretical representation of strictly positive real matrices is as follows.

Theorem 1-5 [14, 18] (Strictly Positive Real Lemma). Let $\{A \ B \ C \ D\}$ be a minimal realization of $Z(s)$. Then $Z(s)$ is strictly positive real if and only if there exist positive definite symmetric matrices P and Q and real matrices L and W_0 such that

$$A^T P + PA = -L^T L - Q \tag{1-20a}$$

$$PB = C^T - L^T W_0 \tag{1-20b}$$

$$D^T + D = W_0^T W_0 \tag{1-20c}$$

If we define $W(s)$ as $W(s) = W_0 + L[sI - A]^{-1}B$, it follows from (1-20) and $P^T = P$ that

$$Z*(s)^T + Z(s) = W*(s)^T W(s) + B^T[s*I - A^T]^{-1}\{2 \ \text{Re}[s]P + Q\}[sI - A]^{-1}B \tag{1-21a}$$

and

$$Z(-s)^T + Z(s) = W(-s)^T W(s) + B^T[-sI - A^T]^{-1}Q[sI - A]^{-1}B \tag{1-21b}$$

Therefore, if $P > 0$ and $Q > 0$ $Z*(s)^T + Z(s) > 0$ for $\text{Re}[s] \geq 0$. ∎

The concept of strictly positive realness was used by Popov in the proof of the hyperstability theorem [19-21]. Popov proved that if the transfer function matrix of the linear block of a certain nonlinear feedback control system is strictly positive real, the system is asymptotically stable. Theorem 1-5 (the strictly positive real lemma) for scalar functions is called the Kalman-Yacubovich lemma. This lemma appeared in the proof of the absolute stability of the nonlinear control problem called the Lur'e problem [30, 31].

1-7 IMPEDANCE MATRICES OF
TWO-ELEMENT NETWORKS

Let $Z_{RC}(s)$ denote the impedance matrix of RC multiport networks consisting of resistors, capacitors, and ideal transformers. Let $Z_{RL}(s)$ and $Z_{LC}(s)$ denote the impedance matrices of RL multiport networks and LC multiport networks, respectively, where RL multiport networks consist of resistors, inductors, and ideal transformers and LC multiport networks consist of inductors, capacitors, and ideal transformers. We call $Z_{RC}(s)$, $Z_{RL}(s)$, and $Z_{LC}(s)$ an RC impedance matrix, an RL impedance matrix, and an LC impedance matrix, respectively. These matrices are all positive real and symmetric because they are the impedance matrices of passive reciprocal networks. $Z_{LC}(s)$, especially, is a symmetric lossless positive real matrix. Therefore, these matrices satisfy both Theorems 1-1 and 1-5 (the positive real lemma) or Theorem 1-4 (the lossless positive real lemma). But as will be shown later, for the case of $Z_{RC}(s)$ and $Z_{RL}(s)$, Theorems 1-1 and 1-5 can be unified into a single theorem.

Now we make some preparations to derive minimal realizations of $Z_{RC}(s)$, $Z_{RL}(s)$, and $Z_{LC}(s)$. The partial fraction expansions of these three matrices become as follows:

$$Z_{RC}(s) = \frac{A_0}{s} + A_\infty + \sum_{i=1}^{q} \frac{A_i}{s + \sigma_i} \qquad (1\text{-}22)$$

$$Z_{RL}(s) = sA_\infty + A_0 + \sum_{i=0}^{q} \frac{sA_i}{s + \sigma_i} \qquad (1\text{-}23)$$

$$Z_{LC}(s) = sA_\infty + A_0 + \sum_{i=0}^{q} \frac{sA_i}{s^2 + \omega_i^2} \qquad (1\text{-}24)$$

where σ_i and ω_i are positive real numbers and A_∞, A_0, and A_i are all $p \times p$ nonnegative definite symmetric matrices. Let us note here Eq. (1-22). Let

$$\text{rank } A_0 = n_0 \le p \qquad (1\text{-}25a)$$

$$\text{rank } A_i = n_i \le p \qquad (1\text{-}25b)$$

and let $\tilde{B}_i$ be a real matrix such that

$$\tilde{B}_i^T \tilde{B}_i = A_i, \quad \tilde{B}_i \in R^{n \times p} \qquad (1\text{-}26)$$

Furthermore, let O_0 be a $p \times p$ null matrix and I_i be an identity matrix of order n_i. Then a minimal realization of $Z_{RC}(s)$ is given by the following $\{\tilde{A} \ \tilde{B} \ \tilde{C} \ D\}$:

$$\tilde{A} = -[O_0 + \sigma_1 I_1 + \sigma_2 I_2 + \cdots + \sigma_q I_q] \le 0 \tag{1-27a}$$

$$\tilde{B} = [B_1^T \ B_2^T \ \cdots \ B_q^T]^T \tag{1-27b}$$

$$\tilde{C} = \tilde{B}^T \tag{1-27c}$$

$$D = A_\infty \tag{1-27d}$$

Equations (1-27) satisfy the following relations:

$$\tilde{A}^T = \tilde{A} \le 0$$

$$\tilde{B} = \tilde{C}^T \tag{1-28}$$

$$D^T = D \ge 0$$

From the results above we obtain a system-theoretical representation of $Z_{RC}(s)$ as follows.

Theorem 1-6 [24, 26]. Let $\{A \ B \ C \ D\}$ be a minimal realization of $Z(s)$. Then $Z(s)$ is realizable as an RC impedance matrix if and only if there exists a positive definite symmetric matrix P and the following conditions hold:

$$A^T P = PA \le 0 \tag{1-29a}$$

$$PB = C^T \tag{1-29b}$$

$$D^T = D \ge 0 \tag{1-29c}$$

∎

It should be noted that Eqs. (1-29) in Theorem 1-6 resemble Eq. (1-3) in Theorem 1-1. Equation (1-29a) with $P > 0$ guarantees that A is diagonalizable and all the eigenvalues of A are real and nonpositive. This fact and (1-29b) and (1-29c) guarantee that the residue matrices of the partial fraction expansion of $Z(s) = D + C[sI - A]^{-1}B$ are all nonnegative definite symmetric. Thus if Eqs. (1-29) with $P > 0$ hold, the partial fraction expansion of $Z(s)$ becomes as in Eq. (1-22). Moreover, it follows from (1-29) with $P > 0$ that

$$Z^*(s)^T + Z(s) = D^T + D + B^T[s*I - A^T]^{-1}\{2 \text{ Re}[s]P - (A^T P + PA)\}[sI - A]^{-1}B$$

$$(1\text{-}30)$$

Since $A^T P + PA \le 0$, $D \ge 0$, and $P > 0$, (1-30) implies that $Z^*(s)^T + Z(s) \ge 0$ for $\text{Re}[s] \ge 0$. Therefore, $Z(s)$ is positive real.

Next, the system-theoretical representation of RL impedance matrices becomes as follows.

Theorem 1-7 [25]. Let $\{A \ B \ C \ D\}$ be a minimal realization of $Z(s)$. Then $\overline{Z}(s)$ is realizable as an RL impedance matrix if and only if there exists a positive definite symmetric matrix P and the following conditions hold:

$$A^T P = PA < 0 \tag{1-31a}$$

$$PB = -C^T \tag{1-31b}$$

$$D - CA^{-1}B \ge 0 \tag{1-31c}$$

■

We should note that Eqs. (1-31) resemble Eqs. (1-3) in Theorem 1-1. Equation (1-31a) with $P > 0$ guarantees that A is diagonalizable and all the eigenvalues of A are negative real numbers. This fact and Eqs. (1-31b) and (1-31c) with $P > 0$ guarantee that the partial fraction expansion of $Z(s)$ becomes as (1-23) and A_0 and A_i become nonnegative definite symmetric. It follows from Eqs. (1-31) with $P > 0$ that for $\text{Re}[s] \ge 0$,

$$Z^*(s)^T + Z(s) = 2(D - CA^{-1}B) + B^T[s*I - A^T]^{-1}\{-|s|^2 A^{T-1}(A^T P + PA)A^{-1}$$

$$+ 2 \text{ Re}[s]P\}[sI - A]^{-1}B \ge 0 \tag{1-32}$$

Therefore, $Z(s)$ is positive real.

It is interesting to note that Theorems 1-6 and 1-7 both resemble Theorem 1-1 in form rather than Theorem 1-3 (the positive real lemma) despite $Z_{RC}(s)$ and $Z_{RL}(s)$ being positive real.

The remaining impedance matrix of a two-element network is an LC impedance matrix. Unfortunately, no appropriate theorem corresponding to Theorem 1-6 or 1-7 has been found for $Z_{LC}(s)$. In other words, Theorem 1-4 (the lossless positive real lemma) and Theorem 1-1 cannot easily be combined in one theorem. It should be noted that Theorem 1-4 does not guarantee that $Z(s)$ is symmetric.

Finally, we note that the mathematical properties of RC impedance matrices and RL admittance matrices are equal. Therefore, Theorem 1-6 can be viewed as the system-theoretical representation of RL admittance matrices. Similarly, as the mathematical properties of RL impedance matrices and RC admittance matrices are equal, Theorem 1-7 can be viewed as the system-theoretical representation of RC admittance matrices.

1-8 VOLTAGE TRANSFER FUNCTIONS OF TRANSFORMERLESS RC THREE-TERMINAL NETWORKS

Let $f(s)$ and $g(s)$ be real polynomials defined by

$$f(s) = s^n + a_1 s^{n-1} + a_2 s^{n-2} + \cdots + a_n \tag{1-33}$$

$$g(s) = K(s^m + b_1 s^{m-1} + \cdots + b_m) \tag{1-34}$$

where $K > 0$ and $n \geq m$. In 1952, Fialkow and Gerst gave the necessary and sufficient conditions for a real rational transfer function

$$T(s) = \frac{g(s)}{f(s)} \tag{1-35}$$

to be realizable as a voltage transfer function of a transformerless RC three-terminal network.

The Conditions of Fialkow and Gerst [26]

(i)   The zeros of $f(s)$ are distinct negative numbers.
(ii) The leading coefficients of $g(s)$ and $f(s) - g(s)$ are positive and the zeros of $g(s)$ and $f(s) - g(s)$ are arbitrary, except that they may not be positive.

Conditions (i) and (ii) are equivalent to the following two conditions (i') and (ii') [27].

(i')   The zeros of $f(s)$ are distinct negative numbers.
(ii') There exist nonnegative integers k and ℓ and positive number ϵ such that

$$\mathrm{Re}[F(j\omega)] \geq 0 \tag{1-36}$$

$$\mathrm{Re}[G(j\omega)] \geq 0 \tag{1-37}$$

where

$$F(s) = \frac{g(-s^2) - (-s^2)^k \epsilon}{f(-s^2)} \tag{1-38}$$

$$G(s) = \frac{f(-s^2) - g(-s^2) - (-s^2)^\ell \epsilon}{f(-s^2)} \qquad (1\text{-}39)$$

The inequalities (1-36) and (1-37) means that the real rational functions $F(s)$ and $G(s)$ are generalized positive real functions [28]. If $\{A \ b \ c \ d\}$ denotes a minimal realization of $T(s)$ in (1-35), a minimal realization of $F(s)$ is given by

$$\left\{ \begin{bmatrix} 0 & -A \\ I & 0 \end{bmatrix} \ \begin{bmatrix} -b \\ 0 \end{bmatrix} \ [0 \ c - \epsilon c_k] \ d - \epsilon \delta_{nk} \right\} \qquad (1\text{-}40)$$

and a minimal realization of $G(s)$ is given by

$$\left\{ \begin{bmatrix} 0 & -A \\ I & 0 \end{bmatrix} \ \begin{bmatrix} -b \\ 0 \end{bmatrix} \ [0 \ -c - \epsilon c_\ell] \ 1 - d - \epsilon \delta_{n\ell} \right\} \qquad (1\text{-}41)$$

where δ_{ij} is a Kronecker's delta and c_i is

$$c_i = [0 \ 0 \ \cdots \ 0 \ 1][b \ Ab \ \cdots \ A^{n-1}b]^{-1} A^i \qquad (1\text{-}42)$$

If we apply the generalized positive real lemma of Anderson and Moore [28] to $F(s)$ and $G(s)$, we obtain a system-theoretical representation of the conditions of Fialkow and Gerst, as follows.

Theorem 1-8 [27]. Let $\{A \ b \ c \ d\}$ be a minimal realization of a real rational transfer function $T(s)$. Then $T(s)$ is realizable as a voltage transfer function of a transformerless RC three-terminal network if and only if the following conditions hold.

(i) There exists a positive definite symmetric matrix P such that

$$A^T P = PA < 0 \qquad (1\text{-}43)$$

(ii) There exist a positive number ϵ, nonsingular symmetric matrix P_1 and P_2, and row vectors ℓ_1 and ℓ_2 such that

$$0 \leq \epsilon \delta_{nk} \leq d \leq 1 - \epsilon \delta_{n\ell} \qquad (1\text{-}44)$$

$$\begin{bmatrix} 0 & -A \\ I & 0 \end{bmatrix}^T P_1 + P_1 \begin{bmatrix} 0 & -A \\ I & 0 \end{bmatrix} = -\ell_1^T \ell_1 \qquad (1\text{-}45a)$$

$$P_1 \begin{bmatrix} -b \\ 0 \end{bmatrix} = [0 \quad c - \epsilon c_k]^T - \sqrt{2(d - \epsilon \delta_{nk})} \, \ell_1^T \tag{1-45b}$$

$$\begin{bmatrix} 0 & -A \\ I & 0 \end{bmatrix}^T P_2 + P_2 \begin{bmatrix} 0 & -A \\ I & 0 \end{bmatrix} = -\ell_2^T \ell_2 \tag{1-46a}$$

$$P_2 \begin{bmatrix} -b \\ 0 \end{bmatrix} = [0 \quad -c - \epsilon c_\ell]^T - \sqrt{2(1 - d - \epsilon \delta_{n\ell})} \tag{1-46b}$$

where c_i is a row vector defined by (1-42) and k and ℓ are the integers such that

$$k = \begin{cases} 0 & \text{if } d - cA^{-1}b \neq 0 \\ i & \text{if } d - cA^{-1}b = cA^{-2}b = \cdots = cA^{-i}b = 0 \end{cases} \tag{1-47}$$

$$\text{and } cA^{-(i+1)}b \neq 0$$

$$\ell = \begin{cases} 0 & \text{if } 1 - (d - cA^{-i}b) \neq 0 \\ i & \text{if } 1 - (d - cA^{-1}b) = cA^{-2}b = \cdots = cA^{-i}b = 0 \end{cases} \tag{1-48}$$

$$\text{and } cA^{-(i+1)}b \neq 0$$

■

Equation (1-44) with P positive definite symmetric guarantees that A is diagonalizable and that all the eigenvalues are negative and distinct. Equations (1-45) and (1-46) with P_1 and P_2 nonsingular symmetric imply that $\text{Re}[F(j\omega)] \geq 0$ and $\text{Re}[G(j\omega)] \geq 0$; that is, $F(s)$ in (1-38) and $G(s)$ in (1-39) are generalized positive real functions [28]. The integers k in (1-47) and ℓ in (1-48) coincide with the lowest degrees of the powers of s in $f(s)$ and $f(s) - g(s)$, respectively.

1-9   BOUNDED REAL MATRICES AND LOSSLESS BOUNDED REAL MATRICES

In passive filter synthesis the role of scattering matrices is important. It is well known that the scattering matrices of the networks, which consist of only passive lumped elements, become rational bounded real (BR) matrices. Conversely, any rational BR matrix can be realized as a scattering matrix by a passive network. Moreover, the scattering matrices of the networks, which consist only of inductors, capacitors, ideal transformers, and ideal gyrators, become rational lossless bounded real (LBR) matrices. Definitions of rational BR matrices and rational LBR matrices follow.

Definition 1-5 [10, 22]. Let $S(s)$ be a $p \times p$ matrix of real rational functions of complex variable s. $S(s)$ is bounded real if and only if the following two conditions hold:

(i)   All the elements of $S(s)$ are analytic in $\text{Re}[s] \geq 0$.
(ii) $I - S^*(s)^T S(s)$ is nonnegative definite Hermitian in $\text{Re}[s] \geq 0$, or equivalently, $I - S(-j\omega)^T S(j\omega)$ is nonnegative definite Hermitian for all real ω.   ∎

Definition 1-6 [10, 22]. Let $S(s)$ be a $p \times p$ matrix of real rational functions of complex variable s. $S(s)$ is lossless bounded real if and only if the following two conditions hold:

(i) $S(s)$ is bounded real.
(ii) $I - S(-s)^T S(s) = 0$ for all s.   ∎

System-theoretical representations of rational BR and rational LBR matrices are given as follows.

Theorem 1-9 [10, 22] (Bounded Real Lemma). Let $\{A \ B \ C \ D\}$ be a minimal realization of a $p \times p$ matrix of a real rational function $S(s)$. Then $S(s)$ is bounded real if and only if there exist real matrices L and W_0 and a positive definite symmetric P such that

$$A^T P + PA = -C^T C - L^T L \qquad (1\text{-}49a)$$

$$PB = -C^T D - L^T W_0 \qquad (1\text{-}49b)$$

$$I - D^T D = W_0^T W_0 \qquad (1\text{-}49c)$$

Equation (1-49a) with $P > 0$ guarantees that the real parts of the eigenvalues of A are all negative. If $W(s)$ denotes

$$W(s) = W_0 + L[sI - A]^{-1}B \qquad (1\text{-}50)$$

it follows from (1-49) and $P = P^T$ that

$$I - S^*(s)^T S(s) = W^*(s)^T W(s) + 2 \, \text{Re}[s] B^T [s^*I - A^T]^{-1} P[sI - A]^{-1} B \qquad (1\text{-}51)$$

Therefore, if $P > 0$, $I - S^*(s)^T S(s) \geq 0$ for $\text{Re}[s] \geq 0$.   ■

A system-theoretical representation of rational LBR matrices is as follows.

Theorem 1-10 [10, 22] (Lossless Bounded Real Lemma). Let $\{A \;\; B \;\; C \;\; D\}$ be a minimal realization of a $p \times p$ matrix of a real rational function S(s). Then S(s) is lossless bounded real if and only if there exists a positive definite symmetric matrix P such that

$$A^T P + PA = -C^T C \qquad (1\text{-}52a)$$

$$PB = -C^T D \qquad (1\text{-}52b)$$

$$I - D^T D = 0 \qquad (1\text{-}52c)$$

Equation (1-52a) with $P > 0$ guarantees that the real parts of the eigenvalues of A are all negative. It follows from (1-52) with $P = P^T$ that

$$I - S^*(s)^T S(s) = 2 \, \text{Re}[s] B^T [s^*I - A^T]^{-1} P[sI - A]^{-1} B \qquad (1\text{-}53)$$

Therefore, if $P > 0$, $I - S^*(s)^T S(s) \geq 0$ for $\text{Re}[s] \geq 0$. Furthermore, it follows from (1-52) with P symmetric and nonsingular that [23]

$$P^{-1}(-A^T + C^T D^{T-1} B^T) P = A$$
$$P^{-1}(-C^T D^{T-1}) = B$$
$$(-D^{T-1} B^T) P = C \qquad (1\text{-}54)$$
$$D^{T-1} = D$$

Equations (1-54) imply that minimal realizations of $[S(-s)^T]^{-1}$ and S(s) are similar. This is equivalent to $[S(-s)^T]^{-1} = S(s)$. Hence we have

$$I - S(-s)^T S(s) = 0 \qquad (1\text{-}55)$$
                                                                                      ■

It should be noted that a scalar rational lossless bounded real function is a stable all-pass transfer function. Hence we can call rational LBR matrices, stable all-pass transfer function matrices [23].

1-10 DISCRETE POSITIVE REAL MATRICES AND DISCRETE
STRICTLY POSITIVE REAL MATRICES

In previous sections we stated the system-theoretical representations of functions of continuous-time systems. Beginning with this section we state the system-theoretical representations of functions of discrete-time systems.

Let $Z(z)$ be a $p \times p$ matrix of real rational functions of complex variable z with $Z(\infty) < \infty$. The concept of positive realness for continuous-time systems can be extended to discrete-time systems as follows.

Definition 1-7 [20,21,29]. A $p \times p$ matrix of real rational functions $Z(z)$ is a discrete positive real (DPR) matrix if and only if the following conditions hold:

(i) All the elements of $Z(z)$ are analytic in $|z| > 1$.

(ii) $Z^*(z)^T + Z(z)$ is a nonnegative definite Hermitian for $|z| \geq 1$. ■

Then a system-theoretical representation of DPR matrices is given as follows.

Theorem 1-11 [20,21,29] (Discrete Positive Real Lemma). Let $\{A \ B \ C \ D\}$ be a minimal realization of $Z(z)$. Then $Z(z)$ is a DPR matrix if and only if there exist real matrices L, W_0, and a positive definite symmetric matrix P such that

$$A^T PA - P = -L^T L \tag{1-56a}$$

$$B^T PA + W_0^T L = C \tag{1-56b}$$

$$D^T + D = B^T PB + W_0^T W_0 \tag{1-56c}$$

Equation (1-56a) with $P > 0$ guarantees that all the eigenvalues of A exist in $|z| \leq 1$. If $W(z)$ denotes $W(z) = W_0 + L[zI - A]^{-1}B$, it follows from (1-56) with $P = P^T$ that

$$Z^*(z)^T + Z(z) = W^*(z)^T W(z) + B^T[z^*I - A^T]^{-1}(|z|^2 - 1)P[zI - A]^{-1}B \tag{1-57a}$$

or

$$Z(z^{-1})^T + Z(z) = W(z^{-1})^T W(z) \tag{1-57b}$$

Therefore, if $P > 0$, we have $Z^*(z)^T + Z(z) \geq 0$ for $|z| \geq 1$. ■

Next, discrete strictly positive real matrices are defined as follows.

Definition 1-8 [20, 21, 29]. A $p \times p$ matrix of real rational functions $Z(z)$ is discrete strictly positive real (DSPR) matrix if and only if the following conditions hold:

(i) All the elements of $Z(z)$ are analytic in $|z| \geq 1$.
(ii) $Z*(z)^T + Z(z)$ is a positive definite Hermitian for $|z| \geq 1$, or equivalently,
(iii) $Z(\rho z)$ is a DPR matrix for some $0 < \rho < 1$.   ∎

Let $\{A\ B\ C\ D\}$ be a minimal realization of $Z(z)$. If $Z(\rho z)$ is positive real for some $0 < \rho < 1$, a minimal realization of $Z(\rho z)$ is given by $\{\hat{A}\ B\ \hat{C}\ D\}$, where

$$\hat{A} = \rho^{-1}A, \quad C = \rho^{-1}C \tag{1-58}$$

and by Theorem 1-11 (DPR lemma) there exist a positive definite symmetric matrix P and matrices L and W_0 such that

$$\hat{A}^T P \hat{A} - P = -L^T L \tag{1-59a}$$

$$B^T P \hat{A} + W_0 L = \hat{C} \tag{1-59b}$$

$$D^T + D = B^T P B + W_0^T W_0 \tag{1-59c}$$

Then it follows from (1-58) and (1-59) that

$$A^T P A - P = -(\rho L^T)(\rho L) - (1 - \rho^2)P \tag{1-60a}$$

$$B^T P A + W_0^T (\rho L) = C \tag{1-60b}$$

$$D^T + D = B^T P B + W_0^T W_0 \tag{1-60c}$$

Therefore, a system-theoretical representation of DSPR matrix is given as follows.

Theorem 1-12 [20, 21, 29] (Discrete Strictly Positive Real Lemma). Let $\{A\ B\ C\ D\}$ be a minimal realization of $Z(z)$. Then $Z(z)$ is a DSPR matrix if and only if there exist positive definite symmetric matrices P and Q and matrices L and W_0 such that

$$A^T P A - P = -L^T L - Q \tag{1-61a}$$

$$B^T PA + W_0^T L = C \qquad\qquad (1\text{-}61b)$$

$$D^T + D = B^T PB + W_0^T W_0 \qquad\qquad (1\text{-}61c)$$

Equation (1-61a) with $P > 0$ and $Q > 0$ guarantees that all the eigen-values of A exist in $|z| < 1$. If $W(z)$ denotes $W(z) = W_0 + L[zI - A]^{-1}B$, it follows from (1-61) with P and Q symmetric that

$$Z^*(z)^T + Z(z) = W^*(z)^T W(z) + B^T [z^*I - A^T]^{-1} \{ (|z|^2 - 1)P + Q \} [zI - A]^{-1}B \qquad (1\text{-}62)$$

Therefore, if $P > 0$ and $Q > 0$, we have $Z^*(z)^T + Z(z) > 0$ for $|z| \geq 1$. ∎

The concept of discrete strictly positive realness plays an important role in the proof of the discrete hyperstability theorem given by Popov [20, 21, 29]. Popov proved that if the transfer function of a linear block of a certain discrete nonlinear feedback control system is a discrete strictly positive real matrix, the system is asymptotically stable.

1-11 DISCRETE BOUNDED REAL MATRICES AND DISCRETE LOSSLESS BOUNDED REAL MATRICES

The concepts of discrete bounded realness and discrete lossless bounded realness are extensions of the concept of bounded realness and lossless bounded realness for continuous-time systems to discrete-time systems.

Let $H(z)$ be a $p \times m$ $(p > m)$ matrix of real rational functions of complex variable z. Then the definition of discrete lossless bounded real matrices is as follows.

Definition 1-9 [32]. $H(z)$ is a discrete lossless bounded real matrix if and only if the following three conditions hold:

    (i)   All the elements of $H(z)$ are analytic in $|z| \geq 1$.
    (ii) $I - H^*(z)^T H(z)$ is nonnegative definite Hermitian of $|z| \geq 1$.
    (iii) $I - H(z^{-1})^T H(z) = 0$ for all z. ∎

The definition of discrete bounded real matrices is as follows.

Definition 1-10 [32]. $H(z)$ is a discrete bounded real matrix if and only if the following two conditions hold:

    (i)   All the elements of $H(z)$ are analytic in $|z| \geq 1$.
    (ii) $I - H^*(z)^T H(z)$ is nonnegative definite Hermitian for $|z| \geq 1$. ∎

A system-theoretical representation of discrete lossless bounded real matrices is as follows.

Theorem 1-13 [32]. Let $\{A \ B \ C \ D\}$ be a minimal realization of $H(z)$. $H(z)$ is a discrete lossless bounded real matrix if and only if there exists a positive definite symmetric matrix P such that

$$A^T P A + C^T C = P \tag{1-63a}$$

$$BPB + D^T D = I \tag{1-63b}$$

$$A^T PB + C^T D = 0 \tag{1-63c}$$

Equation (1-63a) with $P > 0$ guarantees that all the eigenvalues of A exist in $|z| < 1$. It follows from (1-63) with $P = P^T$ that

$$I - H(z^{-1})^T H(z) = 0 \tag{1-64}$$

and

$$I - H^*(z)^T H(z) = B^T [z^*I - A^T]^{-1} (|z|^2 - 1) P[zI - A]^{-1} B \tag{1-65}$$

Therefore, if $P > 0$, $I - H^*(z)^T H(z) \geq 0$ for $|z| \geq 1$.   ∎

Now let us decompose the positive definite symmetric matrix P in (1-63) as

$$P = (T^{-1})^T T^{-1} \tag{1-66}$$

Then

$$x_1(n + 1) = A_1 x_1(n) + B_1 u(n)$$
$$\tag{1-67}$$
$$y(n) = C_1 x_1(n) + D_1 u(n)$$

is also a minimal realization of $H(z)$, where

$$A_1 = T^{-1} AT, \quad B_1 = T^{-1} B, \quad C_1 = CT, \quad D_1 = D \tag{1-68}$$

Then it follows from (1-63) and (1-66) that

$$A_1^T A_1 + C_1^T C_1 = I \tag{1-69a}$$

$$B_1^T B_1 + D_1^T D_1 = I \tag{1-69b}$$

$$A_1^T B_1 + C_1^T D_1 = 0 \tag{1-69c}$$

Hence the matrix R defined by

$$R = \begin{bmatrix} A_1 & B_1 \\ C_1 & D_1 \end{bmatrix} \tag{1-70}$$

satisfies

$$R^T R = I \tag{1-71}$$

In view of (1-71) Eq. (1-67) is called an orthogonal realization of $H(z)$, and it is known as a low-sensitivity realization [32-34, 41]. It follows from (1-71) and (1-67) that

$$x_1(n + 1)^T x_1(n + 1) + y(n)^T y(n) = x_1(n)^T x_1(n) + u(n)^T u(n) \tag{1-72}$$

If the input $u(n)$ is finite energy input and the initial state $x_1(0) = 0$, we have [32]

$$\sum_{n=0}^{\infty} y(n)^T y(n) = \sum_{n=0}^{\infty} u(n)^T u(n) \tag{1-73}$$

which is a time-domain expression of Eq. (1-64).

If $H(z)$ is a square matrix and Eqs. (1-63) with P nonsingular and symmetric and det $D \neq 0$ hold, it follows from (1-63) that

$$P^{-1}(A^{-1})^T P = A - BD^{-1}C$$

$$P^{-1}(A^{-1})^T C^T = BD^{-1}$$

$$-B^T(A^{-1})^T P = -D^{-1}C \tag{1-74}$$

$$D^T - B^T(A^{-1})^T C^T = D^{-1}$$

The equations above imply that P is a similarity transformation matrix and

$$H(z^{-1})^T = H(z)^{-1} \tag{1-75}$$

Hence we again obtain

$$I - H(z^{-1})^T H(z) = 0 \tag{1-76}$$

A system-theoretical representation of discrete bounded real matrices follows.

Theorem 1-14 [32]. Let $\{A \ B \ C \ D\}$ be a minimal realization of $H(z)$. Then $H(z)$ is a discrete bounded real matrix if and only if there exist a positive definite symmetric matrix P and matrices L and W_0 such that

$$A^T PA + C^T C + L^T L = P \tag{1-77a}$$

$$B^T PB + D^T D + W_0^T W_0 = I \tag{1-77b}$$

$$A^T PB + C^T D + L^T W_0 = 0 \tag{1-77c}$$

■

By a lemma of Lyapunov, Eq. (1-77a) with $P > 0$ guarantees that all the eigenvalues of A are strictly within the unit circle. If we define $W(z)$ as $W(z) = W_0 + L[zI - A]^{-1}B$, it follows from (1-77) with $P = P^T$ that

$$I - H^*(z)^T H(z) = W^*(z)^T W(z) + B^T [z^*I - A^T]^{-1}(|z|^2 - 1)P[zI - A]^{-1}B \tag{1-78}$$

Therefore, if $P > 0$, we have $I - H^*(z)^T H(z) \geq 0$ for $|z| \geq 1$. Next we decompose the positive definite symmetric matrix P in (1-77) as in (1-66):

$$P = (T^{-1})^T T^{-1} \tag{1-79}$$

Then we obtain from (1-77)

$$A_1^T A_1 + C_1^T C_1 + L_1^T L_1 = I$$

$$B_1^T B_1 + D_1^T D_1 + W_1^T W_1 = I \tag{1-80}$$

$$A_1^T B_1 + C_1^T D_1 + L_1^T W_1 = 0$$

where $\{A_1 \ B_1 \ C_1 \ D_1\}$ are given by (1-68) and

$$L_1 = LT, \qquad W_1 = W_0 \tag{1-81}$$

Now set

$$C_2 = \begin{bmatrix} C_1 \\ L_1 \end{bmatrix}, \quad D_2 = \begin{bmatrix} D_1 \\ W_1 \end{bmatrix} \tag{1-82}$$

$$G(z) = C_2[zI - A_1]^{-1}B_1 + D_2 = \begin{bmatrix} H(z) \\ W(z) \end{bmatrix} \tag{1-83}$$

Then we have from (1-80)

$$A_1^T A_1 + C_2^T C_2 = I$$

$$B_1^T B_1 + D_2^T D_2 = I \tag{1-84}$$

$$A_1^T B_1 + C_2^T D_2 = 0$$

In view of Theorem 1-13, Eqs. (1-84) imply that $G(z)$ in (1-83) is a discrete lossless bounded real matrix and $\{A_1 \; B_1 \; C_2 \; D_2\}$ is an orthogonal realization of $G(z)$. Equation (1-83) shows that a discrete bounded real matrix can always be embedded in a discrete lossless bounded real matrix [32].

1-12 FUTURE PROSPECTS

So far we have stated system-theoretical representations of various network functions. But the discussion was limited to the case of one-dimensional real-coefficient systems. Here we discuss briefly prospects for the future.

Recent progress in theories of multidimensional digital signal processing [35-38] and multirate digital signal processing [39-41] convince us that multidimensional digital systems and complex-coefficient digital systems [42] will play an important role in many engineering applications. In particular, digital all-pass functions (or discrete lossless bounded real functions) are considered as versatile building blocks for the digital signal processing [41]. However, there has been little study of multidimensional discrete all-pass systems or complex all-pass systems from the system-theoretical point of view. Hence we believe that extension of the discrete lossless bounded real lemma for one-dimensional discrete systems to the case of multidimensional discrete systems or complex-coefficient systems will be important and productive. Whereas multivariable positive real functions in network theory were established in 1960 [43], a multivariable or multidimensional positive real lemma has not yet been derived. Hence this problem is also left to the future.

Finally, we note the difficulties we face when we try to drive system-theoretical representations of transfer functions of multidimensional discrete systems. It is known that a multidimensional discrete transfer function is not always minimally realizable [44]; furthermore, a positive definite solution to the associated multidimensional matrix Lyapunov equation does not always exist, even if the transfer function is stable [45]. These facts are serious obstacles. To derive system-theoretical representations of multidimensional discrete transfer functions, we have to find a clever way to avoid these obstacles in the future.

REFERENCES

1. W. Cauer, Theorie der linearen Wechselstromschaltungen, Becker und Erler, Leipzig, East Germany, 1941.

2. S. Darlington, "Synthesis of reactance four poles which produce prescribed insertion loss characteristics including special application in filter design," J. Math. Phys. (Cambridge, Mass.), vol. 18, pp. 257-353 (1939).

3. K. Ogata, State Spece Analysis of Control Systems, Prentice-Hall, Englewood Cliffs, N.J., 1967.

4. L. A. Zadeh and C. A. Desoer, Linear System Theory, McGraw-Hill, New York, 1963.

5. R. A. Rohrer, Circuit Theory: An Introduction to the State-Variable Approach. McGraw-Hill, New York, 1970.

6. E. S. Kuh and R. A. Rohrer, "The state variable approach to network analysis, Proc. IEE, vol. 53, no. 7, pp. 672-686 (July 1965).

7. T. R. Bashkow, "The A matrix, a new network description," IRE Trans. Circuit Theory, vol. CT-4, no. 3, pp. 117-120 (Sept. 1957).

8. D. C. Youla and P. Tissi, "N-port synthesis via reactance extraction, Part I," IEEE Int. Conv. Rec., part 7, pp. 183-205 (1966).

9. B. D. O. Anderson and S. Vongpanitlerd, "Reciprocal passive impedance synthesis via state-space techniques," Proc. 1969 Hawaii Int. Conf. System Science, pp. 171-174, 1969.

10. S. Vongpanitlerd, "Passive reciprocal network synthesis: the state-space approach," Ph.D. dissertation, University of Newcastle, 1970.

11. S. Vongpanitlerd and B. D. O. Anderson, "Passive reciprocal state-space synthesis using a minimal number of resistors," Proc. IEE, vol. 117, no. 5, pp. 903-911 (May 1970).

12. R. Yarlagadda, "Network synthesis: a state space approach," IEEE Trans. Circuit Theory, vol. CT-19, no. 3, pp. 227-232 (May 1972).

13. S. Vongpanitlerd, "Reciprocal lossless synthesis via state-variable techniques," IEEE Trans. Circuit Theory, vol. CT-17, no. 4, pp. 630-632 (Nov. 1970).

14. K. S. Narenda and J. H. Taylor, Frequency Domain Criteria for Absolute Stability, Academic Press, New York, 1973.

15. F. R. Gantmacher, The Theory of Matrices, Vols. 1 and 2, Chelsea, New York, 1959.

16. B. D. O. Anderson, "A system theory criterion for positive real matrices," SIAM J. Control, vol. 5, no. 2, pp. 171-182 (May 1967).

17. B. D. O. Anderson and S. Vongpanitlerd, Network Analysis and Synthesis: A Modern Systems Approach, Prentice-Hall, Englewood Cliffs, N.J., 1972.

18. B. D. O. Anderson, "A simplified viewpoint on hyperstability," IEEE Trans. Autom. Control, vol. AC-13, no. 3, pp. 292-294 (March 1968).

19. V. M. Popov, "The solution to a new stability problem for controlled systems," Autom. Remote Control (USSR), vol. 24, no. 1, pp. 1-23 (1963).

20. V. M. Popov, Hyperstability of Control Systems, Springer-Verlag, Berlin, 1973.

21. I. D. Landau, Adaptive Control: The Model Reference Approach, Marcel Dekker, New York, 1979.

22. B. D. O. Anderson, "Algebraic description of bounded real matrices," Electron. Lett., vol. 2, no. 12, pp. 464-465 (Dec. 1966).

23. N. Matsumoto, "System theory representation for all pass transfer function matrices of discrete time and continuous time systems," Trans. IECE (Japan), vol. J64-A, no. 9, pp. 783-784 (Sept. 1981).

24. N. Matsumoto, N. Hamada, and S. Takahashi, "System representation for RC impedance matrices," Trans. IECE (Japan), vol. 59-A, no. 7, pp. 598-599 (July 1976).

25. N. Matsumoto, N. Hamada, and S. Takahashi, "System theory representations for linear networks," Trans. IECE (Japan), vol. 60-A, no. 8, pp. 709-716 (Aug. 1977).

26. A. Fialkow and I. Gerst, "The transfer function of general two terminal pair RC networks," Quart. Appl. Math., vol. X, no. 2, pp. 113-127 (July 1952).

27. N. Matsumoto and S. Takahashi, "On the realizability of state equations with RC 3-terminal network," IEEE Trans. Circuits Syst., vol. CAS-27, no. 1, pp. 53-57 (Jan. 1980).

28. B. D. O. Anderson and J. B. Moore, "Algebraic structure of generalized positive real matrices," SIAM J. Control, vol. 6, no. 4, pp. 615-624 (1968).

29. L. Hitz and B. D. O. Anderson, "Discrete positive real functions and their application to system stability," Proc. IEE, vol. 116, no. 1, pp. 153-155 (1969).

30. R. E. Kalman, "Lyapunov functions for the problem of Lur'e in automatic control," Proc. Nat. Acad. Sci. USA, vol. 49, no. 2, pp. 201-205 (Feb. 1963).

31. V. A. Yakubovich, "The solution of certain matrix inequalities in automatic control theory," Dokl. Akad. Nauk SSSR, vol. 143, pp. 1304-1307 (1962).

32. P. P. Vaidyanathan, "The discrete-time bounded-real lemma in digital filtering," IEEE Trans. Circuits Syst., vol. CAS-32, no. 9, pp. 918-924 (Sept. 1985).

33. P. P. Vaidyanthan and S. K. Mitra, "Low passband sensitivity digital filters: a generalized viewpoint and synthesis procedures," Proc. IEEE, vol. 72, no. 4, pp. 404-423 (Apr. 1984).

34. P. P. Vaidyanathan, "A unified approach to orthogonal digital filters and wave digital filters, based on LBR two-pair extraction," IEEE Trans. Circuits Syst., vol. CAS-32, no. 7, pp. 673-686 (July 1985).

35. N. K. Bose, Applied Multidimensional Systems Theory, Van Nostrand Reinhold, New York, 1981.

36. D. E. Dudgeon and R. M. Mersereau, Multidimensional Digital Signal Processing, Prentice-Hall, Englewood Cliffs, N.J., 1984.

37. T. Kaczorek, Two-Dimensional Linear Systems, Lecture Notes in Control and Information Sciences, Vol. 68, Springer-Verlag, Berlin, 1985.

38. S. G. Tzafestas (ed.), Multidimensional Systems: Techniques and Applications, Marcel Dekker, New York, 1986.

39. R. E. Crochiere and L. R. Rabiner, Multirate Digital Signal Processing, Prentice-Hall, Englewood Cliffs, N.J., 1983.

40. P. P. Vaidyanathan, "Quadrature mirror filter banks, M-band extensions and perfect-reconstruction techniques," IEEE ASSP Mag., vol. 4, no. 3, pp. 4-20 (July 1987).

41. P. A. Regalia, S. K. Mitra, and P. O. Vaidyanathan, "The digital all-pass filter: a versatile signal processing building block," Proc. IEEE, vol. 76, no. 1, pp. 19-37 (Jan. 1988).

42. T. H. Crystal and L. Ehrman, "The design and applications of digital filters with complex coefficients," IEEE Trans. Audio Electroacoust., vol. AU-16, no. 3, pp. 315-320 (Sept. 1968).

43. H. Ozaki and T. Kasami, "Positive real functions of several variables and their applications to variable networks," IRE Trans. Circuit Theory, vol. 7, no. 9, pp. 251-260 (Sept. 1960).

44. S. Kung, B. C. Lévy, M. Morf, and T. Kailath, "New results in 2-D systems theory, Part II: 2-D state-space models, realization and the notions of controllability, observability, and minimality," Proc. IEEE, vol. 65, no. 6, pp. 945-961 (June 1977).

45. B. D. O. Anderson, P. Agathoklis, E. I. Jury, and M. Mansour, "Stability and the matrix Lyapunov equation for discrete 2-dimensional systems," IEEE Trans. Circuits Syst., vol. CAS-33, no. 3, pp. 261-267 (Mar. 1986).

2

RC Active Circuit Synthesis via State Variable Method

TOHRU TAKAHASHI

SHIN-ICHI TAKAHASHI

Fukuoka Institute of Technology, Fukuoka, Japan

Keio University, Yokohama, Japan

2-1 INTRODUCTION

In this chapter we consider a network synthesis problem. We focus especially on system-theoretical methods for synthesizing an RC active network having a prescribed transfer function or matrix.

From conventional network synthesis it is well known that there exists a method of obtaining networks having a specified impedance matrix, any

positive real one, based on reactance extraction via the state variable
method [1]. This method, being a modernization of classical network theory,
requires multiwinding transformers and hence is not as practical. Another
method of synthesis involves the use of state variable circuits based on the
analog computer simulation of the transfer function, where the state equations are expressed in the canonic or companion form [2]. However, it requires many integrators to provide stable integration.

State equations of a specified network having a desired rational transfer function are naturally interpreted as follows. We consider state equations
of the form

$$\dot{x} = Ax + bu, \qquad \dot{x} = \frac{dx}{dt} \tag{2-1a}$$

$$y = cx \tag{2-1b}$$

where u is the input, y the output, and x the state variable vector. The coefficient matrices A, b, and c are constant with real elements unless specifically noted. Often, we denote the state equations (2-1a) and (2-1b) by a
3-tuple of coefficient matrices (A, b, c) and denote the state equation (2-1a)
by a 2-tuple of coefficient matrices (A, b).

Matrix A in (2-1a), the eigenvalues of which are identical to the characteristic roots, is characterized by network topology and element values,
the b in (2-1a) specifies a relation between the input and the state, and the c
in (2-1b) gives a relation between the output and the state. Thus we can
observe that the realization of the characteristic polynomial of the transfer
function is expressed in terms of (2-1a) and the numerator polynomial is
expressed in terms of (2-1b).

From the synthesis point of view, we propose two problems. The
first one is, from a given characteristic polynomial, to find a realization
$(\bar{A}, \bar{b})$ which has the same structures as the specified network and input topol-
ogies, and whose elements are described by coefficients of the characteristic
polynomial. The network element values can then be determined from the
given characteristic polynomial by direct comparisons of the elements in
both (A, b) and $(\bar{A}, \bar{b})$. The second problem is, from a desired numerator
polynomial, to determine the vector c. Since the result of such a multiplication of a vector by a vector as indicated in (2-1b) is a linear combination
of the components of the state vector, the system output y is obtained by
taking out each state variable and summing them up by the use of a summer
with appropriate scalings for the purpose of realizing the numerator polynomial. Then each element value of the vector c corresponds to each scaling
of the summer.

Our present approach to synthesis can be viewed as an attempt to find
an equivalent minimum realization $(\bar{A}, \bar{b}, \bar{c})$ from a given transfer function
$T(s)$, where elements of $(\bar{A}, \bar{b}, \bar{c})$ are expressed in terms of coefficients of
$T(s)$ or parameters closely related to coefficients of $T(s)$, and $(\bar{A}, \bar{b}, \bar{c})$ can

be identified with (A, b, c) in the state model of the specified network. In the resulting network, multiwinding transformers and integrators are not used; rather, stable passive networks, in some way, provide the integration, and the active elements are summers.

The strategy is to start with a companion-form realization, whose elements are directly related to the coefficients of T(s), and to obtain a coordinate transformation in the state representation, which provides a link between the state equation in the companion form and the state equation corresponding to the state model of the specified network.

Section 2-2 is intended to provide an introductory treatment for realizing a given characteristic polynomial. We present a method for realizing a Hurwitz characteristic polynomial having roots with nonpositive real parts using an RLC ladder network. As the A-matrix in (2-1a) we will utilize the Routh canonical form, which is a tridiagonal form closely related to a ladder network and has entries derived from a Routh array. However, this method has the disadvantage that each capacitive branch voltage and each inductive branch current as the state variables must be accessed and summed up to realize a numerator polynomial. Therefore, we shall not discuss in this section the problem of realizing a numerator polynomial.

We then consider a method of realizing a characteristic polynomial with an RC network. RC networks, especially, with all common-grounded capacitors can easily be used to construct accessible states (i.e., capacitor voltages) and sum them up for the purpose of realizing the numerator polynomial. To do this, in Sec. 2-3, a modern characterization of RC network functions is first performed. Next, we adopt the Cauer I ladder network as the RC network. In addition, we discuss a method for generating a tridiagonal-form realization related to the RC ladder network. The tridiagonal form has entries derived from the generalized Routh array used in the characterization of RC network functions noted above.

Characteristic polynomials of RC networks are restricted to a class of aperiodic damping polynomials having all negative roots. In Sec. 2-4 we consider a method of realizing a general characteristic polynomial by applying state variable feedback to the controllable RC network. In this way we can easily realize the feedback of accessible states and the numerator polynomial by using two summers.

For network synthesis problems of arbitrary transfer function matrices, we can extend the realization method as mentioned in Sec. 2-4. In Sec. 2-5 we outline several realization methods that have been proposed from a system-theoretical point of view.

2-2 INTRODUCTORY TREATMENT OF CHARACTERISTIC POLYNOMIAL REALIZATION

The basic procedure for realizing a characteristic polynomial is first to find a realization $(\bar{A}, \bar{b})$ that has the same structure as the specified network

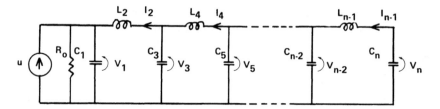

FIGURE 2-1 RLC ladder network.

topology and then give a transformation that provides a link between the
state equation in a companion form and the state equation corresponding to
the state model of the specified network. Once the transformation matrix
is obtained, a constructive procedure is given for determining the values
of network elements from coefficients of the characteristic polynomial.

Our purpose in this section is to realize a characteristic polynomial
with the RLC ladder network shown in Fig. 2-1 by finding the transformation
matrix. The reason that we adopt the RLC ladder network is that the tridiagonal canonical forms termed the Schwarz and Routh canonical forms,
which are closely related to the RLC ladder network [3], are well known in
the field of system theory.

Let a prescribed characteristic polynomial be

$$f(s) = s^n + a_1 s^{n-1} + \cdots + a_{n-1} s + a_n \tag{2-2}$$

and the corresponding state equation in a companion form be

$$\dot{x}_c = A_c x_c + b_c u \tag{2-3}$$

where

$$A_c = \begin{bmatrix} 0 & & & \\ \vdots & & I & \\ 0 & & & \\ -a_n & -a_{n-1} & \cdots & -a_1 \end{bmatrix}, \quad b_c = \begin{bmatrix} 0 \\ \vdots \\ 0 \\ 1 \end{bmatrix} \tag{2-4}$$

and x_c is the state vector, u the input, and I the identity matrix.

On the other hand, the state equation for the RLC ladder network in
Fig. 2-1 is of the form

$$\dot{x}_{LC} = A_{LC} x_{LC} + b_{LC} u \tag{2-5}$$

where

$$A_{LC} = \begin{bmatrix} \dfrac{-1}{C_1 R_0} & \dfrac{1}{\sqrt{C_1 L_2}} & & & & 0 \\ \dfrac{-1}{\sqrt{C_1 L_2}} & 0 & \dfrac{1}{\sqrt{L_2 C_3}} & & & \\ & \dfrac{-1}{\sqrt{L_2 C_3}} & \ddots & \ddots & & \\ & & & & \dfrac{1}{\sqrt{L_{n-1} C_n}} & \\ 0 & & & \dfrac{-1}{\sqrt{L_{n-1} C_n}} & 0 \end{bmatrix}, \quad b_{LC} = \begin{bmatrix} \dfrac{1}{\sqrt{C_1}} \\ 0 \\ \vdots \\ 0 \\ 0 \end{bmatrix} \tag{2-6}$$

and x_{LC} is the state vector with components consisting of each square root of capacitive or inductive energy such that

$$x_{LC} = [\sqrt{C_1} V_1 \ \sqrt{L_2} I_2 \ \cdots \ \sqrt{L_{n-1}} I_{n-1} \ \sqrt{C_n} V_n]^T \tag{2-7}$$

We then consider a similar transformation of the companion-form realization (A_c, b_c) to a tridiagonal-form realization $(\bar{A}_{LC}, \bar{b}_{LC})$ which is identical to the realization (A_{LC}, b_{LC}) and is expressed in terms of the coefficients of the characteristic polynomial f(s). Consider first results derived for f(s) by Parks [4] and Chen and Chu [5]. Parks has related the entries in the Schwarz matrix, which is a representative tridiagonal form, to the Hurwitz determinants and also to the first column of the Routh array; and Chen and Chu have derived a similar transformation of the companion matrix to the Schwarz matrix. We shall state these results in the form of a theorem.

Theorem 2-1. Let the roots of f(s) in (2-2) have negative real parts. Then there exists the Schwarz matrix

$$A_s = \begin{bmatrix} 0 & 1 & & & \\ -\dfrac{C_{n+1,1}}{C_{n-1,1}} & \ddots & \ddots & & 0 \\ & \ddots & \ddots & \ddots & \\ & & -\dfrac{C_{41}}{C_{21}} & 0 & 1 \\ 0 & & & -\dfrac{C_{31}}{C_{11}} & -\dfrac{C_{21}}{C_{11}} \end{bmatrix} \tag{2-8}$$

having f(s) as its characteristic polynomial, where the entries C_{i1} for $i = 1, 2, \ldots, n + 1$ are found from the Routh array:

$$
\begin{matrix}
C_{11}\,(= 1) & C_{12}\,(= a_2) & C_{13}\,(= a_4) & \cdots \\
C_{21}\,(= a_1) & C_{22}\,(= a_3) & \cdots & \\
C_{31} & C_{32} & \cdots & \\
\vdots & \vdots & &
\end{matrix}
\tag{2-9}
$$

and the quotients C_{ij} can be determined from the following Routh algorithm:

$$
C_{ij} = \frac{C_{i-1,1}\,C_{i-2,j+1} - C_{i-2,1}\,C_{i-1,j+1}}{C_{i-1,1}}
\tag{2-10}
$$

In addition, the similar transformation P_c, termed the Chen-Chu transformation satisfying

$$
P_c A_c P_c^{-1} = A_s
\tag{2-11}
$$

is given by

$$
P_c =
\begin{bmatrix}
1 & & & & & & & & \\
0 & & & & & 0 & & & \\
\dfrac{C_{n-1,2}}{C_{n-1,1}} & & & & & & & & \\
0 & & & & & & & & \\
\dfrac{C_{n-3,3}}{C_{n-3,1}} & & & & & & & & \\
0 & & & & & & & & \\
\vdots & & & & 0 & \dfrac{C_{32}}{C_{31}} & 0 & 1 & \\
& \cdots & 0 & \dfrac{C_{23}}{C_{21}} & 0 & \dfrac{C_{22}}{C_{21}} & 0 & 1 &
\end{bmatrix}
\tag{2-12}
$$

∎

Note that $f(s)$ in Theorem 2-1 has zeros with negative real parts and
therefore the entries in the first column of the Routh array are all positive.
It should also be noted that the entries in these canonical forms and transformation can be derived from the coefficients of $f(s)$ by using the Routh
array.

Although the Schwarz matrix A_S is a tridiagonal one, it is not identical
to the A_{LC} in (2-6) derived from the RLC ladder network. We then consider
a similar transformation of the Schwarz matrix A_S to another tridiagonal
matrix $\bar{A}_{LC}$ identical to A_{LC}. The matrix $\bar{A}_{LC}$ is known as the Routh canonical form [3], the entries of which can also be derived from the Routh parameters and is of the form

$$
\bar{A}_{LC} =
\begin{bmatrix}
-\dfrac{C_{21}}{C_{11}} & \sqrt{\dfrac{C_{31}}{C_{11}}} & & & 0 \\[2mm]
-\sqrt{\dfrac{C_{31}}{C_{11}}} & 0 & & & \sqrt{\dfrac{C_{n+1,1}}{C_{n-1,1}}} \\[4mm]
0 & & & -\sqrt{\dfrac{C_{n+1,1}}{C_{n-1,1}}} & 0
\end{bmatrix}
\tag{2-13}
$$

and the similar transformation P_p, termed the Power transformation [6],
satisfying

$$
P_p A_S P_p^{-1} = \bar{A}_{LC}
\tag{2-14}
$$

is given by

$$
P_p =
\begin{bmatrix}
& & & & \sqrt{C_{11}C_{21}} \\
& 0 & & -\sqrt{C_{21}C_{31}} & \\
& & \ddots & & 0 \\
(-1)^{n+1}\sqrt{C_{n1}C_{n+1,1}} & & & &
\end{bmatrix}
\tag{2-15}
$$

Transforming the companion-form realization (A_c, b_c) in (2-4) by the
similar transformation $P = P_p P_c$, we have the Routh canonical-form realization $(\bar{A}_{LC}, \bar{b}_{LC})$, identical to the realization (A_{LC}, b_{LC}), whose entries
are expressed in terms of network elements. The vector $\bar{b}_{LC}$ is then described by

$$\bar{b}_{LC} = Pb_c$$

$$= [\sqrt{C_{11}C_{21}} \quad 0 \quad \cdots \quad 0]^T \qquad (2\text{-}16)$$

Comparing the realization $(\bar{A}_{LC}, \bar{b}_{LC})$ with (A_{LC}, b_{LC}), we observe that the state variables of the realization $(A_{LC}, \bar{b}_{LC})$ are identical to x_{LC} in (2-7) and have the same physical meanings. The network element values can be then determined by direct comparison of the elements in both $(\bar{A}_{LC}, \bar{b}_{LC})$ and (A_{LC}, b_{LC}) as

$$R_0 = 1, \quad C_1 = \frac{1}{C_{21}}$$

$$L_i = \frac{C_{i-1,1}}{C_{i+1,1}} C_{i-1} \quad \text{for } i = 2, 4, \ldots, n-1 \qquad (2\text{-}17)$$

$$C_j = \frac{C_{j-1,1}}{C_{j+1,1}} L_{j-1} \quad \text{for } j = 3, 5, \ldots, n$$

The procedure above, developed for generating an RLC ladder network having a given characteristic polynomial, does not require the roots of the polynomial to be known. The synthesis is performed in a unified way by using the Routh parameters. However, since this synthesis method using an RLC network has the disadvantage that it is difficult to extract and sum up both each capacitor voltage and inductor current, we shall not discuss the problem of realizing the numerator polynomial of a given transfer function.

2-3 REALIZATION OF CHARACTERISTIC POLYNOMIAL WITH AN RC NETWORK

We consider here a method of realizing a given characteristic polynomial with an RC network, where we can easily construct accessible states and sum them up for the purpose of realizing a numerator polynomial. Especially, canonic RC networks with all common-grounded capacitors are of practical use. In addition, it is well known that the Cauer I RC ladder network is representative among these networks.

First, we consider a modern characterization of RC network functions [7]. This can effectively be performed by introducing a system with complex-number coefficients. The system is applicable for the analysis of aperiodic damping systems [8]. In the following discussions, we derive three canonical forms, termed the complex-number companion and the Schwarz and Routh canonical forms, respectively. It will further be shown that the complex-number Routh canonical form, which is also a tridiagonal one, is closely related to the Cauer I RC ladder network. Next we consider a method of realizing an RC impedance function with the RC ladder network [9].

2-3-1 Characterization of RC Impedance Functions

We begin with a theorem [10] useful for characterizing the properties of RC impedance functions.

Theorem 2-2 (Hermite-Biehler). Let $N(s)$ and $D(s)$ be polynomials having real coefficients defined as

$$N(s) = b_1 s^m + b_2 s^{m-1} + \cdots$$

$$D(s) = a_0 s^n + a_1 s^{n-1} + \cdots$$

where $|n - m| \leq 1$ and $b_1 a_0 > 0$. Then the necessary and sufficient condition for all the roots of the following polynomial having complex coefficients,

$$N(s) - jD(s), \quad j \text{ is an imaginary unit}$$

to be located in the upper half of the s-plane (including the real axis) is that all the roots of $N(s)$ and $D(s)$ are real and isolated from each other. ∎

Let a rational function $Z(s)$, which is relatively prime, be given by

$$Z(s) = \frac{N(s)}{D(s)}$$

where

$$N(s) = b_1 s^{n-1} + b_2 s^{n-2} + \cdots + b_n$$

$$D(s) = s^n + a_1 s^{n-1} + \cdots + a_n$$

(2-18)

Then the necessary and sufficient condition for $Z(s)$ to be an RC impedance function is that the following two conditions be satisfied:

Condition 2-1. All the roots of polynomials $N(s)$ and $D(s)$ are real and isolated from each other. ∎

Condition 2-2. The roots of $D(s)$ lie on the negative real axis (including the origin) of the s-plane. ∎

That is, when these conditions are satisfied, all the poles and zeros of Z(s) are on the real axis (including the origin) alternately as pole, zero, pole, zero, ..., the first pole being located closest to the origin. Therefore, the determination of the system theoretical representation of Conditions 2-1 and 2-2 results in a system-theoretical representation of RC impedance functions.

By transforming s into js in the Hermite-Biehler theorem, Condition 2-1 can be replaced by the condition that all the roots of the following polynomial having complex coefficients,

$$N(js) - jD(js) \tag{2-19}$$

lie in the left half (including the imaginary axis) of the s-plane. Let $\hat{f}(s)$ be the polynomial resulting from (2-19) upon normalizing the coefficient of the highest order, that is,

$$\hat{f}(s) = s^n + (-b_1 + ja_1)(-j)^2 s^{n-1} + \cdots + (-b_n + ja_n)(-j)^{n+1} \tag{2-20}$$

Then, for $\hat{f}(s)$ to be in a class of complex-coefficient Hurwitz polynomials is equivalent to Condition 2-1 being satisfied. If $\hat{f}(s)$ is a complex-coefficient Hurwitz polynomial, then, in view of Condition 2-2, the condition corresponds to the condition that coefficients a_i and b_i in (2-18) are positive (constant term a_n is nonnegative), which follows from Descartes's law of signs [11]. We have, then, the following theorem.

Theorem 2-3 [7]. Let the coefficients a_i and b_i of Z(s) in (2-18) be positive (constant term a_n be nonnegative). Then the necessary and sufficient condition for Z(s) to be an RC impedance function is that the polynomial $\hat{f}(s)$ with complex coefficients in (2-20) is a complex-coefficient Hurwitz polynomial.   ∎

We shall continue our discussion with the assumptions that the coefficients a_i and b_i are positive and that $\hat{f}(s)$ is a complex-coefficient Hurwitz polynomial. The following theorem is then useful for a characterization of polynomials having complex coefficients.

Theorem 2-4 [7]. The necessary and sufficient condition for the polynomial $\hat{f}(s)$ in (2-20) to be a complex-coefficient Hurwitz polynomial is that all the real entries, D_{00}, D_{11}, D_{22}, D_{33}, ..., D_{nn} in the first column of the following generalized Routh array [8] are positive:

$$
\begin{array}{llll}
D_{00} & D_{01} & D_{02} & \cdots & D_{0n} \\
D_{11} & D_{12} & \cdots & D_{1n} \\
E_{11} & E_{12} & \cdots & E_{1n} \\
\vdots & \vdots & & \\
D_{nn} & & & \\
E_{nn} & & &
\end{array}
\tag{2-21}
$$

where

$$D_{00} = 1, \quad D_{0i} = (-b_i + ja_i)(-j)^{i+1}$$

and

$$D_{1i} = \begin{cases} \text{Re}(D_{0i}), & i \text{ odd} \\ \\ \text{Im}(D_{0i}), & i \text{ even} \end{cases} \quad \text{for } i = 1, 2, \ldots, n$$

and the generalized Routh algorithms [8] are given by

$$E_{ij} = D_{i-1, j} - (D_{i-1, i-1} D_{i, j+1} / D_{ii})$$

$$D_{ij} = E_{i-1, j} - (E_{i-1, i-1} D_{i-1, j} / D_{i-1, i-1}) \tag{2-22}$$

∎

Note that the D_{ii} for $i = 1, 2, \ldots, n$ are all positive constants since $Z(s)$ is an RC impedance function.

Consider now a set of differential equations having complex coefficients with $\hat{f}(s)$ as its characteristic polynomial. Let a complex-number companion matrix $\hat{A}_c$ that satisfies

$$f(s) = \det(sI - \hat{A}_c)$$

be

$$\hat{A}_c = \begin{bmatrix} 0 & & & \\ \vdots & & I & \\ 0 & & & \\ -D_{0n} & -D_{0, n-1} & \cdots & -D_{01} \end{bmatrix} \tag{2-23}$$

Also, it is known that the complex-number companion matrix $\hat{A}_c$ is similar to the following complex-number Schwarz matrix $\hat{A}_s$:

$$\hat{A}_s = \begin{bmatrix} -\dfrac{E_{nn}}{D_{n-1, n-1}} & 1 & & & & \\ -\dfrac{E_{nn}}{D_{n-2, n-2}} & & \ddots & & 0 & \\ & & & \ddots & & \\ & & & & & 1 \\ 0 & & & & -\dfrac{E_{22}}{D_{00}} & -\dfrac{E_{11}}{D_{00}} \end{bmatrix} \tag{2-24}$$

which is effectively applicable for the stability criteria for complex-coefficient ordinary differential equations [8].

The similar transformation $\hat{P}_c$ satisfying

$$\hat{P}_c \hat{A}_c \hat{P}_c^{-1} = \hat{A}_s \tag{2-25}$$

termed the generalized Chen-Chu transformation [9], is derived in the same manner as the Chen-Chu transformation for real-coefficient systems, and is given by

$$\hat{P}_c = \begin{bmatrix} 1 & & & & \\ \dfrac{D_{n-1,n}}{D_{n-1,n-1}} & 1 & & & \\ \vdots & & \ddots & & \\ \dfrac{D_{1n}}{D_{11}} & \cdots\cdots & \dfrac{D_{12}}{D_{11}} & 1 \end{bmatrix} \tag{2-26}$$

Furthermore, by applying the transformation

$$\hat{P}_p = \begin{bmatrix} & & & & \sqrt{D_{11}D_{00}} \\ & 0 & & -\sqrt{D_{22}D_{11}} & \\ & & \cdots & & \\ & & & 0 & \\ (-1)^{n+1}\sqrt{D_{nn}D_{n-1,n-1}} & & & & \end{bmatrix} \tag{2-27}$$

termed the generalized Power transformation, to the complex-number Schwarz matrix $\hat{A}_s$, the complex-number Routh canonical form matrix $\hat{A}_R$ satisfying

$$\hat{P}_p \hat{A}_s \hat{P}_p^{-1} = \hat{A}_R \tag{2-28}$$

is obtained by

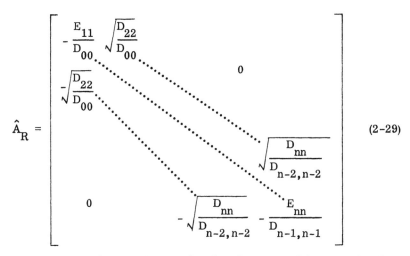

$$\hat{A}_R = \begin{bmatrix} -\dfrac{E_{11}}{D_{00}} & \sqrt{\dfrac{D_{22}}{D_{00}}} & & & 0 \\ -\sqrt{\dfrac{D_{22}}{D_{00}}} & & & & \\ & & & \sqrt{\dfrac{D_{nn}}{D_{n-2,n-2}}} & \\ & & & & -\dfrac{E_{nn}}{D_{n-1,n-1}} \\ 0 & & -\sqrt{\dfrac{D_{nn}}{D_{n-2,n-2}}} & & \end{bmatrix} \qquad (2\text{-}29)$$

It will be shown that the complex-number Routh canonical form is closely related to the Cauer I RC ladder network.

2-3-2 Tridiagonal Matrix Associated with an RC Ladder Network

Based on the characterization of RC impedance functions in the preceding section, we synthesize a given RC impedance function with the Cauer I RC ladder network in Fig. 2-2. Let a strictly proper RC impedance function without pole-zero cancellations be

$$Z(s) = \frac{b_1 s^{n-1} + \cdots + b_n}{s^n + a_1 s^{n-1} + \cdots + a_n} \qquad (2\text{-}30)$$

where $a_i > 0$ and $b_i > 0$ for $i = 1, 2, \ldots, n$ and the corresponding companion-form realization be (A_c, b_c, c_c), where

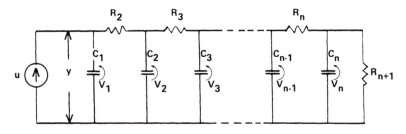

FIGURE 2-2 RC ladder network. (From Ref. 9.)

$$A_c = \begin{bmatrix} 0 & & \\ \vdots & & I \\ \vdots & & \\ 0 & & \\ -a_n & \cdots & -a_1 \end{bmatrix}, \quad b_c = [0 \cdots 0 \ 1]^T, \quad c_c = [b_n \cdots b_1]$$

$$(2\text{-}31)$$

On the other hand, the state model for the RC ladder network in **Fig. 2-2** is of the symmetrically tridiagonal form

$$\dot{x}_{RC} = A_{RC} x_{RC} + b_{RC} u$$

$$y = c_{RC} x_{RC}$$

where

$$A_{RC} = \begin{bmatrix} \dfrac{-1}{C_1 R_2} & \dfrac{1}{R_2\sqrt{C_1 C_2}} & & & 0 \\[2mm] \dfrac{1}{R_2\sqrt{C_1 C_2}} & \dfrac{-1}{C_2}\left(\dfrac{1}{R_2}+\dfrac{1}{R_3}\right) & \dfrac{1}{R_3\sqrt{C_2 C_3}} & & \\[2mm] & \dfrac{1}{R_3\sqrt{C_2 C_3}} & \ddots & & \\[2mm] & & & \ddots & \dfrac{1}{R_n\sqrt{C_n C_{n-1}}} \\[2mm] 0 & & & \dfrac{1}{R_n\sqrt{C_n C_{n-1}}} & \dfrac{-1}{C_n}\left(\dfrac{1}{R_n}+\dfrac{1}{R_{n+1}}\right) \end{bmatrix}$$

$$b_{RC} = [1/\sqrt{C_1} \ \ 0 \cdots 0]^T, \quad c_{RC} = [1/\sqrt{C_1} \ \ 0 \cdots 0]$$

$$(2\text{-}32)$$

and u is the input current, y the output voltage, and x_{RC} the state vector with components consisting of each square root of capacitive energy such that

$$x_{RC} = [\sqrt{C_1}\, V_1 \ \cdots \ \sqrt{C_n}\, V_n]^T \qquad (2\text{-}33)$$

We then consider a similar transformation of the companion–form realization (A_c, b_c, c_c) to a symmetrically tridiagonal-form realization $(\bar{A}_{RC}, \bar{b}_{RC}, \bar{c}_{RC})$ which is identical to the realization (A_{RC}, b_{RC}, c_{RC}) in (2-32).

By the nonsingular matrix

$$\hat{P}_1 = \text{diag}[1 \quad -j \quad \cdots \quad (-j)^{n-1}]$$

the complex-number companion matrix $\hat{A}_c$ derived from the foregoing characterization of $Z(s)$ is similarly transformed as follows:

$$\hat{P}_1^{-1}\hat{A}_c\hat{P}_1 = \begin{bmatrix} 0 \\ -c_c \end{bmatrix} - jA_c \overset{\Delta}{=} \tilde{A}_c \tag{2-34}$$

where the matrix A_c and vector c_c are as defined in (2-31).

The complex-number Routh canonical matrix $\hat{A}_R$ in (2-29) is similarly transformed by the nonsingular matrix

$$\hat{P}_2 = (-1)^{n-1} \text{diag}[(-j)^{n-1} \quad \cdots \quad -j \quad 1]$$

to another tridiagonal form satisfying $\hat{P}_2\hat{A}_R\hat{P}_2^{-1} = \tilde{A}_R$ and $\tilde{A}_R$ is given by

$$\tilde{A}_R = \begin{bmatrix} -\dfrac{E_{11}}{D_{00}} & -j\sqrt{\dfrac{D_{22}}{D_{00}}} & & & & 0 \\ -j\sqrt{\dfrac{D_{22}}{D_{00}}} & & \ddots & & & \\ & & & & & -j\sqrt{\dfrac{D_{nn}}{D_{n-2,n-2}}} \\ 0 & & & & -j\sqrt{\dfrac{D_{nn}}{D_{n-2,n-2}}} & -\dfrac{E_{nn}}{D_{n-1,n-1}} \end{bmatrix} \tag{2-35}$$

From this complex-number matrix $\tilde{A}_R$ we can extract a symmetrically tridiagonal form having real entries as follows. We can observe that the matrix $\tilde{A}_R$ is decomposed into the sum of real and imaginary parts such that

$$\tilde{A}_R = A_{RR} - j\tilde{A}_{RC} \tag{2-36}$$

where

$$A_{RR} = \text{diag}[-D_{11} \quad 0 \quad \cdots \quad 0]$$

and

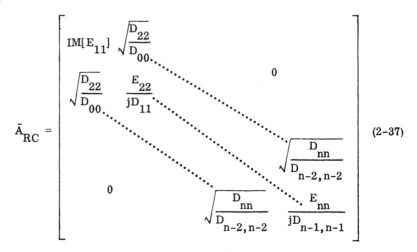

$$(2\text{-}37)$$

The matrices A_{RR} and $\bar{A}_{RC}$ above are obviously proved to have real entries, respectively, by using the following properties [7]:

$$\left. \begin{array}{ll} D_{ij} = \text{pure imaginary,} & i + j \text{ odd} \\ D_{ij} = \text{real,} & i + j \text{ even} \end{array} \right\} i = 1, 2, \ldots, n$$

$$\left. \begin{array}{ll} E_{ij} = \text{real,} & i + j \text{ odd} \\ E_{ij} = \text{pure imaginary,} & i + j \text{ even} \end{array} \right\} i = 2, 3, \ldots, n$$

Through the preceding canonical transformations, it is obvious that the complex-number matrix $\tilde{A}_c$ in (2-34) is similar to the complex-number matrix $\tilde{A}_R$ in (2-36). Thus the similar transformation

$$P = P_2 P_p P_c P_1$$

provides a link between these two matrices such that

$$P\tilde{A}_c P^{-1} = \tilde{A}_R \qquad (2\text{-}38)$$

where

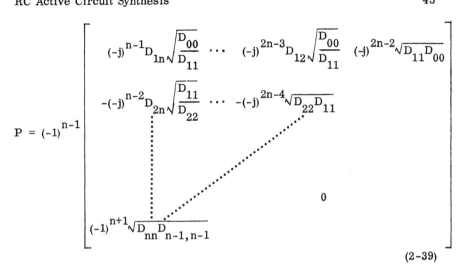

$$(2\text{-}39)$$

Also, the similar transformation P is obviously proved to have real entries by using the properties above. Substituting (2-34) and (2-36) into (2-38), we obtain

$$PA_c P^{-1} = \bar{A}_{RC} \tag{2-40}$$

Then we have

$$Pb_c = [\sqrt{D_{11}} \quad 0 \quad \cdots \quad 0]^T \triangleq \bar{b}_{RC}$$

$$c_c P^{-1} = [\sqrt{D_{11}} \quad 0 \quad \cdots \quad 0] \triangleq \bar{c}_{RC} \tag{2-41}$$

We now have the real similar transformation P between two minimal realizations (A_c, b_c, c_c) and $(\bar{A}_{RC}, \bar{b}_{RC}, \bar{c}_{RC})$ of $Z(s)$. Comparing the symmetrically tridiagonal form realization $(\bar{A}_{RC}, \bar{b}_{RC}, \bar{c}_{RC})$ with the state model (A_{RC}, b_{RC}, c_{RC}) derived from the RC ladder network, we can identify them. Through direct comparisons of each matrix element in both $(\bar{A}_{RC}, \bar{b}_{RC}, \bar{c}_{RC})$ and (A_{RC}, b_{RC}, c_{RC}), the network element values are then evaluated by the entries in the first column of the generalized Routh array as

$$C_1 = \frac{1}{D_{11}}, \quad C_2 = \frac{(\mathrm{Im}\, E_{11})^2}{D_{11} D_{22}} \tag{2-42}$$

$$C_k = \frac{D_{k-1,k-1}}{D_{kk}} \left(\frac{E_{k-1,k-1}}{j\sqrt{D_{k-1,k-1} D_{k-2,k-2}}} \sqrt{C_{k-1}} + \sqrt{\frac{D_{k-2,k-2}}{D_{k-3,k-3}}} \sqrt{C_{k-2}} \right)^2$$

for $k = 3, 4, \ldots, n$

and

$$R_2 = -\frac{D_{11}}{Im(E_{11})}$$

$$R_3 = \frac{jD_{11}^2 D_{22}}{\{(E_{11} - D_{11})(E_{22} - D_{22}) + E_{11}D_{22}\}(E_{11} - D_{11})} \qquad (2\text{-}43)$$

$$R_k = \frac{-(R_{k-2} + R_{k-1})R_{k-1}}{\left(\dfrac{E_{k-1,k-1}E_{k-2,k-2}}{D_{k-1,k-1}D_{k-2,k-2}} + 1\right)R_{k-2} + R_{k-1}} \qquad \text{for } k = 4, \ldots, n+1$$

2-4 REALIZATION OF TRANSFER FUNCTION USING AN RC NETWORK AND SUMMERS

The characteristic polynomial realized with an RC network is restricted to a class of aperiodic damping polynomials. This restriction can be removed by applying state feedback theory [12] for the purpose of realizing general characteristic polynomials.

The method of realizing a general transfer function $T_G(s) = N(s)/D_G(s)$ consists of the following three procedures [9, 13]:

Procedure 2-1. We synthesize an aperiodic damping-type characteristic polynomial $D_{RC}(s)$, which is properly chosen, with an RC network. This implies that we obtain the realization

$$\dot{x} = \bar{A}_{RC}x + \bar{b}_{RC}u$$

satisfying $D_{RC}(s) = \det(sI - \bar{A}_{RC})$.   ∎

Procedure 2-2. We synthesize an RC transfer function $T_{RC}(s) = N(s)/D_{RC}(s)$ having the same numerator polynomial as that of a desired general transfer function and the same characteristic polynomial as used in Procedure 2-1. This implies that we obtain the realization

$$\dot{x} = \bar{A}_{RC}x + \bar{b}_{RC}u$$

$$y = \bar{c}_{RC}x$$

satisfying $T_{RC}(s) = \bar{c}_{RC}(sI - \bar{A}_{RC})^{-1}\bar{b}_{RC}$.   ∎

Procedure 2-3. We synthesize a desired characteristic polynomial by applying the state feedback $u = v - \bar{f}^T x$, where v is the new input and $\bar{f}$ is the feedback gain vector, to the controllable RC active network constructed in Procedure 2-2. This implies that we obtain the realization

$$\dot{x} = (\bar{A}_{RC} - \bar{b}_{RC}\bar{f}^T)x + \bar{b}_{RC}v$$

$$y = \bar{c}_{RC}x$$

satisfying $T_G(s) = \bar{c}_{RC}\{sI - (\bar{A}_{RC} - \bar{b}_{RC}\bar{f}^T)\}^{-1}\bar{b}_{RC}$.   ∎

The notions of Procedures 2-1 and 2-2 are based on the fact that zeros of the transfer function of a scalar system, which is controllable, are invariant despite applying state feedback to the system [12]. The RC network used in Procedure 2-1 can arbitrarily be chosen within the framework of controllable RC networks. From the viewpoint of practical use, it is desirable to choose a canonic RC network with all common-grounded capacitors.

In accordance with the procedures described above, we consider here a method of realizing general transfer functions with an RC active network containing two summers and a Cauer I RC ladder network which is controlable with all common-grounded capacitors [9].

2-4-1 Realization of RC Voltage Transfer Functions

Let an RC voltage transfer function be

$$T_{RC}(s) = \frac{b_0 s^m + b_1 s^{m-1} + \cdots + b_m}{s^n + a_1 s^{n-1} + \cdots + a_n} \triangleq \frac{N(s)}{D_{RC}(s)} \tag{2-44}$$

where the numerator polynomial is identical to that of a desired general transfer function and the characteristic polynomial of which is an aperiodic damping one chosen properly, and let the corresponding companion-form realization be (A_c, b_c, c_c)

Procedure 2-1. For the purpose of realizing the characteristic polynomial $D_{RC}(s)$, we first synthesize an RC impedance function $Z(s) = N_{RC}(s)/D_{RC}(s)$ with a modified RC ladder network in Fig. 2-3. The reason we utilize the RC ladder network is that we want to make the network be driven by the voltage source, which is obtained by the equivalent transformation shown in Fig. 2-3. With no loss of generality, we assume that $R_1 = 1$. Then we must select a numerator polynomial $N_{RC}(s)$ such that $Z(s)$ can be expanded as

                                                    Takahashi and Takahashi

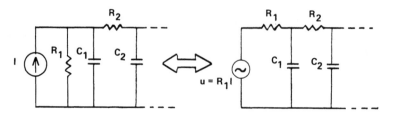

FIGURE 2-3 Modified RC ladder network.

$$Z(s) = \frac{N_{RC}}{(D_{RC} - N_{RC}) + N_{RC}}$$

$$= \frac{1}{\dfrac{1}{1} + \dfrac{1}{N_{RC}/(D_{RC} - N_{RC})}} \tag{2-45}$$

and such that $N_{RC}/(D_{RC} - N_{RC})$ is also an RC impedance function. This expansion corresponds to the ladder network structure in Fig. 2-3. The RC impedance function $N_{RC}/(D_{RC} - N_{RC})$ is synthesized by using the Cauer I RC ladder network according to the method described in Sec. 2-3. The element values of ladder network are then determined by (2-42) and (2-43). Note that D_{ii} and E_{ii} are the entries in the generalized Routh array derived from $N_{RC}/(D_{RC} - N_{RC})$. In this case there is a degree of freedom in selecting $N_{RC}(s)$. One simple selection of $N_{RC}(s)$ is a polynomial $rdD_{RC}(s)/ds$, where $a_n/a_{n-1} \geq r > 0$. All the poles and zeros of $N_{RC}/(D_{RC} - N_{RC})$ are then on the real axis of the s-plane alternately as pole, zero, pole, zero, $\cdots$, the first pole being located closest to the origin.   ■

    Procedure 2-2. To synthesize the RC voltage transfer function $T_{RC}(s)$ with the RC ladder network in Fig. 2-3 as a basic RC network, we transform the companion-form realization (A_c, b_c, c_c) into a symmetrically tridiagonal-form realization $(\bar{A}_{RC}, \bar{b}_{RC}, \bar{c}_{RC})$ by using the transformation matrix P in (2-39). The generalized Routh array derived from $Z(s)$ is expressed in terms of D_{ij} and E_{ij}, which are the entries in the generalized Routh array derived from $N_{RC}/(D_{RC} - N_{RC})$, as follows:

$$D_{00}, \quad D_{01} - jD_{11}, \quad D_{02} - jD_{12}, \quad \cdots, \quad D_{0n} - jD_{1n}$$

$$D_{11}, \quad D_{12}, \quad \cdots, \quad D_{1n}$$

$$E_{11} - jD_{11}, \quad E_{12} - jD_{12}, \quad \cdots, \quad E_{1n} - jD_{1n}$$

$$D_{22}, \quad D_{23}, \quad \cdots, \quad D_{2n}$$

$$\tag{2-46}$$

E_{22}, E_{23}, $\cdots$, E_{2n}

$\vdots$ $\vdots$

D_{nn}

E_{nn}

The matrices $\bar{A}_{RC}$, $\bar{b}_{RC}$, and $\bar{c}_{RC}$ in the symmetrically tridiagonal form realization are then described by

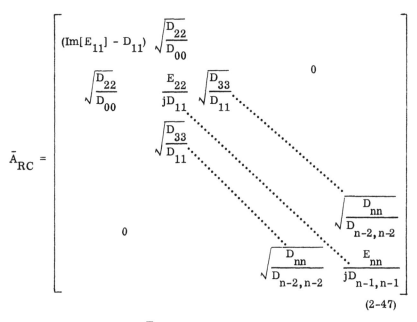

$$\bar{A}_{RC} =$$

(2-47)

$$\bar{b}_{RC} = [\sqrt{D_{11}} \quad 0 \quad \cdots \quad 0]^T$$

$$\bar{c}_{RC} = [0 \quad \cdots \quad 0 M'_{n-m} \quad \cdots \quad M'_n]$$

where

$$M'_{n-m} = \frac{b_0}{\sqrt{D_{n-m,n-m} D_{n-m-1,n-m-1}}}$$

$$M_k' = \frac{1}{\sqrt{D_{kk}D_{k-1,k-1}}}\left[b_{k-n+m} - \sum_{i=n-m}^{k-1} M_i'(-1)^{n+i}(-j)^{2n-k-i}\right.$$

$$\left. \cdot D_{ik}\sqrt{\frac{D_{i-1,i-1}}{D_{ii}}}\right] \text{ for } k = n - m + 1, \ldots, n \qquad (2\text{-}48)$$

Since we define the root of each capacitive energy, $\sqrt{c_i}\, v_i$ for $i = 1, 2, \ldots, n$ as each component of the state variable vector for the realization $(\bar{A}_{RC}, \bar{b}_{RC}, \bar{c}_{RC})$, the output y is given by

$$y = \sum_{k=n-m}^{n} -M_k v_k$$

where

$$M_k = -\sqrt{C_k}\, M_k' \qquad (2\text{-}49)$$

The M_i for $i = 1, 2, \ldots, n$ are the infinite input impedance coefficients. Therefore, the output y is constructed by accessing each capacitor voltage and summing them up by the use of a summer with appropriate scalings. The network resulting from the foregoing synthesis procedures is shown in Fig. 2-4. ∎

2-4-2 Realization of General Voltage Transfer Functions

Consider a general voltage transfer function

$$T_G(s) = \frac{b_0 s^m + b_1 s^{m-1} + \cdots + b_m}{s^n + e_1 s^{n-1} + \cdots + e_n} \triangleq \frac{N(s)}{D_G(s)} \qquad (2\text{-}50)$$

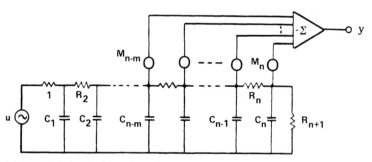

FIGURE 2-4 Realization of $T_{RC}(s)$.

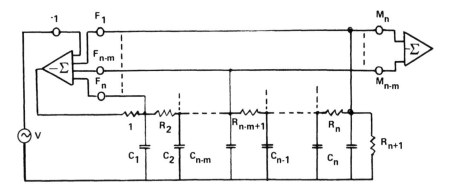

FIGURE 2-5 Realization of general transfer function.

where $D_G(s)$ is a Hurwitz polynomial. We can always decompose $D_G(s)$ into the sum of two polynomials such that

$$D_G(s) = D_{RC}(s) + D_0(s)$$

$$= (s^n + a_1 s^{n-1} + \cdots + a_n) + (f_1 s^{n-1} + \cdots + f_n) \qquad (2\text{-}51)$$

where $D_{RC}(s)$ is an aperiodic damping polynomial.

Using the $D_{RC}(s)$, which is properly chosen, we first synthesize an RC voltage transfer function $T_{RC}(s) = N(s)/D_{RC}(s)$ according to Procedures 2-1 and 2-2 as noted above. After performing these procedures, applying the state feedback to the controllable network in Fig. 2-4, we realize the desired $T_G(s)$ as shown in Fig. 2-5. Then the problem is to determine the values of feedback gains.

Procedure 2-3. Let the state equation in a companion form for $T_{RC}(s)$ be

$$\dot{x} = A_c x + b_c u$$
$$y = c_c x \qquad\qquad (2\text{-}52)$$

Applying the state feedback $u = v - f^T x$ to the system above, where v is the new input and f^T the feedback gain vector such that

$$f^T = [f_n \quad f_{n-1} \quad \cdots \quad f_1] \qquad (2\text{-}53)$$

the system in (2-52) becomes

$$\dot{x} = (A_c - b_c f^T)x + b_c v$$

$$y = c_c x$$

$$(2\text{-}54)$$

Transforming the realization $(A_c - b_c f^T, b_c, c_c)$ into the tridiagonal-form realization $(\bar{A}_{RC} - \bar{b}_{RC}\bar{f}^T, \bar{b}_{RC}, \bar{c}_{RC})$, which corresponds to the realization in Fig. 2-5, by the use of the transformation matrix P in (2-39), we then have

$$\bar{A}_{RC} = PA_c P^{-1}, \quad \bar{b}_{RC} = Pb_{RC}, \quad \bar{c}_{RC} = c_c P^{-1} \qquad (2\text{-}55)$$

and

$$\bar{f}^T = f^T P^{-1} \qquad (2\text{-}56)$$

Setting

$$\bar{f}^T = [\bar{f}_n \quad \bar{f}_{n-1} \quad \cdots \quad \bar{f}_1]$$

we have

$$\bar{f}_n = \frac{f_1}{\sqrt{D_{11}D_{00}}}$$

$$\bar{f}_k = \frac{1}{\sqrt{D_{n-k+1,n-k+1}D_{n-k,n-k}}} \left[f_{n-k+1} - \sum_{i=1}^{n-k} (-1)^{n+i}(-j)^{n-1-i+k} \right.$$

$$\left. \cdot D_{i,n+1-k}\sqrt{\frac{D_{i-1,i-1}}{D_{ii}}}\,\bar{f}_{n-i+1} \right] \quad \text{for } k = n-1, \ldots, 1$$

$$(2\text{-}57)$$

The resulting feedback gains of the network in Fig. 2-5 are then given by

$$F_k = \sqrt{c_{n-k+1}}\,\bar{f}_k \quad \text{for } k = 1, 2, \ldots, n \qquad (2\text{-}58)$$

Sensitivities of transfer characteristics to passive and active elements are easily evaluated by using the transformation matrix P [9]. ∎

Illustrative Example

To illustrate these realization procedures, we realize the third-order Butterworth low-pass voltage transfer function

$$T_G(s) = \frac{N(s)}{D_G(s)} = \frac{1}{s^3 + 2s^2 + 2s + 1}$$

We decompose $D_G(s)$ as

$$D_G(s) = D_{RC}(s) + D_0(s)$$

$$= (s^3 + 3s^2 + 2.51s + 0.51) + (-s^2 - 0.51s + 0.49) \quad (2\text{-}59)$$

According to Procedures 2-1 and 2-2, we realize the RC voltage transfer function

$$T_{RC}(s) = \frac{N(s)}{D_{RC}(s)} = \frac{1}{s^3 + 3s^2 + 2.51s + 0.51}$$

Procedure 2-1. Selecting a polynomial $(s^2 + 1.6s + 0.48)$ as the $N_{RC}(s)$ in (2-45), we realize the RC impedance function $Z(s) = N_{RC}(s)/D_{RC}(s)$. The $Z(s)$ is expanded as

$$Z(s) = \frac{s^2 + 1.6s + 0.48}{s^3 + 3s^2 + 2.51s + 0.51} = \frac{(s + 0.4)(s + 1.2)}{(s + 0.3)(s + 1)(s + 1.7)}$$

$$= \cfrac{1}{\dfrac{1}{1} + \cfrac{1}{\left(\dfrac{s^2 + 1.6s + 0.48}{s^3 + 2s^2 + 0.91s + 0.03}\right)}}$$

Deriving the generalized Routh array from $N_{RC}/(D_{RC} - N_{RC})$, we determine element values of the ladder network from (2-42) and (2-43) as follows:

$$R_1 = 1, \quad R_2 = 2.5, \quad R_3 = 4.32, \quad R_4 = 9.56$$

$$C_1 = 1, \quad C_2 = 0.76, \quad C_3 = 0.44$$

∎

Procedure 2-2. The coefficients of the summer are determined from (2-48) and (2-49) as follows:

$$M_3 = -3.61$$

∎

Procedure 2-3. We determine the feedback gains. From (2-59) we have

$$f_1 = -1, \quad f_2 = -0.51, \quad f_3 = 0.49$$

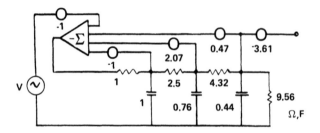

FIGURE 2-6 Realization of $T_G(s)$.

The coefficients of the summer are then determined from (2-57) and (2-58) as follows:

$$F_1 = 0.47, \quad F_2 = 2.07, \quad F_3 = -1$$

The resulting network for $T_G(s)$ is shown in Fig. 2-6.   ∎

    The method of realizing general transfer functions we have described has a degree of freedom in choosing an RC impedance function of the RC network. The realization that gives all the capacitors and all the resistors in the RC ladder network the same value, respectively (which is practically important), is easily obtained by using a proper RC impedance function $Z(s)$ in (2-45). This degree of freedom can also be utilized to obtain a low-sensitivity network by applying optimization techniques [9].

    The procedures mentioned above can also be applied to another RC network structures. However, these RC networks are restricted to a class of controllable networks which contain no cutsets or closed loops consisting of capacitors alone. From the viewpoint of practical use, it is desirable to choose an RC network with all common-grounded capacitors, for example, an RC parallel network of Foster II type which is controllable and also has practically accessible states [13].

2-5 OUTLINE OF REALIZATION METHODS FOR TRANSFER FUNCTION MATRICES VIA STATE VARIABLE METHODS

The design of a multiport network or a multivariable control system often results in a multiport filter or a multivariable compensator, which is described by a transfer function matrix. We focus on a system-theoretical approach that reflects current trends in the research literature for synthesizing transfer function matrices with multiport active RC networks. In this section we outline several studies. For details regarding them, see appropriate sources listed in the references.

Realization of transfer function matrices can easily be accomplished by using multiwinding transformers and active elements [14]. This realization method is, however, a modernization of classical network theory. Work on a synthesis method of multiport active RC networks without using multiwinding transformers and integrators has been performed from a system-theoretical point of view by several researchers [15-21]. In previous works by Stamenkovic [15] and Bichart and Melvin [16], the transfer function matrix has been realized as an admittance matrix of active RC networks, although the realization of voltage transfer function matrices has not been treated. Matsumoto et al. [17] and Shieh et al. [18,19] have extended the state space realization method mentioned in Secs. 2-3 and 2-4 to the case of realizing voltage transfer function matrices.

Matsumoto et al. [17] have proposed a method of synthesizing voltage transfer function matrices with a modified Cauer I RC ladder and summers. This method is based on state feedback theory for multivariable control systems. The movement of zeros of the system, caused by state feedback, is compensated by using input transformation and state feedforward. In this method, however, since for realizing m-input/p-output transfer function matrices with a McMillan degree n, (n + p) summers are used, many summers (i.e., many active elements) are required when the McMillan degree of the system is high. On the other hand, a remedial method [20] has been investigated to overcome this advantage. The features of this method are applicable for the realization of arbitrary transfer function matrices with m inputs and p outputs, and require only (m + p) summers as active elements independent of the McMillan degree.

Shieh et al. [18,19] have proposed elegant methods of transfer function matrix realization without using integrators and multiwinding transformers. They have constructed a block-transformation matrix which transforms a block-state equation from block-companion form to block-tridiagonal form using the conventional or modified matrix Routh algorithm and Routh array. The developed block-state equation in the block-tridiagonal form has been used to realize a square transfer function matrix with a multiport RC ladder network.

In the case where a multivariable system can be decoupled by applying state feedback or output feedback, that is, when the transfer function matrix of the system resulting from the application of state or output feedback can be reduced to a diagonal form, an interesting method of synthesis has been proposed [21]. In this method, based on solving the inverse problem of decoupling, each transfer function of the corresponding scalar system is realized by the synthesis method of a one-port active RC network. Thereafter, a desired transfer function matrix is realized by coupling these scalar systems by again applying state or output feedback.

As mentioned above, we can introduce many results of system theory into the problem of synthesizing multiport active RC networks. There is plenty of room for further development.

REFERENCES

1. B. D. O. Anderson and R. W. Newcomb, "Impedance synthesis via state-space techniques," Proc. IEE, vol. 115, no. 7, pp. 928-936 (1968).

2. L. C. Thomas, "The biquad, Part 1: Some practical design considerations," IEEE Trans. Circuit Theory, vol. CT-18, no. 3, pp. 350-357 (1971).

3. R. Yarlagadda, "An application of tridiagonal matrices to network synthesis," SIAM J. Appl. Math., vol. 16, no. 6, pp. 1146-1162 (1968).

4. P. C. Parks, "A new proof of the Routh-Hurwitz stability criterion using the second method of Lyapunov," Proc. Cambridge Philos. Soc., vol. 58, part 4, pp. 694-702 (1962).

5. C. F. Chen and H. Chu, "A matrix for evaluating Schwarz's form," IEEE Trans. Autom. Control, vol. AC-11, no. 4, pp. 303-305 (1966).

6. H. M. Power, "Canonical form for the matrices of linear discrete-time systems," Proc. IEE, vol. 116, no. 7, pp. 1245-1252 (1969).

7. T. Takahashi, N. Hamada, and S. Takahashi, "System theory considerations on RC driving-point functions," Electron. Commun. Jpn., vol. 59-A, no. 8, pp. 1-7 (1976).

8. N. Hamada and S. Takahashi, "Generalization of Routh algorithm by the second method of Lyanpunov," Electron. Commun. Jpn., vol. 57-A, no. 3, pp. 55-62 (1974).

9. T. Takahashi, N. Hamada, and S. Takahashi, "A state-space realization for transfer functions," IEEE Trans. Circuits Syst., vol. CAS-25, no. 2, pp. 79-88 (1978).

10. F. R. Gantmacher, The Theory of Matrices, Vol. 2, Chelsea, New York, 1971, p. 228.

11. E. I. Jury, Inners and Stability of Dynamic Systems, Wiley, New York, 1974, p. 145.

12. R. W. Brockett, "Poles, zeros and feedback: state space interpretation," IEEE Trans. Autom. Control, vol. AC-10, no. 2, pp. 129-135 (1965).

13. M. Oguchi, T. Takahashi, N. Hamada, and S. Takahashi, "State-variable realization of active RC filters without integrators," Electron. Commun. Jpn., vol. 60-A, no. 11, pp. 37-45 (1977).

14. B. D. O. Anderson and S. Vongpanitlerd, Network Analysis and Synthesis, Prentice-Hall, Englewood Cliffs, N.J., 1973.

15. B. B. Stamenkovic, "Synthesis of n-port active RC network with minimum number of capacitors," Int. J. Circuit Theory Appl., vol. 2, pp. 149-162 (1974).

16. T. A. Bichart and D. W. Melvin, "Synthesis of active RC multiport networks with grounded ports," J. Franklin Inst., vol. 294, no. 5, pp. 289-312 (1972).

17. T. Matsumoto, T. Takahashi, and S. Takahashi, "Synthesis of multiport active RC networks," Electron. Commun. Jpn., vol. 61-A, no. 11, pp. 36-43 (1978).

18. L. S. Shieh and A. Tajvari, "Analysis and synthesis of matrix transfer functions using the new block-state equations in block tridiagonal forms," IEE Proc. Part D, vol. 127, no. 1, pp. 19-31 (1980).

19. L. S. Shieh, S. Yeh, and H. Y. Zhang, "Realization of matrix transfer functions using RC ladders and summers," IEE Proc. Part G, vol. 128, no. 3, pp. 101-110 (1981).

20. K. Hisanaga and S. Takahashi, "Synthesis of voltage transfer function matrix with active RC networks," Proc. IEEE Int. Symp. Circuits and Systems, pp. 610-613, May 1983.

21. J. Tajima, H. Nagase, and S. Takahashi, "Synthesis problem of voltage transfer function matrix," Electron. Commun. Jpn., vol. 61-A, no. 5, pp. 17-25 (1978).

Part II
Interpolation Minimization and Rational Approximation

3
Application of Classical Interpolation Theory

HIDENORI KIMURA

Osaka University, Suita, Osaka, Japan

3-1 INTRODUCTION

Complex function theory has provided the theoretical basis for classical system theory. Many famous results have been established based on complex function theory, such as the realizability condition of Wiener-Payley, the Bode formula concerning the relation between the real and the imaginary parts of minimal phase transfer functions, the Nyquist stability criterion, the Schur-Cohn stability test, and the Fujisawa criterion of ladder network, to name a few. These results, which are concerned primarily with the analytic aspects of the transfer function, do not seem to have received much attention in the far-reaching progress of modern system theory, in which algebraic and/or combinatorial features of systems have been emphasized rather than the analytical aspect. Therefore, the analytical aspects of systems have not been fully exploited beyond the classical results mentioned above, in which only elementary complex function theory in the standard textbooks was involved.

In the late 1960s, however, we had two important contributions in system theory, which created renewed interest in complex function theory at a level beyond university textbooks. One was made by Itakura and Saito, who represented the linear prediction model of speech in terms of the Szegö orthogonal polynomials [1]. The other was by Youla and Saito, who gave a circuit-theoretical proof of the Pick-Nevanlinna interpolation theory [2] in conjunction with the broadband matching problem [3]. The work by Itakura and Saito pioneered a huge amount of subsequent work on statistical signal processing, fast algorithms, and lattice filters, in which classical moment theory originated in the Carathéodory-Féjer problem played an important role. The work by Youla and Saito initiated the application of classical interpolation theory to engineering problems and was elaborated further, from a different point of view, by Belevitch [4] and his school.

In the area of control, dominance of the state space paradigm tended to deemphasize the analytic aspect of systems. Zames was probably the first to address the importance of the analytic aspect of transfer functions in the context of control system design [5]. His initial approach to sensitivity minimization used elementary theory of the Hardy class [6], but later highly advanced operator theory was used to tackle the general optimization problem in the frequency domain [7, 8]. Also, it was found that the problem of robust stabilization had the same mathematical structure as that of sensitivity minimization [9, 10]. It is now widely recognized that the underlying mathematical principle dates back to the classical Pick-Nevanlinna interpolation theory [11, 12].

Model reduction is another area in which the analytic aspects of systems receive considerable attention. The celebrated work by Adamjan et al. [13] made an innovative impact in this area. The results in [13] are based on the mass of literature in classical function theory concerning the generalization of Carathéodory-Féjer problem [14].

As described above, renewed interest in the analytic aspects of systems is generating a new trend in various areas of modern system theory. Mathematics involved in the new analytical trend ranges from elementary function theory to the most advanced operator theory. However, the underlying principles can be traced back to a few fundamental results in classical analysis established about 70 years ago: the Carathédory-Toeplitz characterization of positive functions, the Schur parametrization of bounded functions, and the Pick-Nevanlinna interpolation theory. To become acquainted with these classical results will undoubtedly enhance understanding of the most advanced results in analytic system theory. Fortunately, these classical results are easily understood based on the elementary knowledge of standard complex function theory.

In this chapter we focus on the Pick-Nevanlinna interpolation problem, which is the mathematical basis of H^∞-optimization, robust stabilization, and broadband matching. The first two sections are devoted to a brief survey of the work of Pick and Nevanlinna. Remarks on the connection to the Schur theory will be made occasionally. In Section 3-3 we discuss the algebraic feature of the problem for the purpose of establishing the connection between the preceding two sections. The last section is devoted to a discussion of applications.

In the sequel we made extensive use of the following definitions of the two classes of analytic functions, the class $\mathscr{C}$ (Carathéodory), which is composed of all functions $w = f(z)$ analytic in the unit disk $U = \{z : |z| < 1\}$ and maps U to the right half plane (RHP) $\text{Re}[w] > 0$, and the class $\mathscr{S}$ (Schur), which is composed of all functions $w = f(z)$ analytic in U and maps U to the unit disk $|w| < 1$:

$$\mathscr{C} = \{f(z) : \text{analytic and } \text{Re}[f(z)] \geq 0 \text{ in } U\} \tag{3-1}$$

$$\mathscr{S} = \{f(z) : \text{analytic and } |f(z)| \leq 1 \text{ in } U\} \tag{3-2}$$

A variant $\mathscr{C}_1$ of the class $\mathscr{C}$ defined by

$$\mathscr{C}_1 = \{f(s) : \text{analytic and } \text{Re}[f(s)] \geq 0 \text{ in } \text{Re}[s] > 0\} \tag{3-3}$$

plays an important role in control problems. Also, a variant $\mathscr{S}_1$ of the class $\mathscr{S}$ is defined as

$$\mathscr{S}_1 = \{f(s) : \text{analytic and } |f(s)| \leq 1 \text{ in } \text{Re}[s] > 0\} \tag{3-4}$$

3-2 PICK CRITERION

In 1916, Pick posed the following question [11]: Let $f(z) \in \mathscr{C}$, and assume that $w_i = f(z_i)$, $i = 1, \ldots, n$. What conditions are imposed on the n-pairs of complex numbers (z_i, w_i), $i = 1, \ldots, n$?

This problem is clearly a variant of the celebrated Carathéodory problem: Characterize the class $\mathscr{C}$ in terms of the coefficients c_0, c_1, $\ldots$ of the power series expansion

$$f(z) = \frac{1}{2}c_0 + c_1 z + c_2 z^2 + \cdots \tag{3-5}$$

An elegant solution was given to this problem, which is now called the Carathéodory-Toeplitz theorem.

Carathéodory Toeplitz Theorem. A function $f(z)$ analytic in $U = \{z : |z| < 1\}$ represented in (3-5) belongs to $\mathscr{C}$ if and only if

$$
T_n := \begin{bmatrix} \dfrac{c_0 + \bar{c}_0}{2} & c_1 & \cdots & c_n \\[2mm] \bar{c}_1 & \dfrac{c_0 + \bar{c}_0}{2} & \cdots & c_{n-1} \\[2mm] & & \cdots\cdots\cdots & \\[2mm] \bar{c}_n & \bar{c}_{n-1} & \cdots & \dfrac{c_0 + \bar{c}_0}{2} \end{bmatrix} \geq 0 \qquad (3\text{-}6)
$$

for each n. ∎

It is clear that Pick aimed to derive an analogous condition to (3-6) for his interpolation problem. First, he derived an integral representation of the function $f(z)$ in the disk $|z| < r < 1$ as

$$
f(z) = \frac{1}{2\pi} \int_0^{2\pi} \frac{re^{j\theta} + z}{re^{j\theta} - z} \operatorname{Re}[f(re^{j\theta})]\, d\theta + \frac{j}{2\pi} \int_0^{2\pi} \operatorname{Im}[f(re^{j\theta})]\, d\theta \qquad (3\text{-}7)
$$

This representation is known as the <u>Schwarz's formula,</u> and its derivation is standard [15, p. 168]. From this representation it follows that

$$
\frac{f(z_i) + \overline{f(z_j)}}{r^2 - z_i \bar{z}_j} = \frac{1}{\pi} \int_0^{2\pi} \frac{\operatorname{Re}[f(re^{j\theta})]}{(re^{j\theta} - z_i)(re^{-j\theta} - \bar{z}_j)}\, d\theta
$$

where r is taken sufficiently close to 1 to guarantee $|z_i| < r$ for each i. Thus, for each x_i, $i = 1, \ldots, n$,

$$
\sum_{i,j=1}^n \frac{f(z_i) + \overline{f(z_j)}}{r^2 - z_i \bar{z}_j} x_i \bar{x}_j = \frac{1}{\pi} \int_0^{2\pi} \operatorname{Re}[f(re^{j\theta})] \left| \sum_{i=1}^n \frac{x_i}{re^{j\theta} - z_i} \right|^2 d\theta
$$

Since $f \in \mathscr{C}$, the integrand of the right-hand side is nonnegative. Letting $r \to 1$ verifies the inequality

$$
P := \begin{bmatrix} \dfrac{w_1 + \bar{w}_1}{1 - z_1 \bar{z}_1} & \cdots & \dfrac{w_1 + \bar{w}_n}{1 - z_1 \bar{z}_n} \\[2mm] & \cdots\cdots\cdots & \\[2mm] \dfrac{w_n + \bar{w}_1}{1 - z_n \bar{z}_1} & \cdots & \dfrac{w_n + \bar{w}_n}{1 - z_n \bar{z}_n} \end{bmatrix} \geq 0 \qquad (3\text{-}8)
$$

This is the original solution given by Pick himself, which corresponds to the inequality (3-6). The matrix P defined in (3-8) is usually referred to as the Pick matrix. Now it has been proven that $P \geq 0$ is a necessary condition for the existence of a function $f \in \mathscr{C}$ satisfying the interpolation conditions

$$w_i = f(z_i), \quad i = 1, \ldots, n \tag{3-9}$$

It seems almost obvious that the converse assertion is also valid, that is, if (z_i, w_i), $i = 1, \ldots, n$, satisfy (3-8), there exists a $f \in \mathscr{C}$ which satisfies (3-9). Pick himself did not consider this problem. It is actually far from trivial and an answer was given later by Nevanlinna [12], who derived an algorithm of constructing such f.

Before proceeding to the Nevanlinna construction, we note an essential feature of the condition (3-8) which was fully discussed by Pick. Consider a linear fractional transformation

$$z = \frac{a + b\lambda}{c + d\lambda}, \quad \lambda = \frac{a - cz}{-b + dz} \tag{3-10}$$

from λ to z. This maps a disk on the z-plane to a region on the λ-plane. For instance,

$$z = \frac{1 - \lambda}{1 + \lambda}, \quad \lambda = \frac{1 - z}{1 + z} \tag{3-11}$$

maps the unit disk on the z-plane to the right half plane

$$\lambda + \bar{\lambda} \geq 0$$

It follows that

$$1 - z_i \bar{z}_j = \frac{K(\lambda_i, \bar{\lambda}_j)}{(c + d\lambda_i)(\bar{c} + \bar{d}\bar{\lambda}_j)} \tag{3-12}$$

$$K(\lambda_i, \lambda_j) = (|c|^2 - |a|^2) + (\bar{c}d - \bar{a}b)\lambda_i + (\bar{c}d - \bar{a}b)\bar{\lambda}_j + (|d|^2 - |b|^2)\lambda_i \bar{\lambda}_j$$

Thus the unit disk on the z-plane is mapped into the region

$$K(\lambda, \bar{\lambda}) \geq 0 \tag{3-13}$$

Also, the transformation

$$w = \frac{a_1 + b_1\beta}{c_1 + d_1\beta}, \qquad \beta = \frac{a_1 - c_1 w}{-b_1 + d_1 w} \tag{3-14}$$

maps the right half plane of the w-plane to a region of the β-plane. Similarly, we obtain

$$w_i + \bar{w}_j = \frac{L(\beta_i, \bar{\beta}_j)}{(c_1 + d_1\beta_i)(\bar{c}_1 + \bar{d}_1\bar{\beta}_j)} \tag{3-15}$$

$$L(\beta_i, \bar{\beta}_j) = (a_1\bar{c}_1 + \bar{a}_1 c_1) + (a_1\bar{d}_1 + c_1\bar{b}_1)\bar{\beta}_j + (\bar{c}_1 b_1 + \bar{a}_1 d_1)\beta_i + (b_1\bar{d}_1 + \bar{b}_1 d_1)\beta_i\bar{\beta}_j$$

Thus the right half plane $w + \bar{w} > 0$ is mapped to the region

$$L(\beta, \bar{\beta}) \geq 0 \tag{3-16}$$

From (3-12) and (3-15), it follows that

$$\frac{w_i + \bar{w}_j}{1 - z_i\bar{z}_j} = \frac{(c + d\lambda_i)(\bar{c} + \bar{d}\bar{\lambda}_j)}{(c_1 + d_1\beta_i)(\bar{c}_1 + \bar{d}_1\bar{\beta}_j)} \frac{L(\beta_i, \bar{\beta}_j)}{K(\lambda_i, \bar{\lambda}_j)} \tag{3-17}$$

Therefore, the inequality (3-8) is equivalent to

$$\begin{bmatrix} \dfrac{L(\beta_1, \bar{\beta}_1)}{K(\lambda_1, \bar{\lambda}_1)} & \cdots & \dfrac{L(\beta_1, \bar{\beta}_n)}{K(\lambda_1, \bar{\lambda}_n)} \\ & \cdots\cdots\cdots & \\ \dfrac{L(\beta_n, \bar{\beta}_1)}{K(\lambda_1, \bar{\lambda}_1)} & \cdots & \dfrac{L(\beta_n, \bar{\beta}_n)}{K(\lambda_1, \bar{\lambda}_n)} \end{bmatrix} \geq 0 \tag{3-18}$$

Thus for any function $\beta = f(\lambda)$ which is analytic in the region (3-13) and satisfies (3-9), the interpolation data $\beta_i = f(\lambda_i)$, $i = 1,\ldots,n$, must satisfy the inequality (3-18).

As an example, take $a_1 = c_1 = d_1 = 1$, $b_1 = -1$. Then $\mathrm{Re}[w] \geq 0$ is mapped into the unit disk $|\beta| \leq 1$ by the transformation (3-10). Since $L(\beta_i, \bar{\beta}_j) = 2(1 - \beta_i\bar{\beta}_j)$ in this case, we conclude that a function $f \in \mathcal{B}$ [see (3-2)] satisfies

$$f(z_i) = \beta_i, \qquad i = 1,\ldots,n \tag{3-19}$$

only if

$$P = \begin{bmatrix} \dfrac{1 - \beta_1 \bar{\beta}_1}{1 - z_1 \bar{z}_1} & \cdots & \dfrac{1 - \beta_1 \bar{\beta}_n}{1 - z_1 \bar{z}_n} \\ & \cdots\cdots\cdots & \\ \dfrac{1 - \beta_n \bar{\beta}_1}{1 - z_n \bar{z}_1} & \cdots & \dfrac{1 - \beta_n \bar{\beta}_n}{1 - z_n \bar{z}_n} \end{bmatrix} \geq 0 \qquad (3\text{-}20)$$

Also, we can show analogously that a function $f \in \mathscr{S}_1$ exists satisfying

$$f(\lambda_i) = \beta_i, \qquad \text{Re } \lambda_i > 0, \qquad i = 1, \ldots, n \qquad (3\text{-}21)$$

only if

$$P = \begin{bmatrix} \dfrac{1 - \beta_1 \bar{\beta}_1}{\lambda_1 + \bar{\lambda}_1} & \cdots & \dfrac{1 - \beta_1 \bar{\beta}_n}{\lambda_1 + \bar{\lambda}_n} \\ & \cdots\cdots\cdots & \\ \dfrac{1 - \beta_n \bar{\beta}_1}{\lambda_n + \bar{\lambda}_1} & \cdots & \dfrac{1 - \beta_n \bar{\beta}_n}{\lambda_n + \bar{\lambda}_n} \end{bmatrix} \geq 0 \qquad (3\text{-}22)$$

3-3 NEVANLINNA ALGORITHM

Three years after Pick's work [11], a paper by Nevanlinna [12] came out, in which the converse of the result of Pick was discussed extensively. He worked with $\mathscr{S}$ instead of $\mathscr{C}$ and formulated the problem as follows: What is the necessary and sufficient condition that guarantees the existence of a function $f \in \mathscr{S}$ satisfying

$$\beta_i = f(z_i), \qquad i = 1, \ldots, n \qquad (3\text{-}23)$$

where $|\beta_i| < 1$ and $|z_i| < 1$?

Probably, Nevanlinna did not know the result of Pick at that time because the paper [11] was not quoted in [12] and his approach was totally different from Pick's approach. Instead, the approach of [12] seemed to be strongly influenced by the work of Schur [16], who gave an alternative proof of the Carathéodory-Toeplitz theorem from an essentially different viewpoint. Before stating the Nevanlinna algorithm it will be helpful to mention briefly Schur's approach to the Carathéodory problem.

The Carathéodory problem for the class $\mathscr{S}$ is concerned with the characterization of $\mathscr{S}$ in terms of the power series expansion

$$f(z) = a_0 + a_1 z + a_2 z^2 + \cdots \tag{3-24}$$

If $f \in \mathscr{S}$, then $g = (1 - f)/(1 + f) \in \mathscr{C}$, and vice versa. Therefore, it may be possible to characterize $\mathscr{S}$ by expanding g and using (3-24). Schur took an essentially new approach without appealing to the Carathéodory-Toeplitz characterization (3-6).

If $f \in \mathscr{S}$, it is clear that $|a_0| = |f(0)| \leq 1$. From the maximum modulus theorem, $|a_0| = |f(0)| = 1$ implies that $f(z) \equiv a_0$. To rule out this trivial case, assume that $|a_0| < 1$. The key observation is as follows: $f \in \mathscr{S}$ if and only if

$$g(z) := \frac{f(z) - a_0}{1 - \bar{a}_0 f(z)} \in \mathscr{S} \tag{3-25}$$

This fact is easily seen from the identity

$$1 - |g|^2 = \frac{(1 - |a_0|^2)(1 - |f|^2)}{|1 - \bar{a}_0 f|^2} \tag{3-26}$$

Since $g(0) = 0$, Schwarz's lemma [15, p. 135] implies that $f \in \mathscr{S}$ is equivalent to

$$f_1(z) := \frac{g(z)}{z} = \frac{f(z) - a_0}{z[1 - \bar{a}_0 f(z)]} \in \mathscr{S} \tag{3-27}$$

Thus the problem is reduced to checking whether

$$f_1(z) = \frac{a_1 + a_2 z + \cdots}{1 - \bar{a}_0 (a_0 + a_1 z + \cdots)} = a_0' + a_1' z + \cdots \tag{3-28}$$

$$a_0' = \frac{a_1}{1 - |a_0|^2} \tag{3-29}$$

belongs to $\mathscr{S}$. A necessary condition for $f_1 \in \mathscr{S}$ is $|f_1(0)| = |a_0'| \leq 1$. By the same reasoning as before, $|a_0| = 1$ implies that $f_1(z) \equiv a_0$. If $|a_0'| < 1$, consider a function $f_2 = (f_1 - a_0')/z(1 - \bar{a}_0' f_1)$ and repeat the same procedure.

This procedure is formally represented as

$$f_{\nu+1}(z) = \frac{f_\nu(z) - \tau_\nu}{z[1 - \bar{\tau}_\nu f_\nu(z)]}, \qquad \tau_\nu = f_\nu(0) \tag{3-30}$$

$$f_0(z) = f(z)$$

Based on this recursion, it is concluded that $f \in \mathscr{S}$ if and only if either
(1) $|\tau_\nu| < 1$ for each ν, or (2) $|\tau_\nu| < 1$, $\nu = 0, 1, \ldots, n - 1$ and $|\tau_n| = 1$
for some n. In the latter case, the back substitution of the identity (3-30)
starting from $f_n(z) = \tau_n$ verifies that f must be a rational function of degree n.
This is the core of Schur's recursion. Here the essential idea lies in the
parametrization of the power series (3-24) in terms of the parameters τ_i
whose modulus are not greater than 1. In [16], many important results were
derived, including the celebrated Schur-Cohn stability test based on the
parametrization by τ_i. Here we only note the fundamental result that a
power series (3-24) is in $\mathscr{S}$ if and only if, for each ν,

$$I - \bar{A}_\nu^T A_\nu \geq 0 \tag{3-31}$$

where A_ν is given by

$$A_\nu = \begin{bmatrix} a_0 & a_1 & a_\nu & \cdots & a_\nu \\ 0 & a_0 & a_1 & \cdots & a_{\nu-1} \\ & & \cdots\cdots & & \\ 0 & 0 & 0 & & a_0 \end{bmatrix}$$

Now we go back to the Nevanlinna algorithm. The basic idea of Nevanlinna was to use Schwarz's lemma at an arbitrary point.

Schwarz's Lemma. Assume that $f(z) \in \mathscr{S}$ and $\beta_1 = f(z_1)$, $|z_1| < 1$.
Then for any z satisfying

$$\left| \frac{z - z_1}{1 - \bar{z}_1 z} \right| = r \quad (0 < r < 1)$$

the following inequality holds:

$$\left| \frac{f(z) - \beta_1}{1 - \bar{\beta}_1 f(z)} \right| \leq r$$

∎

The special case $z_1 = \beta_1 = 0$ in the lemma above is usually referred to
as the Schwarz lemma in textbooks of complex function theory.
A direct application of this lemma proves that if $\beta_1 = f(z_1)$ and $f \in \mathscr{S}$
then

$$f_1(z) := \frac{1 - \bar{z}_1 z}{z - z_1} \frac{f(z) - \beta_1}{1 - \bar{\beta}_1 f(z)} \in \mathscr{S} \tag{3-32}$$

Solving the identity above with respect to f yields

$$f(z) = \frac{B_1(z)f_1(z) + \beta_1}{\bar{\beta}_1 B_1(z)f_1(z) + 1}, \qquad B_1(z) = \frac{z - z_1}{1 - \bar{z}_1 z} \tag{3-33}$$

This expression is of the same form as (3-27) if f and a_0 in (3-27) are identified with $B_1(z)f_1(z)$ and $-\beta_1$, respectively. Since $B_1 \in \mathcal{S}$, (3-33) implies
that $f \in \mathcal{S}$. Therefore, $f_1 \in \mathcal{S}$ if and only if $f \in \mathcal{S}$. In the inverse relation
(3-33), $f(z_1) = \beta_1$ is always satisfied irrespective of the selection of f_1, be-
cause $B_1(z_1) = 0$. From the relations (3-23) and (3-32), it is obvious that
the interpolation conditions (3-23) hold if and only if $f_1(z_1)$ satisfies

$$f_1(z_i) = \beta_i^{(2)}, \qquad i = 2, \ldots, n \tag{3-34}$$

$$\beta_i^{(2)} = \frac{1 - \bar{z}_1 z_i}{z_i - z_1} \frac{\beta_i - \beta_1}{1 - \bar{\beta}_1 \beta_i} \tag{3-35}$$

Thus the problem has been reduced to a simpler one, in which the number
of interpolation conditions is one less than the original problem. If a $f_1(z)$
solves the reduced problem, f(z) given by (3-33) solves the original problem.
    Obviously, the inequalities

$$|\beta_i^{(2)}| < 1, \qquad i = 2, \ldots, n \tag{3-36}$$

must be satisfied if the reduced problem (3-34) is solvable. If $|\beta_m^{(2)}| = 1$ for
some m, the maximum modulus theorem implies that $f_1(z) \equiv \beta_m^{(2)}$ is the only
function that satisfies $f_1(z_m) = \beta_m^{(2)}$. Therefore, $\beta_2^{(2)} = \beta_3^{(2)} = \cdots = \beta_n^{(2)}$ must
hold in this case for the solvability of the original interpolation problem. In
that case, substitution of $f_1(z) = \beta_2^{(2)}$ in (3-33) yields

$$f(z) = \frac{\beta_2^{(2)} B_1(z) + \beta_1}{1 + \bar{\beta}_1 \beta_2^{(2)} B_1(z)} \tag{3-37}$$

which is the only function in $\mathcal{S}$ satisfying (3-23). If $|\beta_i^{(2)}| < 1$, $i = 2, \ldots, n$,
we can apply the same procedure to the reduced interpolation problem (3-34).
    The procedure above is repeated to check the solvability of the inter-
polation problem (3-23). Actually, this amounts to calculating the Nevanlinna
parameters $\rho_1, \rho_2, \ldots, \rho_n$, defined later, based on which a solution is computed by a series of back substitutions like (3-33). Thus the algorithm of
solving the interpolation problem (3-23) is divided into the two parts. The

first, called the <u>forward algorithm</u>, is to check the solvability of the problem by calculating the Nevanlinna parameters. The second, the <u>backward algorithm</u>, is to compute the solution based on the Nevanlinna parameters.

Forward Algorithm

  (i) Initialization

$$\beta_i^{(1)} = \beta_i, \quad i = 1, \ldots, n$$

  (ii) Assume that $\{\beta_k^{(k)}, \beta_{k+1}^{(k)}, \ldots, \beta_n^{(k)}\}$ are given such that $|\beta_i^{(k)}| < 1$, $k < i < n$ hold. If $k = n$, stop; otherwise, compute

$$\beta_i^{(k+1)} = \frac{1 - \bar{z}_k z_i}{z_1 - z_k} \frac{\beta_i^{(k)} - \beta_k^{(k)}}{1 - \bar{\beta}_k^{(k)} \beta_i^{(k)}}, \quad i = k + 1, \ldots, n \qquad (3\text{-}38)$$

  (iii) If $|\beta_i^{(k+1)}| < 1$, $k + 1 \le i \le n$, go to (ii). Otherwise, stop.

After carrying out this algorithm, we have either a complete array

$$
\begin{array}{cccccc}
\beta_1^{(1)} & \beta_2^{(1)} & \beta_3^{(1)} & \cdots & \beta_n^{(1)} & \\
& \beta_2^{(2)} & \beta_3^{(2)} & \cdots & \beta_n^{(2)} & \\
& & & \cdots\cdots\cdots & & \\
& & & & \beta_n^{(n)} &
\end{array} \qquad (3\text{-}39)
$$

satisfying $|\beta_i^{(k)}| < 1$, $k < i < n$; $1 \le k \le n$, or an incomplete array

$$
\begin{array}{ccccc}
\beta_1^{(1)} & \beta_2^{(1)} & \ldots\ldots & \beta_n^{(1)} & \\
& \beta_2^{(2)} & \ldots\ldots & \beta_n^{(2)} & \\
& & \ldots\ldots\ldots & & \\
& & \beta_k^{(k)} & \cdots & \beta_n^{(k)}
\end{array} \qquad (3\text{-}40)
$$

where $|\beta_i^{(k)}| \ge 1$ for some $k \le i \le n$. It should be noted that for each row of

these arrays, there corresponds a reduced interpolation problem characterized by

$$f_k(z_i) = \beta_i^{(k)}, \quad i = k, \ldots, n \qquad\qquad (3\text{-}41)$$

The parameters $\rho_j := \beta_j^{(j)}$, $k = 1, \ldots, n$, play important roles in the
recursion (3-38). Actually, it is easily seen that the completed array (3-39)
is obtained if and only if

$$|\rho_j| < 1, \quad j = 1, \ldots, n \qquad\qquad (3\text{-}42\text{a})$$

In the case of incomplete array (3-40), the reduced problem (3-41) is solv-
able if and only if

$$|\rho_j| < 1, \quad j = 1, \ldots, k-1$$

$$|\rho_k| = 1, \quad \rho_k = \beta_k^{(k)} = \beta_{k+1}^{(k)} = \cdots = \beta_n^{(k)} \qquad\qquad (3\text{-}42\text{b})$$

because $f_k(z) \equiv \rho_k$ is the only function in $\mathscr{S}$ satisfying (3-41). In the case
(3-42a), we exhaust all the interpolation constraints at the step n, and hence
$f_{n+1}(z)$ can be chosen freely from $\mathscr{S}$. A series of back substitutions like
(3-33) will give a solution to the original interpolation problem (3-23). In
the case (3-42b), we simply take $f_k(z) = \rho_k$ and use the same back substitution.

Backward Algorithm

    (i) Initialization. Take an arbitrary $f_{n+1}(z) \in \mathscr{S}$ for the case (3-42a)
        and $f_k(z) = \rho_k$ for the case (3-42b).
    (ii) Recursion

$$f_\nu(z) = \frac{B_\nu(z) f_{\nu+1}(z) + \rho_\nu}{\bar{\rho}_\nu B_\nu(z) f_{\nu+1}(z) + 1}, \qquad B_\nu(z) = \frac{z - z_\nu}{1 - \bar{z}_\nu z} \qquad (3\text{-}43)$$

    $f_1(z) = f(z)$ is in $\mathscr{S}$ and satisfies (3-23).

The inequalities (3-42) give a solvability criterion of the Pick-
Nevanlinna problem. In the case (3-42a), infinitely many solutions exist
because $f_{n+1}(z)$ can be chosen freely from $\mathscr{S}$, while in the case (3-42b), the
Pick-Nevanlinna problem has the unique solution which is a rational function
of degree $k-1$.

The recursion (3-43) can be written as

$$f_{\nu}(z) = \frac{n_{\nu}(z)}{d(z)}, \quad \begin{bmatrix} n_{\nu}(z) \\ d(z) \end{bmatrix} = \theta_{\nu}(z) \begin{bmatrix} f_{\nu+1}(z) \\ 1 \end{bmatrix} \tag{3-44}$$

$$\theta_{\nu}(z) = \frac{1}{\sqrt{1 - |\rho_{\nu}|^2}} \begin{bmatrix} 1 & \rho_{\nu} \\ \bar{\rho}_{\nu} & 1 \end{bmatrix} \begin{bmatrix} B_{\nu}(z) & 0 \\ 0 & 1 \end{bmatrix} \tag{3-45}$$

Since $B_{\nu}(z) \bar{B}_{\nu}(\bar{z}^{-1}) = 1$, we notice that

$$\bar{\theta}_{\nu}^T(\bar{z}^{-1}) \begin{bmatrix} 1 & 0 \\ 0 & -1 \end{bmatrix} \theta_{\nu}(z) = \begin{bmatrix} 1 & 0 \\ 0 & -1 \end{bmatrix} \tag{3-46}$$

This implies that $\theta_{\nu}(z)$ is <u>J-unitary</u>. The recursion (3-44) yields the representation of the solution $f(z)$ in the case (3-42a) as

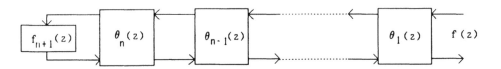

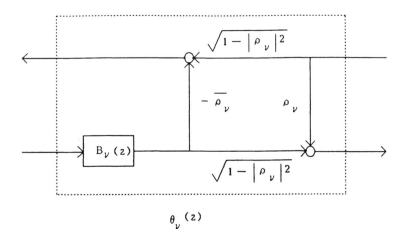

$$\theta_{\nu}(z)$$

FIGURE 3-1 Lattice structure.

$$f(z) = \frac{\theta_{11}(z)f_{n+1}(z) + \theta_{12}(z)}{\theta_{21}(z)f_{n+1}(z) + \theta_{22}(z)}$$

$$\begin{bmatrix} \theta_{11}(z) & \theta_{12}(z) \\ \theta_{21}(z) & \theta_{22}(z) \end{bmatrix} = \theta_1(z)\,\theta_2(z) \cdots \theta_n(z)$$

$$(3\text{-}47)$$

where $f_{n+1}(z)$ is an arbitrary function in $\mathscr{S}$. The representation (3-47) is described schematically in Fig. 3-1, which is a generalization of the lattice filter.

One notices the close resemblance between the Nevanlinna algorithm described above and the Schur algorithm. Nevanlinna's criterion (3-42) is essentially different from Pick's criterion (3-8). The relation between these two criteria is the subject of the next section.

3-4 CHOLESKY FACTORIZATION OF PICK MATRIX

In the previous sections, two different criteria (3-8) and (3-42) were given for the existence of a function in $\mathscr{S}$ satisfying (3-23). These two look very different in many respects, and a question naturally arises: How is one related to the other? The answer to the question is simple but very interesting: that is, the Nevanlinna algorithm discussed in Sec. 3-3 is essentially equivalent to the Cholesky factorization of Pick matrix (3-8). The verification of this fact is the subject of this section.

We start with the identity

$$(1 - z_i \bar{z}_1)(1 - z_1 \bar{z}_j) = (1 - \bar{z}_1 z_1)(1 - z_i \bar{z}_j) + (z_i - z_1)(\bar{z}_j - \bar{z}_1) \qquad (3\text{-}48)$$

Replacing z's in (3-48) by β's yields

$$(1 - \beta_i \bar{\beta}_1)(1 - \beta_1 \bar{\beta}_j) = (1 - \bar{\beta}_1 \beta_1)(1 - \beta_i \bar{\beta}_j) + (\beta_i - \beta_1)(\bar{\beta}_j - \bar{\beta}_1) \qquad (3\text{-}49)$$

Based on these identities, it is not difficult to see that

$$(1 - \beta_i \bar{\beta}_j)(1 - \beta_1 \bar{\beta}_1)(1 - z_i \bar{z}_1)(1 - z_1 \bar{z}_j) - (1 - \beta_i \bar{\beta}_1)(1 - \beta_1 \bar{\beta}_j)(1 - z_i \bar{z}_j)(1 - z_1 \bar{z}_1)$$

$$= (1 - \beta_i \bar{\beta}_1)(1 - \beta_1 \bar{\beta}_j)(z_i - z_1)(\bar{z}_j - \bar{z}_1) - (\beta_i - \beta_1)(\bar{\beta}_j - \bar{\beta}_1)(1 - z_i \bar{z}_1)(1 - z_1 \bar{z}_j)$$

$$(3\text{-}50)$$

Let p_{ij} be the (i, j) element of the Pick matrix (3-20), that is,

$$p_{ij} = \frac{1 - \beta_i \bar{\beta}_j}{1 - z_i \bar{z}_j}, \quad i, j = 1, \ldots, n \tag{3-51}$$

Also, let p_{ij} be the corresponding expression for the reduced problem (3-34), that is,

$$p'_{ij} = \frac{1 - \beta_i^{(2)} \bar{\beta}_j^{(2)}}{1 - z_i \bar{z}_j}, \quad i, j = 2, \ldots, n$$

$$= \frac{1}{1 - z_i \bar{z}_j} \left(1 - \frac{1 - \bar{z}_1 z_i}{z_i - z_1} \frac{1 - z_1 \bar{z}_j}{\bar{z}_j - \bar{z}_1} \frac{\beta_i - \beta_1}{1 - \bar{\beta}_1 \beta_i} \frac{\bar{\beta}_j - \bar{\beta}_1}{1 - \beta_1 \bar{\beta}_j} \right) \tag{3-52}$$

Using the identity (3-50), we have

$$u_i p'_{ij} \bar{u}_j = p_{ij} - p_{i1} p_{11}^{-1} p_{1j}, \quad 2 \leq i \leq n, \quad 2 \leq j \leq n \tag{3-53}$$

$$u_i := \frac{1 - \beta_i \bar{\beta}_1}{\sqrt{1 - |\beta_1|^2}} \frac{z_i - z_1}{1 - \bar{z}_1 z_i} \tag{3-54}$$

Here, we used the assumption $|\beta_1| < 1$. The relation (3-53) implies that the Pick matrix

$$P' = (p'_{ij}), \quad 2 \leq i \leq n, \quad 2 \leq j \leq n \tag{3-55}$$

for the reduced problem (3-34) is congruent to a Schur complement of the original Pick matrix P in (3-20). More precisely, writing

$$P = \begin{bmatrix} p_{11} & \bar{v}^T \\ v & Q \end{bmatrix}$$

we obtain

$$\begin{bmatrix} \dfrac{1}{\sqrt{p_{11}}} & 0 \\ \dfrac{-v}{p_{11}} & I_{n-1} \end{bmatrix} P \begin{bmatrix} \dfrac{1}{\sqrt{p_{11}}} & \dfrac{-\bar{v}^T}{p_{11}} \\ 0 & I_{n-1} \end{bmatrix} = \begin{bmatrix} 1 & 0 \\ 0 & UP'\bar{U}^T \end{bmatrix}$$

where $U = \text{diag}(u_2 \ u_3 \ \cdots \ u_n)$. The relation above is exactly the first step of the Cholesky factorization of P. The Pick matrix P' of the reduced problem can be reduced further if the (1, 1) element of P' is positive, that is,

$$\frac{1 - \beta_2^{(2)} \bar{\beta}_2^{(2)}}{1 - |z_2|^2} > 0 \tag{3-56}$$

The condition above is equivalent to $|\beta_2^{(2)}| = |\rho_2| < 1$. Thus the Cholesky factorization is completed if and only if (3-42a) holds. Now it is clear that P > 0 is equivalent to (3-42a). If the condition (3-42b) is satisfied, then, at the kth step of the Cholesky factorization, the Schur complement vanishes. Hence we conclude that $P \geq 0$ in this case. Now we have established the complete link between the Pick criterion in terms of Pick matrix and the solvability criterion in terms of the Nevanlinna parameters.

3-5 APPLICATIONS

This section is devoted to the application of classical interpolation theory to circuits and control.

3-5-1 Broadband Matching

Consider the simple circuit of Fig. 3-2, consisting of a source E with internal resistance r_g and load r_ℓ. The current I is given by

$$I = \frac{E}{r_g + r_\ell}$$

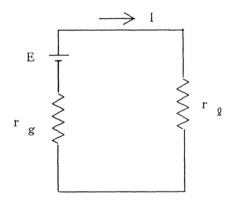

FIGURE 3-2 A simple electrical circuit.

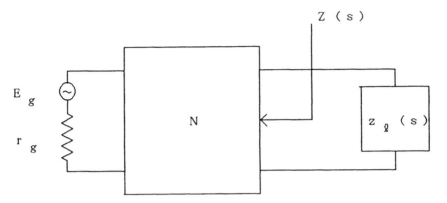

FIGURE 3-3 Impedance matching.

Therefore, the power absorbed by the load is calculated to be

$$P = r_\ell I^2 = \frac{r_\ell}{(r_g + r_\ell)^2} E^2$$

$$= \left[1 - \left(\frac{r_g - r_\ell}{r_g + r_\ell} \right)^2 \right] \frac{E^2}{4r_g}$$

(3-57)

This implies that the maximum power is absorbed by the load when the load is "matched" to the source (i.e., $r_\ell = r_g$) and the maximum absorbed power is given by $P_{max} = E^2/4r_g$.

In the case where the load impedance is frequency dependent and is fixed, we can construct a network N which transfers the maximum power from the source to the load over a wide frequency band range. This is the problem of broadband matching [3, 17, 18]. The problem is depicted schematically in Fig. 3.3. The network N, which is usually called an equalizer, is assumed to be lossless and is specified by the output impedance Z(s).

The ratio of the power P absorbed by the load impedance $z_\ell(s)$ to the maximum power P_{max} absorbed by the matched impedance at a frequency ω is calculated to be

$$\frac{P}{P_{max}} = 1 - \left| \frac{Z(j\omega) - \bar{z}_\ell(j\omega)}{Z(j\omega) + z_\ell(j\omega)} \right|^2$$

(3-58)

This generalizes the identity (3-57), in which $z_\ell(s)$ is constant (resistive). The derivation of (3-58) is found in [17].

Let

$$u(s) = \frac{Z(s) - \bar{z}_{\ell}(-\bar{s})}{Z(s) + z_{\ell}(s)} \tag{3-59}$$

Our task is to choose a passive impedance $Z(s)$ for a given $z_{\ell}(s)$ such that $|u(j\omega)|$ is as small as possible for a given frequency range. It is well known that a function $Z(s)$ represents an impedance of a passive network if and only if $Z \in \mathcal{C}_1$ [see (3-3) for the definition of $\mathcal{C}_1$]. If we can choose $Z(s) = \bar{z}_{\ell}(-\bar{s})$, $u(s) \equiv 0$ and $P = P_{max}$ for each ω. This ideal situation cannot be realized for nonconstant $z_{\ell}(s)$, because $\bar{z}_{\ell}(-\bar{s})$ is not in $\mathcal{C}_1$.

To represent the constraint imposed on the realizable $u(s)$ in (3-59), let $\{\mu_1, \mu_2, \ldots, \mu_m\}$ be the set of poles of $z_{\ell}(s)$. It is clear from the form (3-59) that $s = -\bar{\mu}_i$, $i = 1, \ldots, m$, are the unstable poles of $u(s)$. Define

$$b(s) = \left[\frac{(s - \mu_1)(s - \mu_2) \cdots (s - \mu_m)}{(s + \bar{\mu}_1)(s + \bar{\mu}_2) \cdots (s + \bar{\mu}_m)} \right]^{-1} \tag{3-60}$$

Then it is obvious that

$$v(s) = b(s)u(s) \tag{3-61}$$

is stable (i.e., analytic in $\mathrm{Re}[s] \geq 0$). Since $b(s)\bar{b}(-\bar{s}) = 1$, we have

$$1 - v(s)\bar{v}(-\bar{s}) = 1 - u(s)\bar{u}(-\bar{s})$$

$$= \frac{[Z(s) + \bar{Z}(-\bar{s})][z_{\ell}(s) + \bar{z}_{\ell}(-\bar{s})]}{[Z(s) + z_{\ell}(s)][\bar{Z}(-\bar{s}) + \bar{z}_{\ell}(-\bar{s})]}$$

Since $\mathrm{Re}[Z(j\omega)] > 0$ and $\mathrm{Re}[z_{\ell}(j\omega)] > 0$, we have $|v(j\omega)| \leq 1$. This implies that

$$v \in \mathcal{S}_1 \tag{3-62}$$

Let λ_i, $i = 1, \ldots, n$, be the numbers satisfying

$$z_{\ell}(\lambda_i) + \bar{z}_{\ell}(-\bar{\lambda}_i) = 0, \qquad \mathrm{Re}[\lambda_i] > 0$$

Due to (3-59), $u(\lambda_i) = 1$, and hence

$$v(\lambda_i) = b(\lambda_i), \qquad i = 1, \ldots, n \tag{3-63}$$

for any choice of $Z(s)$. The conditions (3-62) and (3-63) must be satisfied by $v(s)$.

Conversely, if we can choose $v \in \mathscr{S}_1$ satisfying (3-63), we can find $Z(s)$ according to the formula

$$Z(s) = \frac{[z_\ell(s) + \bar{z}_\ell(-\bar{s})]b(s)}{b(s) - v(s)} - z_\ell(s)$$

which is obtained by solving the relation (3-59) with respect to $Z(s)$. Therefore, the problem is reduced to finding $v(s) \in \mathscr{S}_1$ satisfying (3-63). This is exactly a Nevanlinna-Pick interpolation problem. Actually, we would like to find $u(s)$ of the form (3-59) in which $\max_\omega |u(j\omega)|$ is as small as possible. This is equivalent to finding $v(s)$ which is analytic in $\mathrm{Re}[s] \geq 0$ and $\max_\omega |v(j\omega)|$ is as small as possible. The Nevanlinna-Pick interpolation theory will immediately give a complete answer to this problem.

Example 3-1

The load impedance $z_\ell(s)$ given in Fig. 3-4 is calculated to be

$$z_\ell(s) = R_1 + \frac{1}{(1/R_2) + sC} = \frac{R_1 + R_2 + sCR_1R_2}{1 + sCR_2}$$

Hence $\bar{z}_\ell(-\bar{s})$ has an unstable pole at $s = -\bar{\mu}_1 = 1/CR_2$. On the other hand, the equation

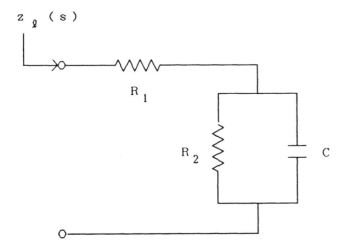

FIGURE 3-4 A simple electric circuit.

$$z_\ell(s) + \bar{z}_\ell(-\bar{s}) = 0$$

has a solution $\lambda_1 = \sqrt{(1 + R_2/R_1)}/CR_2$. We see that

$$\rho_1 = b(\lambda_1) = \frac{\lambda_1 - (1/CR_2)}{\lambda_1 + (1/CR_2)} = \frac{\sqrt{1 + R_2/R_1} - 1}{\sqrt{1 + R_2/R_1} + 1}$$

Our task is to find a function $v(s) \in \mathcal{S}_1$ such that $v(\lambda_1) = \rho_1$.

According to the Nevanlinna algorithm, such a function is given by

$$v(s) = \frac{b_1(s)f_2(s) + \rho_1}{\rho_1 b_1(s)f_2(s) + 1}, \qquad b_1(s) = \frac{s - \lambda_1}{s + \lambda_1} \qquad (3\text{-}64)$$

where $f_2(s)$ is an arbitrary function in $\mathcal{S}_1$. This formula is based on (3-33), where $B_1(z)$ in (3-33) is replaced by $b_1(s)$ in (3-64) because we are working with $\mathcal{S}_1$ instead of $\mathcal{S}$.

3-5-2 Robust Stabilization

In the design of control systems, we are faced with the uncertainty of the model of the plant originated from various sources, such as identification error, the nonlinearity of the plant dynamics, the simplification of the model for the purpose of controller design, the variation of the plant parameters during the operation, and so on. Robust stabilization is concerned with the synthesis of a stabilizing controller that tolerates such model uncertainties.

Consider the closed-loop system of Fig. 3-5, where $p(s)$ denotes the transfer function of the single-input, single-output plant and $c(s)$ the transfer function of the controller. We assume that $p(s)$ belongs to a class of transfer functions which represents the uncertainty of the plant dynamics. A natural class of describing the uncertainty is the set of rational functions $p(s)$ satisfying

    (i) $p(s)$ has the same number of the unstable poles as that of $p_0(s)$.

    (ii) $|p(j\omega) - p_0(j\omega)| \le |r(j\omega)| \quad \forall \omega$.

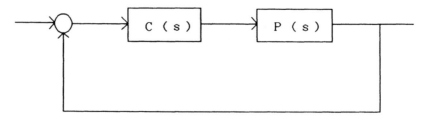

FIGURE 3-5 A unity feedback system.

In the characterization above, $p_0(s)$ denotes the nominal model of the plant dynamics and $r(s)$ characterizes the uncertainty of the nominal plant model. We can assume, without loss of generality, that $r(s)$ is stable. The set of $p(s)$ satisfying (i) and (ii) is denoted by $C(p_0(s), r(s))$, which is usually referred to as the class of <u>additively perturbed plants</u>. A controller $c(s)$ that stabilizes the closed-loop system of Fig. 3-5 for each $p(s) \in C(p_0(s), r(s))$ is called a <u>robust stabilizer of</u> $C(p_0(s), r(s))$. Our problem is to find a condition on $p_0(s)$ and $r(s)$ for which a robust stabilizer of $C(p \in (s), r(s))$ exists. We shall show that this problem is reduced to a classical interpolation problem.

For a given controller $c(s)$, we define

$$q(s) = \frac{c(s)}{1 + p_0(s)c(s)} \tag{3-65}$$

This function was introduced by Zames and Francis [19] to characterize the class of stabilizing controller. They showed that the closed-loop system of Fig. 3-5 is stable for $p(s) = p_0(s)$ if and only if

(i) $q(s)$ is stable.

(ii) $1 - p_0(s)q(s)$ has the zeros at the unstable poles of $p_0(s)$, multiplicity included.

Taking the set of vector loci of $C(p_0(s), r(s))$ and using the Nyquist stability criterion, one can easily show that a stabilizing controller $c(s)$ for the nominal plant $p_0(s)$ is also a robust stabilizer of $C(p_0(s), r(s))$ if and only if

(iii) $|r(j\omega)q(j\omega)| < 1 \quad \forall \omega$ \hfill (3-66)

This inequality was apparently first derived by Doyle [20]. Therefore, we conclude that there exists a robust stabilizer of $C(p_0(s), r(s))$ if and only if there exists a stable $q(s)$ satisfying (ii) and (iii).

The problem above is reduced to the interpolation problem discussed in previous sections [9]. Let $\lambda_1, \lambda_2, \ldots, \lambda_n$ be unstable poles of the nominal plant $p_0(s)$ satisfying $\text{Re}[\lambda_i] > 0$, $i = 1, \ldots, n$. We assume that λ_i are all distinct. Define

$$B(s) = \frac{(s - \lambda_1)(s - \lambda_2) \cdots (s - \lambda_n)}{(s + \bar{\lambda}_1)(s + \bar{\lambda}_2) \cdots (s + \bar{\lambda}_n)} \tag{3-67}$$

From the assumption

$$\tilde{p}_0(s) = p_0(s) B(s)$$

is stable, because all the unstable poles λ_i of $p_0(s)$ are canceled out by the zeros of $B(s)$. Since $q(s)$ has unstable zeros at λ_i, we see that

$$\tilde{q}(s) = \frac{q(s)}{B(s)}$$

is stable. Write

$$\psi(s) = r(s)\tilde{q}(s)$$

Since $r(s)$ is stable, $\psi(s)$ is stable. From $p_0(s)q(s) = \tilde{p}_0(s)\tilde{q}(s) = (\tilde{p}_0(s)/r(s))\psi(s)$, condition (ii) is written as

$$\psi(\lambda_i) = \beta_i, \quad i = 1, \ldots, n \tag{3-68}$$

$$\beta_i = \frac{r(\lambda_i)}{\tilde{p}_0(\lambda_i)} \tag{3-69}$$

Also, since $|B(j\omega)| = 1$ for each ω, condition (iii) can be written as

$$|\psi(j\omega)| < 1 \tag{3-70}$$

The problem of finding a stable $\psi(s)$ satisfying (3-68) and (3-70) is exactly the interpolation problem discussed in previous sections, that is, the problem of finding a function $\psi \in \mathcal{S}_1$ satisfying the interpolation condition (3-68). The necessary and sufficient condition for the existence of such ψ is given in terms of the Pick matrix in (3-22). Characterizing all the robust stabilizers of $C(p_0(s), r(s))$ is equivalent to parametrizing all the solutions of the interpolation problem above, which has been discussed extensively in the previous sections.

Example 3-2

Consider the problem of robust stabilization of the class $C(p_0(s), r(s))$ with

$$p_0(s) = \frac{1}{s^2 - 2s + 2}, \quad r(s) = \frac{b}{s + 1}$$

We shall find the maximum of the uncertainty band b for which a robust stabilizer exists. Since $\lambda_1 = 1 + j$ and $\lambda_2 = 1 - j$, we have

$$B(s) = \frac{s^2 - 2s + 2}{s^2 + 2s + 2}, \quad \tilde{p}_0(s) = \frac{1}{s^2 + 2s + 2}$$

Due to (3-69), we have

$$\beta_1 = \frac{4}{5} b(3 + j), \qquad \beta_2 = \frac{4}{5} b(3 - j)$$

It is straightforward to see that (3-22) holds if and only if

$$0 < b < 0.289578$$

3-6 CONCLUSIONS

This chapter has been devoted to a brief historical review of classical interpolation theory and its applications to circuits and control. The algebraic
aspects of the problem have been exploited fully to establish the analytic
structure of the problem. The parallel between the Nevanlinna-Pick and the
Carathéodory-Féjer interpolation problems has been demonstrated. It is
expected that the reader can taste the flavor of the elegance and the trans-
parency of the really classic (not old-fashioned) theory that has been an
active source of research for more than 70 years and is being revived in
modern engineering.

This chapter has been confined to the classical results. Many important
extensions of these results have been omitted. We mention here only that the
extension to matrix cases is especially important for applications in control
theory [21, 22]. We have also omitted H^∞-control theory, one of the most
important applications of classical interpolation theory [23].

REFERENCES

1. F. Itakura and S. Saito, "Dilital filtering techniques for speech analysis
   and synthesis," Proc. 7th Int. Congr. Acoustics, Budapest, Paper
   25-C-1, pp. 261-264 (1971).

2. D. C. Youla and M. Saito, "Interpolation with positive real functions,"
   J. Franklin Inst., vol. 284, pp. 77-108 (1967).

3. R. M. Fano, "Theoretical limitations on the broadband matching of
   arbitrary impedances," J. Franklin Inst., vol. 249, pp. 57-83 (1960).

4. V. Belevitch, Classical Network Theory, Holden-Day, San Francisco,
   1968.

5. G. Zames, "Feedback and optimal sensitivity: model reference transformation, multiplicative seminorms, and approximate inverses," IEEE
   Trans. Autom. Control, vol. AC-26, pp. 301-320 (1981).

6. P. L. Duren, Theory of H^p Spaces, Academic Press, New York, 1970.

7. B. A. Francis and G. Zames, "On H^∞-optimal sensitivity theory for SISO feedback systems," IEEE Trans. Autom. Control, vol. AC-29, pp. 9-16 (1984).

8. B. A. Francis, J. W. Helton, and G. Zames, "H^∞-optimal controllers for linear multivariable systems," IEEE Trans. Autom. Control, vol. AC-29, pp. 888-900 (1984).

9. H. Kimura, "Robust stabilizability for a class of transfer functions," IEEE Trans. Autom. Control, vol. AC-29, pp. 788-793 (1984).

10. M. Vidyasagar and H. Kimura, "Robust controllers for uncertain linear multivariable systems," Automatica, vol. 22, pp. 85-94 (1986).

11. G. Pick, "Uber die beschränkungen analytischer Funktionen, welche durch vorgegebene Funktionswerte bewirkt werden," Math. Ann., vol. 77, pp. 7-23 (1916).

12. R. Nevanlinna, "Uber beschrankte Funktionen die in Gegebenen Punkten vorgeschreibene Funktionswerte bewirkt werden," Ann. Acad. Sci. Fenn., Ser. A, vol. 13 (1919).

13. V. M. Adamjan, D. Z. Arov, and M. G. Krein, "Analytic properties of Schmidt pairs for a Hankel operator and the generalized Schur-Takagi problem," Math. USSR-Sb., vol. 1t, pp. 31-71 (1971).

14. C. Carathéodory, "Uber den Variabilitätsbereich der Koeffizienten von Potenzreihen, die gegebene Werte nicht annehmen," Math. Ann., vol. 64, pp. 95-115 (1907).

15. L. V. Ahlfors, Complex Analysis, McGraw-Hill, New York, 1979.

16. I. Schur, "Über die Potenzreihen, die im inneren des Einheitskreises beschränkt sind," Part I, J. Reine Angew. Math., vol. 147, pp. 205-232 (1917); Part II, vol. 148, pp. 122-145 (1918).

17. D. C. Youla, "A new theory of broad-band matching," IEEE Trans. Circuit Theory, vol. CT-11, pp. 30-50 (1964).

18. J. W. Helton, "Broadbanding: gain equalization directly from data," IEEE Trans. Circuit Syst., vol. CAS-28, pp. 1125-1137 (1981).

19. G. Zames and B. A. Francis, "Feedback, minimax sensitivity, and optimal robustness," IEEE Trans. Autom. Control, vol. AC-28, pp. 585-601 (1983).

20. J. C. Doyle, "Synthesis of robust controllers and filters," Proc. IEEE Conf. Decision and Control, San Antonio, pp. 109-114, 1983.

21. P. R. Delsarte, Y. Genin, and Y. Kamp, "The Nevanlinna-Pick problem for matrix-valued functions," SIAM J. Appl. Math., vol. 36, pp. 177-187 (1979).

22. H. Kimura, "Directional interpolation approach to H^∞-optimization and robust stabilization," IEEE Trans. Autom. Control, vol. AC-32, pp. 1085-1093 (1987).

23. B. A. Francis, A Course in H^∞ Control Theory, Springer-Verlag, New York, 1987.

4

System-Theoretical Consideration of Multipoint Padé Approximation

KAZUMI HORIGUCHI

Kinki University, Kowakae,
Higashiosaka, Japan

4-1 INTRODUCTION

In many areas of circuit and system theory, it is necessary to approximate a function with desired characteristics by means of a rational function with appropriate degrees. One of the most useful approaches to this subject is Padé approximations [1]. Padé approximation is a well-known classical approximation that has been used for a variety of purposes, including network synthesis [2], filter design [3], system modeling [4], model reduction [5], stability analysis [6] and control system design [7]. Padé approximation is a method of finding a rational function that partially fits the power series expansion of a desired function at the origin, and the Padé problem is defined as follows.

Suppose that we are given $N + 1$ complex numbers $\{f_0, f_1, \ldots, f_N\}$ which represent a desired function $f(z)$, so that

$$f(z) = f_0 + f_1 z + \cdots + f_N z^N + O(z^{N+1}) \tag{4-1}$$

Find a rational function

$$h[m/n](z) = \frac{b(z)}{a(z)} = \frac{b_0 + b_1 z + \cdots + b_m z^m}{a_0 + a_1 z + \cdots + a_n z^n} \tag{4-2}$$

where $a_0 \neq 0$ and $m + n = N$, which interpolates $\{f_0, f_1, \ldots, f_N\}$, that is, satisfies

$$f(z) - h[m/n](z) = O(z^{N+1}) \tag{4-3}$$

$h[m/n](z)$ is said to be an $[m/n]$ Padé approximant of $f(z)$. In system theory, this problem may be interpreted as a partial realization problem based on finite Markov parameters $\{f_0, f_1, \ldots, f_N\}$ [8].

Normally, $h[m/n](z)$ can be obtained by immediate calculation. Multiplying (4-3) by $a(z)$ gives

$$f(z)a(z) - b(z) = O(z^{N+1}) \tag{4-4}$$

In (4-4), equating the coefficients of z^0, z^1, $\ldots$, z^m and z^{m+1}, $\ldots$, z^{m+n}, respectively, we find that

$$\sum_{k=0}^{i} f_k a_{i-k} - b_i = 0, \qquad 0 \leq i \leq m \tag{4-5a}$$

$$\begin{bmatrix} f_{m-n+1} & \cdots & f_{m-1} & f_m & \vdots & f_{m+1} \\ & & & f_{m+1} & \vdots & \\ & & & & \vdots & \\ f_{m-1} & & & & \vdots & f_{m+n-1} \\ f_m & f_{m+1} & & f_{m+n-1} & \vdots & f_{m+n} \end{bmatrix} \begin{bmatrix} a_n \\ \vdots \\ \vdots \\ a_1 \\ a_0 \end{bmatrix} = 0 \tag{4-5b}$$

where $a_k = 0$; $k > n$ and $f_k = 0$; $k < 0$. Thus there exists an $[m/n]$ Padé approximant of $f(z)$ if and only if

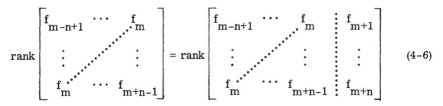

$$\text{rank} \begin{bmatrix} f_{m-n+1} & \cdots & f_m \\ \vdots & & \vdots \\ f_m & \cdots & f_{m+n-1} \end{bmatrix} = \text{rank} \begin{bmatrix} f_{m-n+1} & \cdots & f_m & \vdots & f_{m+1} \\ \vdots & & \vdots & \vdots & \vdots \\ f_m & \cdots & f_{m+n-1} & \vdots & f_{m+n} \end{bmatrix} \qquad (4\text{-}6)$$

and then the coefficients of $a(z)$ and $b(z)$ are determined by linear equations
(4-5a) and (4-5b), respectively. Note that $a(z)$ has at most degree n (i.e.,
$\deg[a(z)] \leq n$ and $\deg[b(z)] \leq m$); for example, in the case

$$a_{q+1} = \cdots = a_n = 0$$

$$b_{p+1} = \cdots = b_m = 0$$

$h[m/n](z)$ is reduced to a $[p/q]$ Padé approximant (i.e., $h[m/n](z) = h[p/q](z)$). The existence of an $[m/n]$ Padé approximant and the actual degrees of its denominator and numerator can be examined via a C-table [1, Part I, Sec. 1.4], which is an array of determinants of Hankel matrices $H_n[m]$:

$$H_n[m] = \begin{cases} 1, & n = 0 \\ \begin{bmatrix} f_{m-n+1} & \cdots & f_m \\ \vdots & & \vdots \\ f_m & \cdots & f_{m+n-1} \end{bmatrix}, & n \geq 1 \end{cases} \qquad (4\text{-}7)$$

If $\det H_n[m] \neq 0$, which will be seen to hold in most practical cases, there
exists an $[m/n]$ Padé approximant with $\deg[a(z)] \leq n$ and $\deg[b(z)] \leq m$
(either holds equality) and it is given by solving (4-5). Thus normally we
may obtain a rational function $h[m/n](z)$ which approximates a desired function $f(z)$ about $z = 0$ via very simple procedures.

The purpose of this chapter is to investigate multipoint Padé approximation [1, Part II, Sec. 1.1], which is an extension of Padé approximation,
from the system theoretical viewpoint. Multipoint Padé approximation is a
method of finding a rational function which partially fits the power series
expansions of a desired function at various points, and the multipoint Padé
problem is defined as follows.

Suppose that we are given r distinct points w_i; $1 \leq i \leq r$ and $N + 1$ complex numbers $\{f_0^{(i)}, f_1^{(i)}, \ldots, f_{N(i)}^{(i)}\}$; $1 \leq i \leq r$ with $\sum_{i=1}^{r} \{N(i) + 1\} = N + 1$,
which represent a desired function $f(z)$, so that

$$f(z) = f_0^{(i)} + f_1^{(i)}(z - w_i) + \cdots + f_{N(i)}^{(i)}(z - w_i)^{N(i)} + O((z - w_i)^{N(i)+1}), \quad 1 \le i \le r$$

$$(4\text{-}8)$$

Find a rational function $h[m/n](z)$ in (4-2) which interpolates $\{f_0^{(i)}, f_1^{(i)}, \ldots,$ $f_{N(i)}^{(i)}\}$; $1 \le i \le r$, that is, satisfies

$$f(z) - h[m/n](z) = O((z - w_i)^{N(i)+1}), \quad 1 \le i \le r \qquad (4\text{-}9)$$

$h[m/n](z)$ is said to be an $\lfloor m/n \rfloor$ multipoint Pade approximant of $f(z)$.

When $r = 1$ and $w_1 = 0$, the problem above is reduced to a Padé problem; moreover, when $N(i) = 0$; $1 \le i \le r(= N + 1)$, and when $N(i) = 1$; $1 \le i \le r$, the problem above is equivalent to a rational interpolation problem (Cauchy interpolation problem [9]) and to a rational Hermite interpolation problem [10], respectively. Hence a multipoint Padé problem is posed as a general interpolation problem, which includes each of these problems as a special case. In system theory, this problem is also interpreted as partial realization problem based on finite input-output responses (see Sec. 4-4). If the finite data $\{f_0^{(i)}, f_1^{(i)}, \ldots, f_{N(i)}^{(i)}\}$; $1 \le i \le r$ represent the desired function sufficiently and without redundancy, we may obtain via simple procedures a rational approximant with small error and low degree.

Section 4-2 is a preliminary section which contains the definition of divided differences and generalized Hankel matrices, which are used throughout this chapter, and the formulation of a multipoint Padé problem (Problem 4-1) and associated problem (Problem 4-2). In Section 4-3 we present three types of expression of the solution to a multipoint Padé problem on the assumption that it exists. The existence of an $\lfloor m/n \rfloor$ multipoint Padé approximant and its degree are discussed in Sec. 4-4. This section is also devoted to a consideration of the minimal partial realization problem (Problem 4-3), which is concerned with finding a system with minimal degree interpolating the given input-output data. In addition to the theoretical results on these problems (Theorems 4-1, 4-2, and 4-3), practical procedures for examining the existence of an approximant and to find it if it exists (Algorithm 4-1), and for finding a system of minimal partial realization (Algorithm 4-2), are presented. In Sec. 4-5, multipoint Padé approximants are connected with continued fraction expansions. Here a class of multipoint Padé approximants is represented as a regular C-type continued fraction, and based on the representation, a simple numerical procedure to construct the approximant successively is derived (Algorithm 4-3).

4-2 PRELIMINARIES

In this section we introduce divided differences and generalized Hankel matrices associated with a function $f(z)$ and a set of points $\{z_0, z_1, \ldots, z_N\}$

which play important roles in the following sections. Using divided differences, we can formulate a multipoint Padé problem in more suitable fashion for treatment together with the associated problem.

Definition 4-1. For $f(z)$ and $\{z_0, z_1, \cdots, z_N\}$, divided differences are defined as follows [1, Part II, Sec. 1.1]:

$$f[z_0] = f(z_0)$$

$$f[z_0, z_1, \cdots, z_k] = \begin{cases} \dfrac{f[z_0, \cdots, z_{j-1}, z_{j+1}, \cdots, z_k] - f[z_0, \cdots, z_{i-1}, z_{i+1}, \cdots, z_k]}{z_i - z_j}, \\ \hspace{6cm} z_i \neq z_j \\[4mm] \dfrac{1}{k!} \dfrac{d^k}{dz^k} f(z_0), \quad z_0 = z_1 = \cdots = z_k \end{cases}$$

$$(4\text{-}10)$$

∎

The value of $f[z_0, z_1, \cdots, z_k]$ is independent of the order of the points $z_0, z_1, \cdots, z_k$; that is, for any permutations of the points, say $z_0', z_1', \cdots, z_k'$,

$$f[z_0, z_1, \cdots, z_k] = f[z_0', z_1', \cdots, z_k'] \tag{4-11}$$

To simplify notation, let us define f_{ij}:

$$f_{ij} = \begin{cases} 0, & i > j \\ f[z_i, \cdots, z_j], & i \leq j \end{cases} \tag{4-12}$$

Using divided differences, $f(z)$ can be represented as follows [1, Part II, Sec. 1.1]:

$$f(z) = f[z_0] + f[z_0, z_1](z - z_0) + \cdots + f[z_0, z_1, \cdots, z_N] \prod_{k=0}^{N-1} (z - z_k)$$

$$+ f[z_0, z_1, \cdots, z_N, z] \prod_{k=0}^{N} (z - z_k) \tag{4-13}$$

Formula (4-13) also holds for any permutations of $z_0, z_1, \cdots, z_N$.

Definition 4-2. Generalized Hankel matrices are composed of divided differences in the following manner [11]:

$$
H_n[z_1, \ldots, z_{m+n}] = \begin{cases} 1, & n = 0 \\[2em] \begin{bmatrix} f_{n,m+1} & \cdots & f_{n,m+n} \\ f_{n-1,m+1} & \cdots & f_{n-1,m+n} \\ \vdots & & \vdots \\ f_{1,m+1} & \cdots & f_{1,m+n} \end{bmatrix}, & n \geq 1 \end{cases} \qquad (4\text{-}14)
$$

$\blacksquare$

In the description of $H_n[z_1, \ldots, z_{m+n}]$, the subscript n stands for the size of the matrix, and $[z_1, \ldots, z_{m+n}]$ indicates that the (n, n) element of the matrix equals the divided difference $f[z_1, \ldots, z_{m+n}]$. Due to property (4-11), it is easily shown that for any permutations of $z_1, \ldots, z_{m+n}$, say $z_1', \ldots, z_{m+n}'$,

$$
\det H_n[z_1, \ldots, z_{m+n}] = \det H_n[z_1', \ldots, z_{m+n}'] \qquad (4\text{-}15)
$$

When $z_1 = \cdots = z_{m+n} = 0$ then since

$$
f_{ij} = \frac{1}{(j-1)!} \frac{d^{j-1}}{dz^{j-1}} f(0) = f_{j-1}
$$

the definition of generalized Hankel matrices (4-14) is reduced to that of regular Hankel matrices (4-7).

Now we reformulate the multipoint Padé problem mentioned in Sec. 4-1 using divided differences to simplify the notation and treatment. Corresponding to the r distinct points $\{w_1, \ldots, w_r\}$ and their degrees of multiplicity $N(1), \ldots, N(r)$, let us set $N + 1$ points $\{z_0, z_1, \ldots, z_N\}$ so that

$$
z_{L(i-1)} = z_{L(i-1)+1} = \cdots = z_{L(i)-1} = w_i, \qquad 1 \leq i \leq r \qquad (4\text{-}16a)
$$

where $L(0) = 0$ and $L(i) = \sum_{k=1}^{i}\{N(k) + 1\}$; $1 \leq i \leq r$. In (4-16a) if $L(i-1) < L(i) - 1$ ($N(i) > 0$), those points are said to be confluent points. Then the data $\{f_0^{(i)}, f_1^{(i)}, \ldots, f_{N(i)}^{(i)}\}$; $1 \leq i \leq r$ are represented as

$$f_0^{(i)} = f[z_{L(i-1)}] = f_{L(i-1), L(i-1)}$$

$$f_1^{(i)} = f[z_{L(i-1)}, z_{L(i-1)+1}] = f_{L(i-1), L(i-1)+1}$$

$$\cdot \ \cdot \ \cdot \ \cdot \ \cdot \ \cdot \ \cdot \ \cdot \ \cdot \ \cdot \ \cdot \ \cdot \qquad , \quad 1 \leq i \leq r \qquad (4\text{-}16\text{b})$$

$$f_{N(i)}^{(i)} = f[z_{L(i-1)}, \cdots, z_{L(i)-1}] = f_{L(i-1), L(i)-1}$$

Hence applying (4-13), we can write (4-9) as

$$f(z) - h[m/n](z) = (f - h)[z_{L(i-1)}, \cdots, z_{L(i)-1}, z] \prod_{k=0}^{N(i)} (z - z_{L(i-1)+k}), \quad 1 \leq i \leq r$$

$$(4\text{-}17)$$

Here (4-17) can be unified with respect to i and represented as

$$f(z) - h[m/n](z) = (f - h)[z_0, z_1, \cdots, z_N, z] \prod_{k=0}^{N} (z - z_k) \qquad (4\text{-}18)$$

Finally, the multipoint Padé problem is reformulated as follows.

Problem 4-1

Given the data (4-16) from a desired function f(z), find a rational function h[m/n](z) in (4-2) satisfying

$$f(z) - h[m/n](z) = \epsilon(z) \prod_{k=0}^{N} (z - z_k) \qquad (4\text{-}19)$$

with $\epsilon(z)$ being analytic at $z = z_k$; $0 \leq k \leq N$.

Associated with Problem 4-1, we set up the following problem to get a candidate for the multipoint Padé approximant h[m/n](z).

Problem 4-2

Given the data (4-16) from a desired function f(z), find two polynomials a(z) and b(z) with deg[a(z)] $\leq$ n, deg[b(z)] $\leq$ m and m + n = N satisfying

$$f(z)a(z) - b(z) = e(z) \prod_{k=0}^{N} (z - z_k) \qquad (4\text{-}20)$$

with e(z) being analytic at $z = z_k$; $0 \leq k \leq N$.

Note that if $h[m/n](z) = b(z)/a(z)$ is a solution to Problem 4-1, then $a(z)$ and $b(z)$ are also solutions to Problem 4-2, but the converse is not always true, due to the possibility of inaccessible points (see Sec. 4-4). However, starting with (4-20) we can get some expressions of the solution to Problem 4-1, as shown in the following section.

4-3 EXPRESSIONS OF A MULTIPOINT PADÉ APPROXIMANT

In this section we present three types of expression of an $[m/n]$ multipoint Padé approximant on the assumption that it exists. The investigation on the existence of the approximant is deferred to Section 4-4. We start with (4-20) and derive equations corresponding to (4-5) in the Padé approximation. According to (4-13), $f(z)$ can be represented in the following forms:

$$
f(z) = f_{00} + f_{01}(z - z_0) + \cdots + f_{0N} \prod_{k=0}^{N-1} (z - z_k) + \cdots
$$

$$
= f_{11} + f_{12}(z - z_1) + \cdots + f_{1N} \prod_{k=1}^{N-1} (z - z_k) + \cdots
$$

$$
N = m + n
$$

$$
\cdot \quad \cdot \quad \cdot \quad \cdot \quad \cdot \quad \cdot \quad \cdot \quad \cdot \quad \cdot
$$

$$
= f_{nn} + f_{n,n+1}(z - z_n) + \cdots + f_{nN} \prod_{k=n}^{N-1} (z - z_k) + \cdots
$$

$$
(4\text{-}21)
$$

Let us represent $a(z)$ and $b(z)$ in similar forms:

$$
a(z) = a_{00} + a_{01}(z - z_0) + \cdots + a_{0n} \prod_{k=0}^{n-1} (z - z_k)
$$

$$
(4\text{-}22)
$$

$$
b(z) = b_{00} + b_{01}(z - z_0) + \cdots + b_{0m} \prod_{k=0}^{m-1} (z - z_k)
$$

Substitute (4-21) and (4-22) into (4-20) to get

$$
a_{00} \sum_{j=0}^{\infty} f_{0j} \prod_{k=0}^{j-1} (z - z_k) + a_{01} \sum_{j=1}^{\infty} f_{1j} \prod_{k=0}^{j-1} (z - z_k) + \cdots + a_{0n} \sum_{j=n}^{\infty} f_{nj} \prod_{k=0}^{j-1} (z - z_k)
$$

$$
- \left[b_{00} + b_{01}(z - z_0) + \cdots + b_{0m} \prod_{k=0}^{m-1} (z - z_k) \right] = e(z) \prod_{k=0}^{N} (z - z_k) \qquad (4\text{-}23)
$$

Equating the coefficients of $\Pi_{k=0}^{j-1}$ $(z - z_k)$; $0 \le i \le N$, we find that

$$\sum_{k=0}^{i} f_{ki} a_{0k} - b_{0i} = 0, \qquad 0 \le i \le m \tag{4-24a}$$

$$\begin{bmatrix} f_{n,m+1} & \cdots & f_{1,m+1} & \vdots & f_{0,m+1} \\ \vdots & & \vdots & \vdots & \vdots \\ f_{n,m+n-1} & \cdots & f_{1,m+n-1} & \vdots & f_{0,m+n-1} \\ f_{n,m+n} & \cdots & f_{1,m+n} & \vdots & f_{0,m+n} \end{bmatrix} \begin{bmatrix} a_{0n} \\ \vdots \\ a_{01} \\ a_{00} \end{bmatrix} = 0 \tag{4-24b}$$

where $a_{0k} = 0$; $k > n$ and $f_{ij} = 0$; $i > j$. Thus an $[m/n]$ multipoint Padé approximant is expressed as

$$h[m/n](z) = \frac{b_{00} + b_{01}(z - z_0) + \cdots + b_{0m} \Pi_{k=0}^{m-1}(z - z_k)}{a_{00} + a_{01}(z - z_0) + \cdots + a_{0n} \Pi_{k=0}^{n-1}(z - z_k)}, \qquad a_{00} \ne 0 \tag{4-25}$$

with the coefficients satisfying (4-24), which is similar to (4-2) with (4-5) in Padé approximation. Here, recalling Cramer's rule on linear equations, we can deduce an explicit form of the approximant from (4-24) and (4-25) on the assumption that the coefficient matrix in (4-24b) has full rank [1, Part II, Sec. 1.1].

$$h[m/n](z) = \frac{b[m/n](z)}{a[m/n](z)}$$

$$a[m/n](z) = \det \begin{bmatrix} f_{n,m+1} & \cdots & f_{1,m+1} & f_{0,m+1} \\ \vdots & & & \vdots \\ f_{n,m+n} & \cdots & f_{1,m+n} & f_{0,m+n} \\ \Pi_{k=0}^{n-1}(z - z_k) & \cdots & (z - z_0) & 1 \end{bmatrix} \tag{4-26a}$$

$$b[m/n](z) = det \begin{bmatrix} f_{n,m+1} & \cdots & f_{1,m+1} & f_{0,m+1} \\ \vdots & & \vdots & \vdots \\ f_{n,m+n} & & f_{1,m+n} & f_{0,m+n} \\ \sum_{j=n}^{m} f_{nj} \prod_{k=0}^{j-1}(z-z_k) & \cdots & \sum_{j=1}^{m} f_{1j} \prod_{k=0}^{j-1}(z-z_k) & \sum_{j=0}^{m} f_{0j} \prod_{k=0}^{j-1}(z-z_k) \end{bmatrix}$$

(4-26b)

Next, we derive an expression of $h[m/n](z)$ associated with state space representation in system theory. Let us transform $a[m/n](z)$ and $b[m/n](z)$ in (4-26) according to the following processes, respectively:

$a[m/n](z)$ for $j = 1,2,\ldots,n$:
   (jth column) := (jth column) − [(j + 1)th column]$(z - z_{n-j})$

$b[m/n](z)$ for $j = 1,2,\ldots,n$:
   (jth column) := (jth column) − [(j + 1)th column]$(z - z_{n-j})$

This changes the last row of $b[m/n](z)$ into

$$\left[-f_{n-1,m} \prod_{k=0}^{m}(z-z_k), \ldots, -f_{0m} \prod_{k=0}^{m}(z-z_k), \sum_{j=0}^{m} f_{0j} \prod_{k=0}^{j-1}(z-z_k) \right]$$

For $j = n, \ldots, 2, 1$:

$$[(n+1)th\ column] := [(n+1)th\ column] + (jth\ column) \bigg/ \prod_{k=0}^{n-j}(z-z_k)$$

This changes the last column of $b[m/n](z)$ into

$$\left[f_{n,m+1} \bigg/ \prod_{k=0}^{n-1}(z-z_k), \ldots, f_{n,m+n} \bigg/ \prod_{k=0}^{n-1}(z-z_k), \sum_{j=n}^{m} f_{nj} \prod_{k=n}^{j-1}(z-z_k) \right]^T$$

After the operations above, $a[m/n](z)$ and $b[m/n](z)$ are transformed into

$$a[m/n](z) = det[E - zA] \tag{4-27a}$$

$$b[m/n](z) = \det \begin{bmatrix} E - zA & c^T \Big/ \prod_{k=0}^{n-1} (z - z_k) \\ -b^T \prod_{k=0}^{m} (z - z_k) & \sum_{j=n}^{m} f_{nj} \prod_{k=n}^{j-1} (z - z_k) \end{bmatrix} \qquad (4\text{-}27b)$$

respectively, where

$$E = H_n[z_1, \cdots, z_{m+n}] - H_n[z_0, \cdots, z_{m+n}] \cdot D_n; \quad D_n = \text{diag}[z_{n-1}, \cdots, z_0]$$

$$A = H_n[z_0, \cdots, z_{m+n}], \quad b = [f_{n-1,m} \quad \cdots \quad f_{1m} \quad f_{0m}]^T \qquad (4\text{-}28)$$

$$c = [f_{n,m+1} \quad \cdots \quad f_{n,m+n-1} \quad f_{n,m+n}]$$

From (4-27), an [m/n] multipoint Padé approximant is expressed as

$$h[m/n](z) = \sum_{j=n}^{m} f_{nj} \prod_{k=0}^{j-1} (z - z_k) + \prod_{k=n}^{m} (z - z_k) c(E - zA)^{-1} b \qquad (4\text{-}29)$$

which corresponds to the form referred to as Nuttall's compact form in Padé approximation [1, Part I, Sec. 1.3]. Equation (4-29) has a close relation to state space equations in system theory. As an example, when we interpret z as a delay element of a discrete-time system, the state space equations

$$x(k + 1) = A_p x(k) + b_p u(k)$$
$$y(k) = c_p x(k) + d_p u(k) \qquad (4\text{-}30)$$

whose transfer function equals $h[m/n](z)$ in (4-29), that is,

$$h[m/n](z) = d_p + zc_p(I - zA_p)^{-1} b_p \qquad (4\text{-}31)$$

are given as follows:

(i) In the case $m < n$

$$A_p = E^{-1}A, \qquad b_p = E^{-1}b$$

$$c_p = cE^{-1}A, \qquad d_p = cE^{-1}b \qquad (4\text{-}32a)$$

(ii) In the case $m = n$

$$A_p = E^{-1}A, \qquad\qquad b_p = E^{-1}b$$

$$c_p = c(I - z_n E^{-1}A), \qquad d_p = f_{nn} - z_n cE^{-1}b \qquad\qquad (4\text{-}32b)$$

(iii) In the case $m > n$

$$A_p = \begin{bmatrix} E^{-1}A & E^{-1}bh \\ 0 & F \end{bmatrix}, \qquad b_p = \begin{bmatrix} E^{-1}bj \\ g \end{bmatrix}$$

$$c_p = [c(I - z_m E^{-1}A) \quad (f_{nm} - z_m cE^{-1}b)h + \tilde{c}F] \qquad\qquad (4\text{-}32c)$$

$$d_p = (f_{nm} - z_m cE^{-1}b)j + \tilde{c}g$$

where

$$\begin{array}{c}\begin{bmatrix} g & F \\ j & h \end{bmatrix}\begin{matrix}\updownarrow\ m-n \\ \times \\ \updownarrow\ 1\end{matrix} \\ \overset{\longleftrightarrow}{\underset{1\ \ m-n}{}} \end{array} = \begin{bmatrix} 1 & & & \\ z_n & & 0 & \\ & \ddots & & \\ 0 & & z_n & 1 \end{bmatrix}^{-1}$$

$$\tilde{c} = [f_{nn} \quad f_{n,n+1} \quad \cdots \quad f_{n,m-1}]$$

Note that since $\det E = a[m/n](0)$ [see (4-27a)], the existence of E^{-1} is guaranteed if $a[m/n](0) \neq 0$, or equivalently, if discrete-time system $h[m/n](z)$ is causal. Equations (4-32) provide a solution to the partial realization problem for discrete-time systems treated in Sec. 4-4.

Finally, we present the third expression of an $[m/n]$ multipoint Padé approximant. The approximant $h[m + n/0](z)$ is a polynomial interpolating the given data, which is constructed directly by

$$h[m + n/0](z) = f_{00} + f_{01}(z - z_0) + \cdots + f_{0,m+n} \prod_{k=0}^{m+n-1} (z - z_k) \qquad\qquad (4\text{-}33)$$

Based on $h[m + n/0](z)$, we can construct any $[m/n]$ multipoint Padé approximants by a procedure analogous to that of Padé approximants. It is easily verified that for any $h[m/n](z) = b(z)/a(z)$ $(m + n = N)$,

$$h[m + n/0](z) - \frac{b(z)}{a(z)} = \frac{q(z)}{a(z)} \prod_{k=0}^{N} (z - z_k) \tag{4-34}$$

where $q(z)$ is a polynomial with $\deg[q(z)] \leq n - 1$. Multiply (4-34) by $a(z)/\prod_{k=0}^{N} (z - z_k)$ to get

$$\frac{h[m + n/0](z)}{\prod_{k=0}^{N} (z - z_k)} a(z) - \frac{1}{\prod_{k=0}^{N} (z - z_k)} b(z) = q(z) \tag{4-35}$$

Here we set

$$a(z) = a_0 + a_1 z + \cdots + a_n z^n$$

$$b(z) = b_0 + b_1 z + \cdots + b_m z^m$$

$$q(z) = q_0 + q_1 z + \cdots + q_{n-1} z^{n-1}$$

$$\frac{h[m + n/0](z)}{\prod_{k=0}^{N} (z - z_k)} = h_1 z^{-1} + h_2 z^{-2} + \cdots \tag{4-36}$$

$$\frac{1}{\prod_{k=0}^{N} (z - z_k)} = g_1 z^{-N-1} + g_2 z^{-N-2} + \cdots$$

Then in (4-35), equating the coefficients of $z^{-m-n-1}, \ldots, z^{-n-1}; z^{-n}, \ldots, z^{-1}; z^0, \ldots, z^{n-1}$, we find that

$$\begin{bmatrix} h_{n+1} & \cdots & h_{2n} & \vdots & h_{2n+1} \\ \vdots & & \vdots & \vdots & \vdots \\ h_{n+m} & \cdots & h_{m+2n-1} & \vdots & h_{m+2n} \\ h_{m+n+1} & \cdots & h_{m+2n} & \vdots & h_{m+2n+1} \end{bmatrix} \begin{bmatrix} a_0 \\ \vdots \\ a_{n-1} \\ a_n \end{bmatrix} - \begin{bmatrix} 0 & & & g_1 \\ & & \ddots & \vdots \\ g_1 & \cdots & & g_m \\ g_1 & \cdots & g_m & g_{m+1} \end{bmatrix} \begin{bmatrix} b_0 \\ \vdots \\ b_{m-1} \\ b_m \end{bmatrix} = 0 \tag{4-37a}$$

$$
\begin{bmatrix}
h_1 & \cdots & h_{n-1} & h_n & \vdots & h_{n+1} \\
\vdots & & & h_{n+1} & \vdots & \vdots \\
h_{n-1} & & & & \vdots & h_{2n-1} \\
h_n & h_{n+1} & \cdots & h_{2n-1} & \vdots & h_{2n}
\end{bmatrix}
\begin{bmatrix}
a_0 \\
\vdots \\
a_{n-1} \\
a_n
\end{bmatrix}
= 0
\qquad (4\text{-}37\mathrm{b})
$$

$$
\begin{bmatrix}
h_1 & \cdots & h_{n-1} & h_n \\
& h_1 & \cdots & h_{n-1} \\
& & \ddots & \vdots \\
0 & & & h_1
\end{bmatrix}
\begin{bmatrix}
a_1 \\
a_2 \\
\vdots \\
a_n
\end{bmatrix}
=
\begin{bmatrix}
q_0 \\
q_1 \\
\vdots \\
q_{n-1}
\end{bmatrix}
\qquad (4\text{-}37\mathrm{c})
$$

Thus an [m/n] multipoint Padé approximant is expressed as

$$
h[m/n](z) = \frac{b_0 + b_1 z + \cdots + b_m z^m}{a_0 + a_1 z + \cdots + a_n z^n}
\qquad (4\text{-}38)
$$

with the coefficients satisfying (4-37a) and (4-37b). Equation (4-37b) has the same form as (4-5b), which means that the procedure to calculate multipoint Padé approximants is simply reduced to that of Padé approximants. Observe that if (4-37a), (4-37b), and (4-37c) hold, (4-35) holds identically; that is, the coefficients of z^{-m-n-2}, z^{-m-n-3}, ... are also equivalent on both sides.

In this section we present three types of expression of the [m/n] multi-point Padé approximant: (4-25) with (4-24), which is a generalization of (4-2) with (4-5); (4-29) with (4-28), which is relevant to state space representation; and (4-38) with (4-37), which is analogous to that of a Padé approximant.

4-4 EXISTENCE AND MINIMAL DEGREE OF MULTIPOINT PADÉ APPROXIMANTS

In this section we consider the problem how to examine the existence of an [m/n] multipoint Padé approximant and its degrees of denominator and numerator. Furthermore, we treat the minimal partial realization problem corresponding to the multipoint Padé problem. We present the practical algorithms to find solutions to these problems as well as the theoretical results on them.

The following are fundamental results on the existence of multipoint Padé approximants [12].

Lemma 4-1

(i) Problem 4-2 always has solutions $a(z)$ and $b(z)$ which are given by (4-22) with arbitrary coefficients satisfying (4-24).

(ii) Let $a(z)$ and $b(z)$ be arbitrary solutions to Problem 4-2, and let $\hat{b}(z)/\hat{a}(z)$ be an irreducible form of $b(z)/a(z)$. Then $h[m/n](z) + \hat{b}(z)/\hat{a}(z)$ is unique and it is a solution to Problem 4-1 if and only if $\hat{a}(z)$ and $\hat{b}(z)$ are solutions to Problem 4-2.   ∎

Remark 1

In Lemma 4-1(ii), if $a(z)$ and $b(z)$ have a common factor $(z - z_k)^{M(k)}$ $0 \leq k \leq N$, and $a(z)/(z - z_k)^{M(k)}$ and $b(z)/(z - z_k)^{M(k)}$ are not solutions to Problem 4-2, then an inaccessible point occurs at $z = z_k$; that is, the approximant $\hat{a}(z)/\hat{b}(z)$ has data that cannot be interpolated at $z = z_k$.

The next theorem shows a condition for the existence of a solution to Problem 4-1 in terms of divided differences.

Theorem 4-1. There exists an $[m/n]$ multipoint Padé approximant if and only if for any permutations of the points $z_0, z_1, \cdots, z_{m+n}$ in (4-16a),

$$
\operatorname{rank}
\begin{bmatrix}
f_{n,m+1} & \cdots & f_{1,m+1} \\
\vdots & & \vdots \\
f_{n,m+n-1} & \cdots & f_{1,m+n-1} \\
f_{n,m+n} & \cdots & f_{1,m+n}
\end{bmatrix}
= \operatorname{rank}
\begin{bmatrix}
f_{n,m+1} & \cdots & f_{1,m+1} & \cdot & f_{0,m+1} \\
\vdots & & \vdots & \cdot & \vdots \\
f_{n,m+n-1} & \cdots & f_{1,m+n-1} & \cdot & f_{0,m+n-1} \\
f_{n,m+n} & \cdots & f_{1,m+n} & \cdot & f_{0,m+n}
\end{bmatrix}
$$

$$(4-39)$$
∎

Proof. Necessity is clear because the existence of $h[m/n](z) = b(z)/a(z)$ implies that (4-24b) holds with $a_{00} \neq 0$ for any permutations of $z_0, z_1, \cdots, z_N$.

Sufficiency: The assumption implies that there exist $a(z)$ and $b(z)$ satisfying (4-24) with $a_{00} \neq 0$; that is, there exist $a(z)$ and $b(z)$ such that (4-20) is satisfied and $\hat{b}(z)/\hat{a}(z)$ [irreducible form of $b(z)/a(z)$] has no inaccessible points at $z = z_0$ for any permutations of $z_0, z_1, \cdots, z_N$. Thus the uniqueness of $\hat{b}(z)/\hat{a}(z)$ (Lemma 4-1) claims that $\hat{b}(z)/\hat{a}(z)$ is an $[m/n]$ multipoint Padé approximant.   ∎

Remark 2

Due to properties (4-11) and (4-15), we have only to evaluate (4-39) for the following r permutations of z_0, z_1, $\cdots$, z_{m+n} rather than for all the permutations.

$$z_{L(0)}, \cdots, z_{m+n}$$

$$z_{L(1)}, \cdots, z_{m+n}, z_0, \cdots, z_{L(1)-1}$$

$$\cdot \quad \cdot \quad \cdot \quad \cdot \quad \cdot \quad \cdot \quad \cdot \quad \cdot \quad \cdot \quad \cdot \quad \cdot$$

$$z_{L(r-1)}, \cdots, z_{m+n}, z_0, \cdots, z_{L(r-1)-1}$$

If no confluent points exist, one has to evaluate (4-39) m + n + 1 times, while in the case of Padé approximation: $z_0 = z_1 = \cdots = z_{m+n} = 0$, one has to evaluate it only once.

The next theorem states how to examine the existence and degrees of multipoint Padé approximants by evaluating generalized Hankel matrices.

Theorem 4-2

   (i) If for any permutations of z_0, z_1, $\cdots$, z_{m+n},

$$\det H_n \lfloor z_1, \cdots, z_{m+n} \rfloor \neq 0 \qquad (4\text{-}40a)$$

   then there exists an $\lfloor m/n \rfloor$ multipoint Padé approximant $h\lfloor m/n \rfloor(z) = a(z)/b(z)$ with $\deg[a(z)] \leq n$ and $\deg[b(z)] \leq m$ (either holds equality).
   (ii) If for any permutations of z_0, z_1, $\cdots$, z_{m+n},

$$\det H_n \lfloor z_1, \cdots, z_{m+n} \rfloor = 0 \qquad (4\text{-}40b)$$

   then $h\lfloor m/n \rfloor(z)$ degenerates into $h\lfloor m - 1/n - 1 \rfloor(z)$, so that the results on existence and degrees of an $\lfloor m/n \rfloor$ multipoint Padé approximant are reduced to those of an $\lfloor m - 1/n - 1 \rfloor$ approximant.
   (iii) Otherwise, that is, for some permutations of z_0, z_1, $\cdots$, z_{m+n} (4-40a) holds, and for the other permutations (4-40b) holds, then an $\lfloor m/n \rfloor$ multipoint Padé approximant does not exist. ∎

Proof

   (i) The assumption implies that (4-39) holds, so there exists an $\lfloor m/n \rfloor$ multipoint Padé approximant represented as (4-26). Suppose that $\deg[a(z)] < n$ and $\deg[b(z)] < m$. Then

$$a_{0n} = \det H_n \lfloor z_0, \cdots, z_{m+n} \rfloor = 0 \text{ and } b_{0m} = \det H_{n+1} \lfloor z_0, \cdots, z_{m+n} \rfloor = 0$$

Applying Sylvester's determinant identity [1, Part I, Sec. 1.4],

$$\det H_{n+1}[z_0, \ldots, z_{m+n}] \det H_{n-1}[z_1, \ldots, z_{m+n-1}]$$

$$= \det H_n[z_1, \ldots, z_{m+n-1}] \det H_n[z_0, \ldots, z_{m+n}]$$

$$- \det H_n[z_0, \ldots, z_{m+n-1}] \det H_n[z_1, \ldots, z_{m+n}]$$

we find that

$$\det H_n[z_0, \ldots, z_{m+n-1}] \det H_n[z_1, \ldots, z_{m+n}] = 0$$

which contradicts the assumption (4-40a).

(ii) The assumption implies that $a_{00} = b_{00} = 0$ in (4-26), so $a(z)$ and $b(z)$ have a common factor $(z - z_0)$. Due to the uniqueness of $\hat{b}(z)/\hat{a}(z)$ (Lemma 4-1), we conclude that $h[m/n](z) = h[m - 1/n - 1](z)$.

(iii) Suppose that there exists an $[m/n]$ multipoint Padé approximant $h[m/n](z) = b(z)/a(z)$ [$a(z)$, $b(z)$: coprime] with $\deg[a(z)] = n$ and $\deg[b(z)] \leq m$ or with $\deg[a(z)] \leq n$ and $\deg[b(z)] = m$. Then (4-40a) holds because $\det H_n[z_1, \ldots, z_{m+n}] = 0$ implies that $a(z)$ and $b(z)$ have a common factor $(z - z_0)$. Suppose there also exists an approximant with $\deg[a(z)] = q < n$ and $\deg[b(z)] = p < m$. Then by analogy with (4-24b),

$$\begin{bmatrix} f_{q,p+1} & \cdots & f_{1,p+1} & f_{0,p+1} \\ \vdots & & \vdots & \vdots \\ f_{qm} & \cdots & f_{1m} & f_{0m} \\ \vdots & & \vdots & \vdots \\ f_{q,m+n-1} & \cdots & f_{1,m+n-1} & f_{0,m+n-1} \end{bmatrix} \begin{bmatrix} a_{0q} \\ \vdots \\ a_{01} \\ a_{00} \end{bmatrix} = 0$$

is led for any permutations of $z_0, z_1, \ldots, z_{m+n}$. Since

$$\begin{bmatrix} f_{qm} & \cdots & f_{1m} & f_{0m} \\ \vdots & & \vdots & \vdots \\ f_{q,m+n-1} & \cdots & f_{1,m+n-1} & f_{0,m+n-1} \end{bmatrix}$$

is equal to the last $q + 1$ columns of $H_n[z_0, \ldots, z_{m+n-1}]$,   (4-40b) holds. Thus if an [m/n] multipoint Padé approximant exists, then either (4-40a) or (4-40b) is satisfied, which is equivalent to Theorem 4-2(iii).   ∎

Example 4-1

Suppose that a desired function $f(z)$ is represented as follows:

$$f(z) = 1 + z - 2z^2 + O(z^3)$$

$$f(z) = -1 + (z + 1) + O((z + 1)^2)$$

$$f(z) = -1 + O((z + 2))$$

this is,

$$N = 5, \quad z_0 = z_1 = z_2 = 0, \quad z_3 = z_4 = -1, \quad z_5 = -2$$

$$f[z_0] = 1, \quad f[z_0, z_1] = 1, \quad f[z_0, z_1, z_2] = -2$$

$$f[z_3] = -1, \quad f[z_3, z_4] = 1, \quad f[z_5] = -1$$

Then evaluating the generalized Hankel matrices for each m and n, we have the following table (generalized C-table):

m\n	0	1	2	3	4	5	6
0	*	*	*	*	*	*	*
1	*	*	*	*	X	*	
2	*	*	*	*	*		
3	*	*	O	O			
4	*	*	O				
5	*	*					
6	*						

* : $\det H_n[z_1, \ldots, z_{m+n}] \neq 0$; for any permutations of $z_0, \ldots, z_{m+n}$
O : $\det H_n[z_1, \ldots, z_{m+n}] = 0$; for any permutations of $z_0, \ldots, z_{m+n}$
X : otherwise

From the table, we find that

$$h[5/0](z) = 1 + z - 2z^2 + z^3 + 3z^4 + z^5$$

$$h[3/2](z) = h[2/3](z) = h[1/4](z) = h[1/2](z) = \frac{1 + 2z}{1 + z + z^2}$$

$$h[0/5](z) = \frac{1}{1 - z + 3z^2 + 26z^3 + 28z^4 + 8z^5}$$

are all interpolating the data. As

$$\det H_1[z_0, z_1, z_2, z_3, z_4] = 1, \quad \det H_1[z_3, z_4, z_5, z_0, z_1] = -1$$

$$\det H_1[z_5, z_0, z_1, z_2, z_3] = 0$$

thus

$$h[4/1](z) = h[3/0](z) = 1 + z - 2z^2 - z^3$$

has an inaccessible point at $z = -1$; in fact,

$$h[4/1](-1) = f(-1) = -1 \text{ but } \frac{d}{dz}h[4/1](-1) = 2 \neq \frac{d}{dz}f(-1) = 1$$

Theorem 4-2 provides the extended results of those on Padé approximants. We can get a lot of information from the C-table about [m/n] Padé approximants. In a multipoint Padé problem, however, we have to evaluate some generalized Hankel matrices for each (m,n) (Remark 2), which is much more cumbersome than in the Padé problem, where we only have to evaluate a regular Hankel matrix for each (m,n). Therefore, Theorem 4-2, as well as Theorem 4-1, is not attractive from the practical viewpoint. In fact, to examine the existence of an [m/n] multipoint Padé approximant and to find it if it exists, the following algorithm is recommended.

Algorithm 4-1

(i) Transform the matrix

$$\begin{bmatrix} f_{n,m+1} & \cdots & f_{1,m+1} & \cdot & f_{0,m+1} \\ \vdots & & \vdots & \cdot & \vdots \\ & & & \cdot & \\ f_{nN} & \cdots & f_{1N} & \cdot & f_{0N} \end{bmatrix} \qquad (4\text{-}41)$$

into the following form by row operations.

$$\begin{bmatrix} 0 & \cdots & \cdots & \cdots & \cdots & \cdots & 0 & \cdot & \phi_{0,m+1} \\ \vdots & & & & & & \vdots & \cdot & \vdots \\ 0 & \cdots & \cdots & \cdots & \cdots & \cdots & 0 & \cdot & \\ * & \cdots & * & \phi_{k(q),N-q} & 0 & \cdots & 0 & \cdot & \phi_{0,N-q} \\ \vdots & & & & & & \vdots & \cdot & \vdots \\ * & \cdots & \cdots & \cdots & * & \phi_{k(1),N-1} & 0 & \cdots & 0 & \cdot & \phi_{0,N-1} \\ * & \cdots & \cdots & \cdots & \cdots & * & \phi_{k(0),N} & 0 & \cdots & 0 & \cdot & \phi_{0N} \end{bmatrix}$$

$$(4\text{-}42)$$

where $\phi_{k(i),N-i} \neq 0; \ 0 \leq i \leq q$.

(ii) If $\phi_{0,m+1} = \cdots = \phi_{0,N-q-1} = 0$, solve the equations

$$
\begin{bmatrix}
\phi_{k(q),N-q} & & & \cdot & \phi_{0,N-q} \\
\vdots & & 0 & \cdot & \vdots \\
* & \cdots\cdots \phi_{k(1),N-1} & & \cdot & \phi_{0,N-1} \\
* & \cdots\cdots\cdots * & \phi_{k(0),N} & \cdot & \phi_{0N}
\end{bmatrix}
\begin{bmatrix}
a_{0,k(q)} \\
\vdots \\
a_{0,k(0)} \\
a_{00}
\end{bmatrix}
= 0
$$

$$(4\text{-}43)$$

with $a_{00} = 1$, and set $a_{0j} := 0;\ j \neq k(i)$ to get

$$
a(z) = \sum_{j=0}^{n} a_{0j} \prod_{k=0}^{j-1} (z - z_k)
$$

Otherwise, stop because the approximant does not exist.

(iii) Evaluate $a(z)$ at various values of z_k. If $a(z_k) \neq 0;\ 0 \leq k \leq N$, the approximant exists, which is given by $h[m/n](z) = b(z)/a(z)$ with $b(z)$ satisfying (4-24a). Otherwise, the approximant does not exist.

In the remainder of this section we discuss the partial realization problem corresponding to the multipoint Padé problem. Hereafter in this section we interpret $f(z)$ as the transfer function of a desired discrete-time system where z stands for delay element. Then the data (4-16) are the input-output responses of a desired system at complex frequency points z_0, z_1, $\ldots$, z_N and the [m/n] multipoint Padé approximant represents the discrete-time system with degree $\max\{m,n\}$, which matches the responses specified at z_0, z_1, $\ldots$, z_N. We formulate the minimal partial realization problem as follows.

Problem 4-3

Given the input-output data (4-16) from a desired discrete-time system $f(z)$, where z stands for a delay element, find a system $(A_p\ b_p\ c_p\ d_p)$ with minimal degree which interpolates the data, that is, whose transfer function

$$
h(z) = d_p + z c_p (I - z A_p)^{-1} b_p
$$

$$(4\text{-}44)$$

satisfies (4-19).

In the case of $z_0 = z_1 = \cdots = z_N = 0$ (Padé problem), Problem 4-3 is reduced to the well-known minimal partial realization problem from a finite sequence of Markov parameters: f_0, f_1, $\ldots$, f_N. The next result shows how to determine the minimal degree of the interpolating system, which is an extension of that on the Padé problem.

Theorem 4-3. The minimal degree of the system interpolating the given data is the minimal integer k such that

$$
\text{rank} \begin{bmatrix} f_{k-1,k} & \cdots & f_{1k} \\ \vdots & & \vdots \\ f_{k-1,N} & \cdots & f_{1N} \end{bmatrix} \neq \text{rank} \begin{bmatrix} f_{k-1,k} & \cdots & f_{1k} & \cdot & f_{0k} \\ \vdots & & & & \cdot \\ f_{k-1,N} & \cdots & f_{1N} & \cdot & f_{0N} \end{bmatrix} \tag{4-45a}
$$

for a permutation of z_0, z_1, $\cdots$, z_N, and

$$
\text{rank} \begin{bmatrix} f_{k,k+1} & \cdots & f_{1,k+1} \\ \vdots & & \vdots \\ f_{kN} & \cdots & f_{1N} \end{bmatrix} = \text{rank} \begin{bmatrix} f_{k,k+1} & \cdots & f_{1,k+1} & \cdot & f_{0,k+1} \\ \vdots & & \vdots & \cdot & \vdots \\ f_{kN} & \cdots & f_{1N} & \cdot & f_{0N} \end{bmatrix} \tag{4-45b}
$$

for any permutations of z_0, z_1, $\cdots$, z_N. ∎

This is an immediate result from Theorem 4-1.

According to Theorem 4-3, to determine the minimal degree we have to check whether (4-45b) holds for some permutations of z_0, z_1, $\cdots$, z_N, which is cumbersome and not suitable for computation, as has been mentioned before. For computational convenience, we present a procedure to find a minimal partial realization.

Algorithm 4-2

(i) Set $k := \left\lceil \dfrac{N+1}{2} \right\rceil$.

(ii) Set $m := k$ and $n := k$.

(iii) Find $h_{\lfloor m/n \rfloor}(z) = a(z)/b(z)$ if it exists according to Algorithm 4-1. If $h_{\lfloor m/n \rfloor}(z)$ does not exist, then $k := k + 1$ and go to (ii).

(iv) Find a minimal partial realization of $h_{\lfloor m/n \rfloor}(z)$ by an appropriate realization method.

Remark 3

Algorithm 4-2 always terminates, because the data (4-16) can always be interpolated by the system with degree N [see (4-33)]. Namely, even in the worst case the algorithm finds a system with degree N.

Remark 4

An interesting algorithm has appeared in [13], which directly finds a state space representation of the minimal partial realization from the input-output data. But it does not meet the problem of inaccessible points suffi-ciently. In any case, the characteristic polynomial of the system must be evaluated at the points z_0, z_1, $\cdots$, z_N to examine whether or not inaccessible point occur. Thus a solution to Problem 4-3 cannot be calculated according to such an efficient algorithm as Rissanen's algorithm [14] for minimal partial realization based on finite Markov parameters.

Example 4-2

Suppose that a desired transfer function $f(z)$ is represented as follows:

$$f(z) = 1 + O((z-1)^2)$$

$$f(z) = O((z+1))$$

$$f(z) = 1 + z + O(z^2)$$

that is,

$$N = 4, \quad z_0 = z_1 = 1, \quad z_2 = -1, \quad z_3 = z_4 = 0$$

$$f[z_0] = 1, \quad f[z_0, z_1] = 0, \quad f[z_2] = 0, \quad f[z_3] = 1, \quad f[z_3, z_4] = 1$$

Then, let us find a minimal partial realization via Algorithm 4-2.

(i) Set $k := 2$, $m := 2$, and $n := 2$. Solve an equation:

$$\begin{bmatrix} f_{23} & f_{13} & f_{03} \\ f_{24} & f_{14} & f_{04} \end{bmatrix} \begin{bmatrix} a_{02} \\ a_{01} \\ a_{00} \end{bmatrix} \begin{bmatrix} 1 & -1/2 & 1/4 \\ 0 & -1/2 & 3/4 \end{bmatrix} \begin{bmatrix} a_{02} \\ a_{01} \\ 1 \end{bmatrix} = 0$$

and get a solution:

$$[a_{00} \quad a_{01} \quad a_{02}] = [1 \quad 3/2 \quad 1/2]$$

Evaluate

$$a(z) = 1 + (3/2)(z-1) + (1/2)(z-1)^2$$

at $z = 1$, -1, and 0. $a(1) \neq 0$ but $a(-1) = a(0) = 0$; hence $z = -1$ and 0 are inaccessible points.

(ii) Set $k := 3$, $m := 3$, and $n := 3$. Solve an equation:

$$[f_{34} \quad f_{24} \quad f_{14} \quad f_{04}] \begin{bmatrix} a_{03} \\ a_{02} \\ a_{01} \\ a_{00} \end{bmatrix} = [1 \quad 0 \quad -1/2 \quad 3/4] \begin{bmatrix} a_{03} \\ a_{02} \\ a_{01} \\ 1 \end{bmatrix}$$

and get a solution:

$$[a_{00} \ a_{01} \ a_{02} \ a_{03}] = [1 \ 3/2 \ 0 \ 0]$$

Since $a(z) = 1 + (3/2)(z - 1) \neq 0$ at $z = 1$, -1, and 0, the data can be interpolated by the system with degree 3. Its transfer function is

$$\frac{b(z)}{a(z)} = \frac{1 + (3/2)(z - 1) + (1/2)(z - 1)^2 - (1/2)(z - 1)^2(z + 1)}{1 + (3/2)(z - 1)}$$

In this section we have shown the necessary conditions for an $[m/n]$ multipoint Padé approximant (Theorem 4-1), and the relationship between generalized Hankel matrices and multipoint Padé approximants (Theorem 4-2). A practical procedure to find the approximant if it exists has also been presented (Algorithm 4-1). In addition, we formulated the minimal partial realization problem corresponding to the multipoint Padé problem, and showed how to determine the minimal degree (Theorem 4-3) and how to find a minimal partial realization (Algorithm 4-2).

4-5 CONNECTION WITH CONTINUED FRACTIONS

In this section we represent a restricted class of multipoint Padé approximants as C-type continued fractions under some assumptions, and based on this representation we derive a simple algorithm to calculate the multipoint Padé approximants successively.

In circuit and system theory, continued fractions play important roles, such as for network synthesis [15], stability analysis [16, Sec. 7.4], and model reduction [17, 18]. There are many kinds of continued fractions, and each represents a class of Padé approximants [1, Part I, Chap. 4]. Among them the C-type (Cauer type) of continued fraction [16, Sec. 7.1], which is the most popular and simplest, can be used to represent a class of multipoint Padé approximants with slight modifications. The C-type continued fraction for multipoint Padé approximants is [16, Sec. 5.5]

$$h(z) = \frac{b(z)}{a(z)} = \frac{c_0}{1} + \frac{c_1(z - z_0)}{1} + \cdots + \frac{c_N(z - z_{N-1})}{1} \tag{4-46a}$$

or

$$h(z) = \frac{b(z)}{a(z)} = c_0 + \frac{c_1(z - z_0)}{1} + \cdots + \frac{c_N(z - z_{N-1})}{1} \tag{4-46b}$$

The degrees of $a(z)$ and $b(z)$ in (4-46a) or (4-46b) are, respectively,

$$\deg[a(z)] = \left[\frac{N+1}{2}\right], \quad \deg[b(z)] = \left[\frac{N}{2}\right] \tag{4-47a}$$

$$\deg[a(z)] = \left[\frac{N}{2}\right], \quad \deg[b(z)] = \left[\frac{N+1}{2}\right] \tag{4-47b}$$

where $\lfloor \cdot \rfloor$ on the right-hand side indicate "Gaussian." Hence the $[m/n]$ multipoint Padé approximant with $|m - n| \leq 1$ can be represented by either (4-46a) or (4-46b) under some assumptions, as will be shown later.

Next we consider how to interpolate the data (4-16) via the continued fraction of (4-46a) or (4-46b). We start by defining

$$h^{(i)}(z) = \frac{b^{(i)}(z)}{a^{(i)}(z)} = \frac{c_0}{1} + \frac{c_1(z - z_0)}{1} + \cdots + \frac{c_i(z - z_{i-1})}{1} \tag{4-48a}$$

$$0 \leq i \leq N$$

$$h^{(i)}(z) = \frac{b^{(i)}(z)}{a^{(i)}(z)} = c_0 + \frac{c_1(z - z_0)}{1} + \cdots + \frac{c_i(z - z_{i-1})}{1} \tag{4-48b}$$

With respect to $a^{(i)}(z)$ and $b^{(i)}(z)$, the following three term recurrence relations hold [16, Sec. 2.1]:

$$a^{(i+1)}(z) = a^{(i)}(z) + c_{i+1}(z - z_i)a^{(i-1)}(z)$$

$$b^{(i+1)}(z) = b^{(i)}(z) + c_{i+1}(z - z_i)b^{(i-1)}(z) \qquad 0 \leq i \leq N - 1 \tag{4-49}$$

$$a^{(0)}(z) = 1, \ a^{(-1)}(z) = 1, \ b^{(0)}(z) = c_0, \ b^{(-1)}(z) = 0 \tag{4-50a}$$

$$a^{(0)}(z) = 1, \ a^{(-1)}(z) = 0, \ b^{(0)}(z) = c_0, \ b^{(-1)}(z) = 1 \tag{4-50b}$$

Our present aim is to determine c_i's, so that

$$f(z) - h^{(i)}(z) = e^{(i)}(z) \prod_{k=0}^{i} (z - z_k), \quad 0 \leq i \leq N \tag{4-51}$$

holds with $e^{(i)}(z)$ being analytic at z_k, $0 \leq k \leq i$. Suppose that (4-51) holds; then due to the uniqueness of multipoint Padé approximants,

$$h^{(2j)}(z) = h[j/j](z) = \frac{b[j/j](z)}{a[j/j](z)}, \quad h^{(2j+1)}(z) = h[j/j + 1](z) = \frac{b[j/j + 1](z)}{a[j/j + 1](z)}$$

$$\tag{4-52a}$$

$$h^{(2j)}(z) = h[j/j](z) = \frac{b[j/j](z)}{a[j/j](z)}, \quad h^{(2j+1)}(z) = h[j+1/j](z) = \frac{b[j+1/j](z)}{a[j+1/j](z)}$$

$$(4\text{-}52b)$$

Recall that $a[m/n](z)$ is represented by (4-26a) and apply Sylvester's determinant identity to it. We have the following three term recurrence relations:

$$a[j/j](z) = \alpha_{2j} a[j-1/j](z) + \beta_{2j}(z - z_{2j-1}) a[j-1/j-1](z)$$

$$a[j/j+1](z) = \alpha_{2j+1} a[j/j](z) + \beta_{2j+1}(z - z_{2j}) a[j-1/j](z)$$

$$\alpha_{2j} = \frac{\det H_j[z_0, \cdots, z_{2j-2}, z_{2j}]}{\det H_j[z_0, \cdots, z_{2j-2}]}, \quad \alpha_{2j+1} = \frac{\det H_{j+1}[z_0, \cdots, z_{2j-1}, z_{2j+1}]}{\det H_j[z_0, \cdots, z_{2j-1}]}$$

$$\beta_{2j} = -\frac{\det H_{j+1}[z_0, \cdots, z_{2j}]}{\det H_j[z_0, \cdots, z_{2j-2}]}, \quad \beta_{2j+1} = -\frac{\det H_{j+1}[z_0, \cdots, z_{2j+1}]}{\det H_j[z_0, \cdots, z_{2j-1}]}$$

$$(4\text{-}53a)$$

$$a[j/j](z) = \alpha_{2j} a[j/j-1](z) + \beta_{2j}(z - z_{2j-1}) a[j-1/j-1](z)$$

$$a[j+1/j](z) = \alpha_{2j+1} a[j/j](z) + \beta_{2j+1}(z - z_{2j}) a[j/j-1](z)$$

$$\alpha_{2j} = \frac{\det H_j[z_0, \cdots, z_{2j-2}, z_{2j}]}{\det H_{j-1}[z_0, \cdots, z_{2j-2}]}, \quad \alpha_{2j+1} = \frac{\det H_j[z_0, \cdots, z_{2j-1}, z_{2j+1}]}{\det H_j[z_0, \cdots, z_{2j-1}]}$$

$$\beta_{2j} = -\frac{\det H_j[z_0, \cdots, z_{2j}]}{\det H_{j-1}[z_0, \cdots, z_{2j-2}]}, \quad \beta_{2j+1} = -\frac{\det H_{j+1}[z_0, \cdots, z_{2j+1}]}{\det H_j[z_0, \cdots, z_{2j-1}]}$$

$$(4\text{-}53b)$$

Then (4-52) implies that

$$a[j/j](z) = \gamma_{2j} a^{(2j)}(z), \quad a[j/j+1](z) = \gamma_{2j+1} a^{(2j+1)}(z) \qquad (4\text{-}54a)$$

$$a[j/j](z) = \gamma_{2j} a^{(2j)}(z), \quad a[j+1/j](z) = \gamma_{2j+1} a^{(2j+1)}(z) \qquad (4\text{-}54b)$$

with appropriate constants γ_i's. Set

$$\gamma_i = \alpha_i \alpha_{i-1} \cdots \alpha_1 \qquad (4\text{-}55)$$

and substitute (4.54) into (4.53) to get

$$a^{(i+1)}(z) = a^{(i)}(z) + \frac{\beta_{i+1}}{\alpha_{i+1}\alpha_i} a^{(i-1)}(z) \qquad (4\text{-}56a)$$

A similar argument works for the numerator b(z) and leads to

$$b^{(i+1)}(z) = b^{(i)}(z) + \frac{\beta_{i+1}}{\alpha_{i+1}\alpha_i} b^{(i-1)}(z) \qquad (4\text{-}56b)$$

Comparing (4-49) and (4-56), we conclude that $c_{i+1} = \beta_{i+1}/(\alpha_{i+1}\alpha_i)$; that is,

$$c_0 = f[z_0], \qquad c_1 = -\frac{f[z_0, z_1]}{f[z_1]}$$

$$c_{2j} = -\frac{\det H_{j-1}[z_0, \cdots, z_{2j-3}]\, \det H_{j+1}[z_0, \cdots, z_{2j}]}{\det H_j[z_0, \cdots, z_{2j-3}, z_{2j-1}]\, \det H_j[z_0, \cdots, z_{2j-2}, z_{2j}]}$$

$$j = 1, 2, \ldots$$

$$c_{2j+1} = -\frac{\det H_j[z_0, \cdots, z_{2j-2}]\, \det H_{j+1}[z_0, \cdots, z_{2j+1}]}{\det H_j[z_0, \cdots, z_{2j-2}, z_{2j}]\, \det H_{j+1}[z_0, \cdots, z_{2j-1}, z_{2j+1}]}$$

$$(4\text{-}57a)$$

$$c_0 = f[z_0], \qquad c_1 = f[z_0, z_1]$$

$$c_{2j} = -\frac{\det H_{j-1}[z_0, \cdots, z_{2j-3}]\, \det H_j[z_0, \cdots, z_{2j}]}{\det H_{j-1}[z_0, \cdots, z_{2j-3}, z_{2j-1}]\, \det H_j[z_0, \cdots, z_{2j-2}, z_{2j}]}$$

$$j = 1, 2, \ldots$$

$$c_{2j+1} = -\frac{\det H_{j-1}[z_0, \cdots, z_{2j-2}]\, \det H_{j+1}[z_0, \cdots, z_{2j+1}]}{\det H_j[z_0, \cdots, z_{2j-2}, z_{2j}]\, \det H_j[z_0, \cdots, z_{2j-1}, z_{2j+1}]}$$

$$(4\text{-}57b)$$

Equations (4-57) give an explicit representation of c_i's in terms of divided differences calculated by the given data. It is required that c_i's be all nonzero finite values, so that $h^{(i)}(z)$'s are all well defined. Thus

$$\det H_j[z_1, \cdots, z_{2j-1}] = 0; \text{ for any permutations of } z_0, \cdots, z_{2j-1}$$

$$(4\text{-}58a)$$

$$\det H_j[z_1, \cdots, z_{2j}] = 0; \text{ for any permutations of } z_0, \cdots, z_{2j}$$

$$j = 1, 2, \ldots$$

$$\det H_j[z_1, \cdots, z_{2j}] = 0; \text{ for any permutations of } z_0, \cdots, z_{2j}$$

$$(4\text{-}58b)$$

$$\det H_j[z_1, \ldots, z_{2j+1}] = 0; \quad \text{for any permutations of } z_0, \ldots, z_{2j+1}$$

must hold. Conversely, under the assumption (4-58), $h^{(i)}(z)$ with the parameters c_i's in (4-57) satisfies (4-51); that is, $h^{(i)}(z)$ coincides with the multipoint Padé approximant interpolating the given data at $z_0, z_1, \ldots, z_i$. Finally, we can represent the [m/n] multipoint Padé approximants with $|m - n| \leq 1$ as the continued fraction (4-46a) or (4-46b) if and only if either (4-58a) or (4-58b) holds [11].

It should be noted that c_i in (4-57) depends only on the data at the points $z_0, z_1, \ldots, z_i$, which enable us to construct the multipoint Padé approximants successively through continued fractions. The algorithm to calculate the approximant based on (4-46a) is as follows [a similar algorithm based on (4-46b) can also be derived]:

Algorithm 4-3

(i) Initialization

$$c_0 = f[z_0], \quad c_1 = - \frac{f[z_0, z_1]}{f[z_1]}$$

$$a_0^{(0)} = 1, \quad a_0^{(1)} = 1 - c_1 z_0, \quad a_1^{(1)} = c_1$$

$$b_0^{(0)} = c_0, \quad b_0^{(1)} = c_0$$

$$a_{00}^{(0)} = 1, \quad a_{00}^{(1)} = 1, \quad a_{01}^{(1)} = c_1$$

$$b_{00}^{(0)} = c_0, \quad b_{00}^{(1)} = c_0$$

(ii) Iteration. For $i = 1, 2, \ldots, N - 1$

$$\sigma_{i+1} = \sum_{j=0}^{\left[\frac{i+1}{2}\right]} a_{0j}^{(i)} f[z_j, \ldots, z_i, z_{i+1}]$$

$$\tau_{i+1} = \sum_{j=0}^{\left[\frac{1}{2}\right]} a_{0j}^{(i-1)} f[z_j, \ldots, z_{i-1}, z_{i+1}]$$

(4-59)

$$c_{i+1} = - \frac{\sigma_{i+1}}{\tau_{i+1}}$$

(4-60)

$$a_j^{(i+1)} = a_j^{(i)} + c_{i+1}(a_{j-1}^{(i-1)} - z_i a_j^{(i-1)}), \qquad 0 \le j \le \left[\frac{i}{2}\right] + 1$$

$$b_j^{(i+1)} = b_j^{(i)} + c_{i+1}(b_{j-1}^{(i-1)} - z_i b_j^{(i-1)}), \qquad 0 \le j \le \left[\frac{i+1}{2}\right]$$

(4-61)

$$a_{0j}^{(i+1)} = a_{0j}^{(i)} + c_{i+1}\{a_{0,j-1}^{(i-1)} + (z_j - z_i)a_{0j}^{(i-1)}\}, \qquad 0 \le j \le \left[\frac{i}{2}\right] + 1$$

$$b_{0j}^{(i+1)} = b_{0j}^{(i)} + c_{i+1}\{b_{0,j-1}^{(i-1)} + (z_j - z_i)b_{0j}^{(i-1)}\}, \qquad 0 \le j \le \left[\frac{i+1}{2}\right]$$

(4-62)

(iii) Termination

$$h(z) = \frac{b(z)}{a(z)} = \frac{b_0^{(N)} + b_1^{(N)} + \cdots + b_m^{(N)} z^m}{a_0^{(N)} + a_1^{(N)} + \cdots + a_n^{(N)} z^n}; \qquad m = \left[\frac{N}{2}\right], \quad n = \left[\frac{N+1}{2}\right]$$

In the algorithm above, $a_j^{(i)}$, $b_j^{(i)}$, $a_{0j}^{(i)}$, and $b_{0j}^{(i)}$ are, respectively, the coefficients of $h^{(i)}(z)$, so that

$$h^{(i)}(z) = \frac{\sum_{j=0}^{\left[\frac{i}{2}\right]} b_j^{(i)} z^j}{\sum_{j=0}^{\left[\frac{i+1}{2}\right]} a_j^{(i)} z^j} = \frac{\sum_{j=0}^{\left[\frac{i}{2}\right]} b_{0j}^{(i)} \Pi_{k=0}^{j-1} (z - z_k)}{\sum_{j=0}^{\left[\frac{i+1}{2}\right]} a_{0j}^{(i)} \Pi_{k=0}^{j-1} (z - z_k)}$$

(4-63)

To verify (4-59) and (4-60), which are main part of the algorithm, we shall show that

$$\det H_{j+1}[z_0, \cdots, z_{2j}] = \gamma_{2j-1} \sigma_{2j}$$

$$\det H_{j+1}[z_0, \cdots, z_{2j+1}] = \gamma_{2j} \sigma_{2j+1}$$

$$\det H_j[z_0, \cdots, z_{2j-2}, z_{2j}] = \gamma_{2j-2} \tau_{2j}$$

(4-64)

$$\det H_{j+1}[z_0, \cdots, z_{2j-1}, z_{2j+1}] = \gamma_{2j-1} \tau_{2j+1}$$

It follows from (4-26a) and (4-54a) that

$$
a[j - 1/j] = \det \begin{bmatrix} f_{jj} & \cdots & f_{1j} & f_{0j} \\ \vdots & & \vdots & \vdots \\ f_{j,2j-1} & \cdots & f_{1,2j-1} & f_{0,2j-1} \\ \prod_{k=0}^{j-1}(z-z_k) & \cdots & (z-z_0) & 1 \end{bmatrix}
$$

$$
= \sum_{k=0}^{j} \gamma_{2j-1} a_{0k}^{(2j-1)} \prod_{i=0}^{k-1}(z-z_i)
$$

Considering the equation above, we expand

$$
\det H_{j+1}[z_0, \ldots, z_{2j}] = \det \begin{bmatrix} f_{jj} & \cdots & f_{1j} & f_{0j} \\ \vdots & & \vdots & \vdots \\ f_{j,2j-1} & \cdots & f_{1,2j-1} & f_{0,2j-1} \\ f[z_j,\ldots,z_{2j}] & \cdots & f[z_1,\ldots,z_{2j}] & f[z_0,\ldots,z_{2j}] \end{bmatrix}
$$

by its last row to get

$$
\det H_{j+1}[z_0, \ldots, z_{2j}] = \sum_{k=0}^{j} \gamma_{2j-1} a_{0k}^{(2j-1)} f[z_k,\ldots,z_{2j}]
$$

which is equivalent to the first one in (4-64). The remaining three equations in (4-64) are also shown via similar arguments. Thus from (4-57a) we conclude that

$$
c_{2j} = -\frac{1}{\alpha_{2j-1}} \frac{\gamma_{2j-1}\sigma_{2j}}{\gamma_{2j-2}\tau_{2j}} = -\frac{\sigma_{2j}}{\tau_{2j}}
$$

$$
j = 1, 2, \ldots
$$

$$
c_{2j+1} = -\frac{1}{\alpha_{2j}} \frac{\gamma_{2j}\sigma_{2j+1}}{\gamma_{2j-1}\tau_{2j+1}} = -\frac{\sigma_{2j+1}}{\tau_{2j+1}}
$$

[recall (4-55)], which is just equivalent to (4-60). Also, (4-61) and (4-62) are easily verified using the relations in (4-49).

In general, the following properties are important and desirable for the algorithm of calculating the multipoint Padé approximants based on continued fractions.

     (i) It is simple and demands a small amount of computation.

    (ii) It allows confluent points.

  (iii) It can find the approximants with arbitrary degrees.

  (iv) It achieves the objective approximant $[h^{(N)}(z)$ in the above] independent of the existence of the intermediate approximans $[h^{(i)}(z)$'s
in the above].

   (v) It can detect whether or not the objective approximant exists.

  (vi) It is numerically stable; that is, small changes in the data lead to
small changes in the approximant.

The algorithm we have presented here is not satisfactory with respect
to properties (iii) to (vi). Some advanced algorithms that have some of these
properties are dealt with in [1, Part II, Sec. 1.1].

4-6 CONCLUSIONS

In this chapter we have investigated multipoint Padé approximation from a
system theoretical viewpoint. Three types of expression of the $[m/n]$ multipoint Padé approximant—(4-25) with (4-24), (4-29) with (4-28), and (4-38)
with (4-37)—have been derived on the assumption that it exists. How to examine the existence and degrees of $[m/n]$ multipoint Padé approximant has been
shown in Theorems 4-1 and 4-2, which are the extended results of those on
Padé approximation. A minimal partial realization problem has been formulated as a problem to find a discrete-time system of minimal degree that
interpolates the given input-output data (4-16) with z interpreted as a delay
element. Theorem 4-3 states how to determine the minimal degree, which
is also an extended result of that on Padé approximation, that is, minimal
partial realization based on a finite sequence of Markov parameters. In the
multipoint Padé problem, due to the possibility of inaccessible points occur-
ring, the conditions for the existence [Eq. (4-39) or (4-40)] and the minimal
degree [Eq. (4-45)] must be checked for some permutations of the interpolation points, which is much more cumbersome than in the Padé problem.
Therefore, the results in Theorems 4-1, 4-2, and 4-3 are not very attractive
from a practical viewpoint. In Sec. 4-4 more practical procedures (Algorithms 4-1 and 4-2) have also been presented, by which we can find the [m/n]
multipoint Padé approximant if it exists and the minimal partial realization
with a small amount of computation. If no inaccessible points occur for $[j/j]$
and $[j/j + 1]$ or $[j/j]$ and $[j + 1/j]$ multipoint Padé approximants $(j = 0, 1, \ldots)$,
that is, conditions (4-58a) or (4-58b) hold, the approximant $h[m/n](z)$ with
$|m - n| \le 1$ can be represented by the continued fraction (4-46a) or (4-46b).
Based on this representation, a simple numerical procedure (Algorithm 4-3)
to permit successive updating of the approximant has been derived.

    Multipoint Padé approximation is the most general method available to
find a rational function that interpolates the desired data and is sufficiently

powerful in many fields of circuit and system theory. However, it should be noted that except for the property of interpolating the data, the approximant is not guaranteed to have any kinds of properties (which might be expected), such as being analytic in $|z| \leq 1$ or being bounded in $|z| \leq 1$. If one requires that the approximant have some properties, they must be embedded in the data to be interpolated beforehand. Conversely, if the data satisfy some conditions, one can always obtain an approximant that both has the required properties and interpolates the data. One of the most popular and important cases for circuit and system theory is that in Nevanlinna-Pick interpolation theory [19, 20].

REFERENCES

1. G. A. Baker, Jr., and P. R. Graves-Morris, Padé Approximants, Part I: Basic Theory, Part II: Extensions and Applications, Addison-Wesley, London, 1981.

2. K. Morimoto, N. Matsumoto, N. Hamada, and S. Takahashi, "Matrix Padé approximants and multiport networks," Trans. IECE (Japan), vol. 61-A, no. 4, pp. 28-36 (1978).

3. R. Hastings-James and S. K. Mehra, "Extensions of the Padé-approximant technique for the design of recursive digital filters," IEEE Trans. Acoust. Speech Signal Process., vol. ASSP-25, no. 6, pp. 501-509 (1977).

4. A. J. Tether, "Construction of minimal linear state-variable models from input-output data," IEEE Trans. Autom. Control, vol. AC-15, no. 4, pp. 427-436 (1970).

5. Y. Shamash, "Linear system reduction using Padé approximation to allow retention of dominant modes," Int. J. Control, vol. 21, no. 2, pp. 257-272 (1975).

6. K. Horiguchi, S. Hayashi, and N. Hamada, "New sets of stability criteria as a generalization of Markov stability theorem," Int. J. Control, vol. 43, no. 5, pp. 1581-1591 (1986).

7. S. S. Lamba and G. Michaeilesco, "Comment on suboptimal control using Padé approximations techniques," IEEE Trans. Autom. Control, vol. AC-27, no. 1, pp. 279-280 (1982).

8. R. E. Kalman, "On partial realizations, transfer functions, and canonical forms," Acta Tech. Scand., vol. 31, pp. 9-32 (1979).

9. J. Meinguet, "On the solubility of the Cauchy interpolation problem," in Approximation Theory, ed. A. Talbot, Academic Press, London, 1970, pp. 137-163.

10. V. Belevitch, "Interpolation matrices," Philips. Res. Rep., vol. 25, pp. 337-369 (1970).

11. K. Horiguchi and N. Hamada, "System theoretical consideration on N-point Padé approximation," Trans. IECE (Jpn.), vol. J-68-A, no. 3, pp. 263-270 (1985) (in Japanese).

12. G. Claessens, "On the Newton-Padé approximation problem," J. Approx. Theory, vol. 22, pp. 150-160 (1978).

13. D. R. Audley, "A method of constructing minimal approximate realizations of linear input-output behavior," Automatica, vol. 13, pp. 409-415 (1977).

14. J. Rissanen, "Recursive identification of linear systems," SIAM J. Control, vol. 9, no. 3, pp. 420-430 (1971).

15. L. Weinberg, Network Analysis and Synthesis, McGraw-Hill, New York, 1962, Chap. 9.

16. W. B. Jones and W. J. Thron, Continued Fractions, Analytic Theory and Applications, Addison-Wesley, London, 1980.

17. D. J. Wright, "The continued fraction representation of transfer functions and model simplification," Int. J. Control, vol. 18, no. 3, pp. 449-454 (1973).

18. Y. Katsube, K. Horiguchi, and N. Hamada, "System reduction by continued-fraction expansion about $s = j\omega_i$," Electron. Lett., vol. 21, no. 16, pp. 678-680 (1985).

19. Ph. Delsarte, Y. Genin, and Y. Kamp, "On the role of the Nevanlinna-Pick problem in circuit and system theory," Circuit Theory Appl., vol. 9, pp. 177-187 (1981).

20. K. Horiguchi and N. Hamada, "Multipoint Padé approximation and its application to design of all-pass type digital filter," 7th European Conf. Circuits, Systems, and Theory, vol. 2, pp. 407-410, 1985.

5
Algebraic Stability Criteria and Their Application to Model Reduction

NOZUMU HAMADA

Keio University, Yokohama, Japan

5-1 INTRODUCTION

Model reduction is a current issue in the search for a simplified model that adequately approximates a given complex system. The principal aims in simplifying a system are:

(i) To reduce computational efforts in the analysis and simulation of a higher-order system
(ii) To reduce computational efforts in the design of a control system
(iii) To simplify implementation of a controller

The method of model reduction of scalar systems has been discussed extensively [1-4], grouped into several techniques. The well-known reduction methods of transfer function are based on (1) the dominant-pole approach [5], (2) Padé approximation [6-8], (3) continued fraction expansion [9-13], (4) the truncation of a particular form that relies on stability conditions such as Routh and Hurwitz [14-18, 31], (5) Hankel norm approximation [19], (6) error minimization, and so on. In the state space domain, the reduction methods are based on (7) balanced realization [20], (8) theory of aggregation [21,22], (9) perturbation methods, and so on. One of the important requirements for the reduction method is that an approximant of a stable system be stable.

To preserve stability, some reduction methods, such as the methods in [14-18], use the stability criteria and the related canonical representation for a given transfer function. The basic idea is stated as follows. At first, a canonical representation of a given transfer function is obtained. In this procedure, stability conditions are expressed explicitly in the realization of the denominator polynomial. Then a reduced-order model can be obtained by truncating the canonical form such that it preserves stability.

Another important aspect of model reduction is how the characteristic of an original system is approximated by a reduced-order system. One popular error criterion between the original and the reduced model is the squared integral or squared sum of the error. In fact, there is an approach to reducing the original system by minimizing such an error criterion. With regard to computational cost, this type of error minimization approach is labor-intensive. On the other hand, Padé approximation uses another criterion to measure the error between the original system and its approximant. The coefficients of Laurent expansion of the original transfer function and reduced model are equated as much as possible. The Padé procedure, even though simple to apply, does not always give a stable reduced system for a given stable higher-order system. Routh approximation [14] is a representative approach to overcoming this defect of the Padé approximation; it preserves both stability and a number of expansion coefficients in a Laurent series of the original system.

Topics dealt with in this chapter and Chapter 7 include two extensions of Routh approximations. Extended Routh approximations of scalar continuous-time and discrete-time systems are treated in this chapter. Another extension of Routh approximations for multiinput/multioutput continuous-time systems is described in Chapter 7.

Routh Approximation

Let us review briefly the conventional Routh approximation techniques of Hutton and Friedland [14]. Routh's stability criterion is a key to simplifying a system. The denominator polynomial f(s) of a system is decomposed into its even and odd parts:

$$f(s) = h(s^2) + sg(s^2) = \sum_{i=0}^{n} a_{n-i} s^i$$

Their ratio, either $sg(s^2)/h(s^2)$ or $h(s^2)/sg(s^2)$, is expanded into the following continued fraction:

$$\frac{1}{\alpha_1 s} + \frac{1}{\alpha_2 s} + \cdots + \frac{1}{\alpha_n s} \qquad (5\text{-}1)$$

This presentation is called Cauer's continued fraction expansion when all coefficients α_k ($k = 1, \ldots, n$) are positive. The existence of Cauer's continued fraction for the polynomial $f(s)$ is the necessary and sufficient condition that the polynomial $f(s)$ be stable [24], or equivalently, that rational function (5-1) is an LC driving point function (reactance function [32]). The expansion coefficients α_k are computed using the Routh array.

The Routh representation of a stable transfer function via Cauer's continued fraction is the canonical representation, written as

$$H(s) = b_0 + \frac{F_{1,n}(s)}{1 + F_{1,n}(s)} [\beta_1 + F_{2,n}(s)[\beta_2 + \cdots + F_{n,n}(s)[\beta_n] \cdots]] \qquad (5\text{-}2)$$

where

$$F_{i,n}(s) = \frac{1}{\alpha_i s} + \cdots + \frac{1}{\alpha_n s}$$

The state space realization can be derived from the Routh representation (5-2). It is called a Routh realization and its system matrix is given by the symmetric Schwartz form [23].

The mth-order Routh approximation of $H(s)$ is given by truncating Routh representation as follows:

$$H_m(s) = b_0 + \frac{F_{1,m}(s)}{1 + F_{1,m}(s)} [\beta_1 + F_{2,m}(s)[\beta_2 + \cdots + F_{m,m}(s)[\beta_m] \cdots]] \qquad (5\text{-}3)$$

where

$$F_{i,m}(s) = \frac{1}{\alpha_i s} + \frac{1}{\alpha_{i+1} s} + \cdots + \frac{1}{\alpha_m s}$$

The truncation procedure from (5-2) to (5-3) corresponds to state aggregation of the original Routh realization.

Remark 1

To calculate the coefficients of the reduced transfer function from those of the original system, we can use an effective tabular algorithm called an alpha-beta table [14].

Remark 2

Routh approximation assures the stability of a reduced system for any stable higher-order systems.

Remark 3

The first $m + 1$ Markov parameters of the Routh approximant are exactly the same as those of the original system. This means that the Routh approximation preserves the high-frequency characteristics of the original system. When we hope to obtain a good approximation in the low-frequency range, as usually required in the case of approximating a controlled system, suitable low-high frequency transformation is used together with the reduction process.

Remark 4

The energy E_i of impulse response $h_i(t)$ of the Routh approximant $H_i(s)$ [where $H_n(s) = H(s)$], defined as

$$E_i = \int_0^{+\infty} h_i(t)^2 \, dt \tag{5-4}$$

converges monotonically to the impulse response energy of the original system. That is,

$$E_1 \leq E_2 \leq \cdots \leq E_n \tag{5-5}$$

5-2 STABILITY CRITERIA BY CONTINUED FRACTIONS WITH TWO-POINT EXPANSION

5-2-1 Continuous-Time System

The basic idea underlying the extension of conventional Routh approximation is to generalize the Routh stability criterion. A new form of the Hurwitz test used here is stated as the next theorem [25].

Theorem 5-1. If an nth-order polynomial $f(s) = h(s^2) + sg(s^2)$ with order n (= 2p or 2p + 1) is Hurwitz, the ratio $sg(s^2)/h(s^2)$ can always be expanded in the following two types of continued fraction.

Type C1

$$\frac{sg(s^2)}{h(s^2)} = \alpha_0 s + \cfrac{1}{\alpha_1 s} + \cdots + \cfrac{1}{\alpha_{2r} s} + \cfrac{1}{\alpha_{2r+1} s + \alpha_{2r+2} s^{-1}} + \cdots + \cfrac{1}{\alpha_{2p-1} s + \alpha_{2p} s^{-1}}$$

$$(5\text{-}6)$$

Type C2

$$\frac{sg(s^2)}{h(s^2)} = \alpha_0^* s + \cfrac{1}{\alpha_1^* s} + \cdots + \cfrac{1}{\alpha_{2r}^* s} + \cfrac{1}{\alpha_{2r+1}^* s^{-1} + \alpha_{2r+2}^* s} + \cdots + \cfrac{1}{\alpha_{2p-1}^* s^{-1} + \alpha_{2p}^* s}$$

$$(5\text{-}7)$$

where r is an arbitrary integer between 0 and p, and

$$\alpha_0 = \alpha_0^* = 0 \text{ for } n = 2p, \quad \alpha_0 = \alpha_0^* > 0 \text{ for } n = 2p + 1$$

$$\alpha_i > 0, \quad \alpha_i^* > 0 \quad (1 \le i \le 2p)$$

$$(5\text{-}8)$$

Conversely, for any r $(0 \le r \le p)$, if either of the continued fraction expansions (5-6) and (5-7) with the conditions (5-8) exists, the polynomial $f(s) = h(s^2) + sg(s^2)$ is Hurwitz. ∎

Proofs of this theorem and of some sets of stability conditions in relation to Markov's stability criterion are given in [25]. From circuit-theoretical points of view, expansion (5-6) is the mixed form of Cauer's first and third forms [32], and expansion (5-7) is the mixture of Cauer's second and third forms. Thus two continued fraction expansions are relevant to the ladder realizations of a reactance function $sg(s^2)/h(s^2)$. In the particular case of r = p, type C1 becomes Cauer's expansion (5-1) used in conventional Routh approximation.

5-2-2 Discrete-Time System

The z-domain analogy of the stability criterion, Theorem 5-1, is discussed here. For a denominator polynomial of a given transfer function,

$$f_n(z) = \sum_{i=0}^{n} a_{n-i} z^i$$

$$(5\text{-}9)$$

The tangent function $\rho_n(z)$ is defined as [33]

$$\rho_n(z) = \frac{f_n(z) - f_n^*(z)}{f_n(z) + f_n^*(z)} \qquad\qquad (5\text{-}10)$$

where $f_n^*(z)$ is the reciprocal polynomial of $f_n(z)$, that is,

$$f_n^*(z) = \sum_{i=0}^{n} a_i z^i = z^n f_n(z^{-1})$$

A real polynomial $f_n(z)$ can be represented as

$$f_n(z) = \frac{1}{2} S_n(z) + \frac{1}{2} A_n(z)$$

where

$$S_n(z) = f_n(z) + f_n^*(z) = S_n^*(z)$$

$$A_n(z) = f_n(z) - f_n^*(z) = -A_n^*(z)$$

More precisely, we have

$$f_{2p+1}(z) = \frac{1}{2} A_{2p+1}(z) + \frac{1}{2}(z + 1) S_{2p}(z)$$

$$f_{2p}(z) = \frac{1}{2} S_{2p}(z) + \frac{1}{2}(z + 1) A_{2p-1}(z)$$

where relations $S_{2i+1}(z) = (z + 1) S_{2i}(z)$ and $A_{2i}(z) = (z + 1) A_{2i-1}(z)$ are used.

The necessary and sufficient condition for the polynomial $f_n(z)$ to be stable is that all the zeros of $S_n(z)$ and $A_n(z)$ be on the unit circle in the z-plane and that they be simple and alternate with one another [26].

The ratio of $A_n(z)$ and $S_n(z)$, $\rho_n(z) = A_n(z)/S_n(z)$ and its inverse $\rho_n^{-1}(z)$ derived from stable $f_n(z)$, are called the z-domain reactance function. There are several continued fraction expansions which correspond to Cauer's forms in the s-domain.

The z-domain analogy of Cauer's continued fraction in the s-domain was proposed by Davis [27], Bistritz [28], and Ismail [29, 36]. This is introduced by applying LDI (lossless digital integrator) transformation to Cauer's continued fraction in the s-domain. The LDI transformation

$$s = \frac{z^{\frac{1}{2}} - z^{-\frac{1}{2}}}{2}$$

proposed by Bruton [34] for synthesizing digital filter and switched capacitor filters is the conformal mapping that maps the unit circle in the z-domain $C = \{z \mid |z| = 1\}$ on the interval of the imaginary axis $J = \{s \mid s = j\omega, \omega \in (-1, 1)\}$.

Of course, the well-known bilinear transformation $s = (z - 1)/(z + 1)$ can derive z-domain continued fraction expansions, which provide schemes to test stability. The terms $(z - 1)/(z + 1)$ and $(z + 1)/(z - 1)$, however, do not make an appropriate analogy to the meaning of s and s^{-1} in continuous-time systems. Cauer's first form in the z-domain expansion, called the discrete Cauer I form (DCI) [30], is given by

$$n = 2p: \ z(z + 1)^{-1} \rho_{2p}(z) = \frac{zA_{2p-1}(z)}{S_{2p}(z)}$$

$$= \frac{1}{c_1(1 - z^{-1})} + \frac{1}{c_2(z - 1)} + \cdots + \frac{1}{c_{2p}(z - 1)} \qquad (5\text{-}11)$$

For other forms, such as the discrete Cauer second and third forms (DCII, DCIII), consult [30].

Taking LDI transformations of (5-6) and (5-7) yields two types of z-domain continued fraction.

Type D1

(i) $n = 2p$

$$z(z + 1)^{-1} \rho_{2p}(z) = \frac{zA_{2p-1}(z)}{S_{2p}(z)}$$

$$= \frac{1}{c_1(1 - z)^{-1}} + \frac{1}{c_2(z - 1)} + \cdots + \frac{1}{c_{2r}(z - 1)}$$

$$+ \frac{1}{c_{2r+1}(1 - z^{-1}) + c^*_{2r+1}/(z - 1)} + \frac{1}{c_{2r+2}(z - 1) + c^*_{2r+2}/(1 - z^{-1})} + \cdots$$

$$(5\text{-}12)$$

(ii) $n = 2p + 1$

$$(z + 1)\rho_{2p+1}(z) = \frac{A_{2p+1}(z)}{S_{2p}(z)} = c_0(z - 1) + \text{right-hand side of (5-12)}$$

Type D2

(i) $n = 2p$

$$(z + 1)^{-1}\rho_{2p}(z) = \frac{A_{2p-1}(z)}{S_{2p}(z)}$$

$$= \frac{1}{c_1^*/(1 - z^{-1})} + \frac{1}{c_2^*/(z - 1)} + \cdots + \frac{1}{c_{2r}^*/(z - 1)}$$

$$+ \frac{1}{c_{2r+1}^*/(1 - z^{-1}) + c_{2r+1}(z - 1)} + \frac{1}{c_{2r+2}^*/(z - 1) + c_{2r+2}(1 - z^{-1})} + \cdots$$

$$(5-13)$$

(ii) $n = 2p + 1$

$$(z + 1)\rho_{2p+1}(z) = \frac{A_{2p+1}(z)}{S_{2p}(z)} = \frac{1}{c_0^*/(z - 1)} + \text{right-hand side of (5-13)}$$

Remark 1

Type D1 (Type D2) continued fraction (5-12) [(5-13)] is the mixture of DC I [(DC II)] and DC III.

Remark 2

Integer r is an arbitrary number satisfying $0 \leq r \leq [n/2]$ ([ ] = Gauss number). When $r = 0$, both type D1 of (5-12) and type D2 of (5-13) become DC III. When $r = [n/2]$, type D1 of (5-12) becomes DC I of (5-11) and type D2 of (5-13) becomes DC II.

Unlike the s-domain case, the positivity of all c_i, c_i^* in (5-12) or (5-13) is not sufficient for the stability of $f(z)$. In the expansion (5-11), Bistritz [28] derives additional conditions with which positivity of all c_i and c_i^* imply stability of $f(z)$.

For the derivation of the necessary and sufficient condition of $f(z)$ in terms of expansions (5-12), we define the intermediate polynomials $f_j(z)$ and $f_k(z)$ for $n = 2p$ in the expansion of (5-12):

$$z(z + 1)^{-1}\rho_j(z) = \frac{z[f_j(z) - f_j^*(z)]}{(z + 1)[f_j(z) + f_j^*(z)]} \qquad \begin{array}{l} 0 \leq i \leq r - 1 \\ j = 2(p - i) \end{array}$$

$$= \frac{1}{c_{2i+1}(1 - z^{-1})} + \frac{1}{c_{2i+2}(z - 1)} + \cdots \qquad (5-14)$$

$$z(z + 1)^{-1}\rho_k(z) = \frac{z[f_k(z) - f_k^*(z)]}{(z + 1)[f_k(z) + f_k^*(z)]} \qquad \begin{array}{l} 0 \le i \le p - r - 1, \\ k = 2(p - r - i) \end{array}$$

$$= \frac{1}{c_{2r+1+i}(1 - z^{-1}) + c_{2r+1+i}^*/(z - 1)} + \cdots \qquad (5\text{-}15)$$

Theorem 5-2 [37]. A given 2pth-order polynomial $f(z)$ is stable if and only if the expansion (5-12) exists and

$$c_i > 0, \quad c_j^* > 0 \quad (i = 1, 2, \ldots, p + r; \; j = 2r + 1, \ldots, p + r)$$

are satisfied and the polynomials defined by (5-14) and (5-15) satisfy

$$f_j(-1) > 0, \quad j = 2p, 2(p - 1), \ldots, 2(p - r + 1)$$

$$f_k(-1) > 0, \quad k = 2(p - r), 2(p - r - 1), \ldots, 2 \qquad (5\text{-}16)$$

In the case of type D2 expansion, intermediate polynomials $f_j(z)$, $f_k(z)$ for $n = 2p$ are defined as

$$(z + 1)^{-1}\rho_j(z) = \frac{z[f_j(z) - f_j^*(z)]}{(z + 1)[f_j(z) + f_j^*(z)]} \qquad \begin{array}{l} 0 \le i \le r \\ j = 2(p - i + 1) \end{array}$$

$$= \frac{1}{c_{2i-1}/(1 - z^{-1})} + \frac{1}{c_{2i}/(z - 1)} + \cdots \qquad (5\text{-}17)$$

$$(z + 1)^{-1}\rho_k(z) = \frac{[f_k(z) - f_k^*(z)]}{(z + 1)[f_k(z) + f_k^*(z)]} \qquad \begin{array}{l} 0 \le i \le p - r - 1 \\ k = 2(p - r - i) \end{array}$$

$$= \frac{1}{c_{2r+1+i}^*/(1 - z^{-1}) + c_{2r+1+i}(z - 1)} + \cdots \qquad (5\text{-}18)$$

$\blacksquare$

Theorem 5-3 [37]. A given 2pth polynomial $f(z)$ is stable if and only if the expansion (5-13) exists and

$$c_i > 0, \quad c_j^* > 0 \quad (i = 1, 2, \ldots, p + r; \; j = 2r + 1, \; p + r)$$

and the polynomials defined by (5-17) and (5-18) satisfy

$$f_j(-1) > 0, \qquad j = 2p, 2(p-1), \ldots, 2(p-r+1)$$

$$f_k(-1) > 0, \qquad k = 2(p-r), 2(p-r-1), \ldots, 2 \tag{5-19}$$

In the case of odd degree, $f_{2p+1}(z)$ is stable if and only if $c_0 = (a_0 - a_{2p+1})/(a_0 + a_{2p+1})$ is positive and $f_{2p}(z) = \{1 - c_0(z - z^{-1})\} S_{2p}(z) + (1 + z^{-1}) A_{2p+1}(z)$ is an even-order stable polynomial, where $2f_{2p+1}(z) = A_{2p+1}(z) + (z+1) S_{2p}(z)$. ∎

5-3 EXTENDED ROUTH APPROXIMATION

This section is devoted to dealing with the extended Routh approximation in both continuous- and discrete-time systems. But owing to limited space, original and reduced systems are restricted to even-order systems.

5-3-1 Continuous-Time System

Let us consider a linear time-invariant continuous system with the transfer function

$$H(s) = \frac{\sum_{i=0}^{n} b_i s^{n-i}}{\sum_{i=0}^{n} a_i s^{n-i}} \tag{5-20}$$

If a transfer function $H(s)$ is asymptotically stable, it can always be expanded into the following two types of canonical forms for any integer $r(0 \le r \le p)$:

Type CE1

$$H(s) = b_0 + \frac{F_1(s)}{1 + F_1(s)} [\beta_1 + F_2(s)[\cdots + F_{2r}(s)[\beta_{2r} + F_{2r+2}(s)$$

$$\cdot [(\beta_{2r+1} + \beta_{2r+2} s^{-1}) + F_{2r+4}(s)[\cdots + F_{2p}(s)(\beta_{2p-1} + \beta_{2p} s^{-1})] \cdots]$$

$$\tag{5-21}$$

Type CE2

$$H(s) = b_0 + \frac{F_1^*(s)}{1 + F_1^*(s)} [\beta_1^* s^{-1} + F_2^*(s)[\cdots + F_{2r}^*(s)[\beta_{2r}^* s^{-1} + F_{2r+2}^*(s)$$

$$\cdot [(\beta_{2r+1}^* s^{-1} + \beta_{2r+2}^*) + F_{2r+4}^*(s)[\cdots + F_{2p}^*(s)(\beta_{2p-1}^* s^{-1} + \beta_{2p}^*)] \cdots]$$

$$\tag{5-22}$$

where $F_i(s)$ and F_i^* are defined by the continued fraction expansions

$$F_i(s) = \frac{1}{\alpha_i s} + \cdots + \frac{1}{\alpha_{2r} s} + \frac{1}{\alpha_{2r+1} s + \alpha_{2r+2} s^{-1}} + \cdots + \frac{1}{\alpha_{2p-1} s + \alpha_{2p} s^{-1}}$$

$$(i = 0, 1, \ldots, 2r)$$

$$(5\text{-}23)$$

$$F_{2j}(s) = \frac{1}{\alpha_{2j-1} s + \alpha_{2j} s^{-1}} + \cdots + \frac{1}{\alpha_{2p-1} s + \alpha_{2p} s^{-1}}, \qquad (j = r + 1, \ldots, p)$$

$$F_i^*(s) = \frac{1}{\alpha^* s^{-1}} + \cdots + \frac{1}{\alpha_{2r}^* s^{-1}} + \frac{1}{\alpha_{2r+1}^* s^{-1} + \alpha_{2r+2}^* s} + \cdots + \frac{1}{\alpha_{2p-1}^* s^{-1} + \alpha_{2p}^* s}$$

$$(i = 0, 1, \ldots, 2r)$$

$$(5\text{-}24)$$

$$F_{2j}^*(s) = \frac{1}{\alpha_{2j-1}^* s^{-1} + \alpha_{2j}^* s} + \cdots + \frac{1}{\alpha_{2p-1}^* s^{-1} + \alpha_{2p}^* s} \qquad (j = r + 1, \ldots, p)$$

and $\alpha_0 = \alpha_0^*$, $\beta_0 = \beta_0^*$, $F_0^*(s) = F_0(s)$. Similar formulations can be given for odd-order systems. To calculate the parameters (α_i, β_i) and (α_i^*, β_i^*) from the coefficients (a_i, b_i) of the transfer function (5-20), we can use an effective tabular algorithm which is an extension of the alpha-beta table in Routh approximation, as shown in Table 5-1.

The reduced systems with order m (= 2q; 0 < q < p) can be obtained by truncating the last 2(p - q) terms in the expansion (5-21) and (5-22) and replacing $F_i(s)$ with $F_{i,2q}(s)$ and $F_i^*(s)$ with $F_{i,2q}^*(s)$. That is,

Type CR1

$$H_{2q}(s) = b_0 + \frac{F_{1,2q}(s)}{1 + F_{1,2q}(s)}[\beta_1 + F_{2,2q}(s)[\cdots + F_{2r,2q}(s)[\beta_{2r}$$

$$+ F_{2r+2,2q}(s)[(\beta_{2r+1} + \beta_{2r+2} s^{-1}) + F_{2r+4,2q}(s)[\cdots$$

$$+ F_{2q,2q}(s)(\beta_{2q-1} + \beta_{2q} s^{-1})]$$

$$(5\text{-}25)$$

where

TABLE 5-1 Calculation Algorithm of (α_i, β_i) from (a_i, b_i)

(a) Type CE-1

From a_i to α_i

	a_0^0	a_1^0	a_2^0	$\cdots\cdots$	$a_{[n+\frac{1}{2}]}^0$
α_1	a_0^1	a_1^1	a_2^1	$\cdots\cdots$	
α_2	a_0^2	a_1^2	a_2^2	$\cdots\cdots$	
	$\cdot$	$\cdot$	$\cdot$		
	$\cdot$	$\cdot$	$\cdot$		

where

α-table

$$\begin{cases} a_i^0 = a_{2i} & (0 \le i \le \nu) \\[2mm] a_i^1 = a_{2i+1} & (0 \le i \le \nu - 1) \quad (n = 2\nu) \\[4mm] a_i^0 \quad a_{2i+1} & (0 \le i \le \nu) \\[2mm] a_i^1 = a_{2i+2} - \alpha_0 a_{2i+3} & (0 \le i \le \nu - 1) \quad (n = 2\nu + 1) \end{cases}$$

and

$$\alpha_0 = a_0/a_1 \quad (n = 2\nu + 1)$$

$$\alpha_i = a_0^{i-1}/a_0^i, \quad a_k^{2i+1} = a_{k+1}^{i-1} - \alpha_i a_{k+1}^i \quad (1 \le i \le 2\gamma + 1 : 0 \le k \le \nu - 1 - [i/2])$$

$$\begin{cases} \alpha_{2i} = a_{\nu-i}^{2i}/a_{\nu-i}^{2i-1}, \quad a_k^{2i+1} = a_k^{2i} - \alpha_{2i} a_k^{2i-1} & (\gamma+1 \le i \le \nu - 1 : \\[3mm] \alpha_{2i+1} = a_0^{2i-1}/a_0^{2i+1}, \quad a_k^{2i+2} = a_{k+1}^{2i-1} - \alpha_{2i+1} a_{k+1}^{2i+1} & 0 \le k \le \nu - i - 1) \end{cases}$$

$$\alpha_{2\nu} = a_0^{2\nu}/a_0^{2\nu-1}$$

From (a_i, b_i) to β_i

β_1	b_0^1	b_1^1	b_2^1	$\cdots\cdots$
β_2	b_0^2	b_1^2	b_2^2	$\cdots\cdots$
	$\cdot$	$\cdot$	$\cdot$	
	$\cdot$	$\cdot$	$\cdot$	

where

β-table

$$
\begin{cases}
b_i^1 = b_{2i+1} & (0 \le i \le \nu) \\[2ex]
b_i^2 = b_{2i+2} & (0 \le i \le \nu - 1) \qquad (n = 2\nu) \\[2ex]
b_i^1 = b_{2i+2} & (0 \le i \le \nu) \\[2ex]
b_i^2 = a_{2i+3} - \beta_0 a_{2i+4} & (0 \le i \le \nu - 1) \qquad (n = 2\nu + 1)
\end{cases}
$$

and

$$\beta_0 = b_1 / a_1 \qquad (n = 2\nu + 1)$$

$$\beta_i = b_0^i / a_0^i, \quad b_k^{i+2} = b_{k+1}^i - \beta_i a_{k+1}^i \quad (1 \le i \le 2\gamma : 0 \le k \le \nu - 1 - [i + \tfrac{1}{2}])$$

$$
\begin{cases}
\beta_{2i-1} = b_0^{2i-1}/a_0^{2i-1}, \quad b_k^{2i-1} = b_{k+1}^{2i-1} - \beta_{2i-1} a_{k+1}^{2i-1} & (\gamma + 1 \le i \le \nu - 1 : \\[2ex]
\beta_{2i} = b_{\nu-i}^{2i}/a_{\nu-i}^{2i-1}, \quad b_k^{2i+1} = b_k^{2i} - \beta_{2i} a_k^{2i-1} & \quad 0 \le k \le \nu - i - 1)
\end{cases}
$$

$$\beta_{2\nu-1} = b_0^{2\nu-1}/a_0^{2\nu-1}, \quad \beta_{2\nu} = b_0^{2\nu}/a_0^{2\nu-1}$$

(b) Type CE-2

Initial setting for α- and β-tables are given by

$$
\begin{cases}
a_i^0 = a_{2^\nu - 2i} & (0 \le i \le \nu) \\[2ex]
a_i^1 = a_{2^\nu - 2i - 1} & (0 \le i \le \nu - 1) \qquad (n = 2\nu) \\[2ex]
a_i^0 = a_{2^\nu - 2i} - \alpha_0 a_{2^\nu - 2i+1} & (0 \le i \le \nu) \\[2ex]
a_i^1 = a_{2^\nu - 2i} - \alpha_0 a_{2^\nu - 2i+1} & (0 \le i \le \nu - 1) \qquad (n = 2\nu + 1)
\end{cases}
$$

$$
\begin{cases}
b_i^1 = b_{2^\nu - 2i} & (0 \le i \le \nu) \\[2ex]
b_i^2 = b_{2^\nu - 2i - 1} & (0 \le i \le \nu - 1) \qquad (n = 2\nu) \\[2ex]
b_i^1 = b_{2^\nu - 2i+1} - \beta_0^* b_{2^\nu - 2i+2} & (0 \le i \le \nu) \\[2ex]
b_i^2 = a_{2^\nu - 2i} & (0 \le i \le \nu - 1) \qquad (n = 2\nu + 1)
\end{cases}
$$

And the same algorithm as in Type CE-1 yields α_i^* and β_i^*.

$$F_{i,2q}(s) = \frac{1}{\alpha_i s} + \cdots + \frac{1}{\alpha_{2r} s} + \frac{1}{\alpha_{2r+1} s + \alpha_{2r+2} s^{-1}} + \cdots + \frac{1}{\alpha_{2q-1} s + \alpha_{2q} s^{-1}}$$

$$(i = 0, 1, \ldots, 2r)$$

$$(5\text{-}26)$$

$$F_{2j,2q}(s) = \frac{1}{\alpha_{2j-1} s + \alpha_{2j} s^{-1}} + \cdots + \frac{1}{\alpha_{2q-1} s + \alpha_{2q} s^{-1}}, \quad (j = r+1, \ldots, q)$$

For type CR2 expansion, the same types of formulation are derived. In this case, reduced order is restricted to even numbers. Extended Routh approximations $H_{2q}(s)$ and $H^*_{2q}(s)$ preserve some useful properties, as in conventional Routh approximation.

Stability Preservation

If $H(s)$ is asymptotically stable, each approximant $H_{2q}(s)$ is always asymptotically stable. This is easily verified from Theorem 5-1.

Partial Padé Approximation

The extended Routh approximation achieves a partial two-point Padé approximation at $s = 0$ and $s = \infty$ [35]. The meaning of this statement is expressed as follows. If a given original system $H(s)$ and its reduced system $H_m(s)$ of order m (= 2q) are expanded by a Laurent series at $s = 0$ and $s = \infty$, respectively, as follows:

$$H(s) = \begin{cases} \Sigma h_i s^{-i} \\ \Sigma h^*_i s^i \end{cases} \qquad H_{2q}(s) = \begin{cases} \Sigma \hat{h}_i s^{-i} \\ \Sigma \hat{h}^*_i s^i \end{cases}$$

then the following equalities are satisfied.

Type CR1

$$h_i = \hat{h}_i \quad (0 < i < q - r - 1), \qquad h^*_i = \hat{h}^*_i \quad (0 \le i < q + r) \qquad (5\text{-}27)$$

Type CR2

$$h_i = \hat{h}_i \quad (0 < i < q - r - 1), \qquad h^*_i = \hat{h}^*_i \quad (0 \le i < q - r) \qquad (5\text{-}28)$$

These equalities mean that extended Routh approximations can match a number of Taylor series coefficients at $s = 0$ and those at $s = \infty$. The total number of these are $2q + 1$, and the trade-off between expansion coefficients matching at two points is achieved by selecting the integer r. Since a reduced transfer function is obtained for a given r, the extended Routh approximation problem from the 2pth original system to 2qth reduced systems has $2q + 1$ solutions.

With regard to the selection of r, there are no fixed rules. But the properties above provide us with the following information. Type-CR1 expansion is used for matching more coefficients at $s = \infty$ than at $s = 0$. If we choose a larger r, the derived system is likely to be approximated adequately in the high-frequency domain. Type CR2 expansion is useful for coefficients matching at $s = 0$. If a larger r is chosen for type CR2, the approximant is likely to approximate well in the lower-frequency domain. In both types with $r = 0$, the same numbers of coefficients matching at $s = 0$ and $s = \infty$ can be attained.

Impulse Response Energy

The energy E_i of impulse response of the system $H_{2i}(s)$ or $H_{2i}^{*}(s)$ defined by (5-4) monotonically increases by i, which is expressed as

$$E_0 \leq E_2 \leq \cdots \leq E_{2p}$$

The overall procedure of extended Routh approximation can be summarized as follows:

(i) The coefficients (a_i, b_i) $(0 \leq i \leq n)$ of original higher-order transfer functions are given.

(ii) Select a type (type CE1 or type CE2) and an integer r $(0 \leq r \leq p)$.

(iii) Compute (α_i, β_i) or $(\alpha_i^{*}, \beta_i^{*})$ $(0 \leq i \leq n)$ from (a_i, b_i) $(0 \leq i \leq n)$ by the tabular algorithm presented in Table 5-1 [38].

(iv) The coefficients of a reduced mth-order system (a_i, b_i) $(0 \leq i \leq m)$ are computed from (α_i, β_i) or $(\alpha_i^{*}, \beta_i^{*})$ $(0 \leq i \leq m)$ by the inverse recursion formula of the tabular algorithm above.

5-3-2 Discrete-Time System

Consider an nth-order linear discrete-time asymptotically stable system having the transfer function

$$H_n(z) = \frac{g_n(z)}{f_n(z)}$$

For a given denominator polynomial $f_n(z)$ and the tangent function $\rho_n(z)$ derived from it, we then expand type D1 and type D2 continued fractions of (5-12) and (5-13).

According to its type, the transfer function $H_n(z)$ can be expanded into the following canonical forms.

Type DE1 $(n = 2p)$

$$H_{2p}(z) = d_0 + \frac{F_1(z)}{1 + F_1(z)(z+1)} \left\{ d_1 + F_2(z)\left[d_2 + \cdots + F_{2r}(z)\left[d_{2r} \right.\right.\right.$$

$$\left.\left.+ F_{2r+1}(z)\left[\left(d_{2r+1} + \frac{d^*_{2r+1}}{z-1} \right) + \cdots + F_{p+r}(z)\left(d_{p+r} + \frac{d^*_{p+r}}{z-1} \right) \right] \cdots \right]\right\}$$

$$(5\text{-}29)$$

where

$$F_i(z) = \frac{1}{c_i(z-1)} + \frac{1}{c_{i+1}(1 - z^{-1})} + \cdots, \qquad i = 1, \ldots, 2r \ (n = 2p)$$

$$F_i(z) = \frac{1}{c_i(z-1) + c^*_i/(1 - z^{-1})} + \cdots, \qquad i = 2r+1, \ldots, p+r \ (n = 2p)$$

Type DE2 $(n = 2p)$

$$H_{2p}(z) = d^*_0 + \frac{F_1(z)}{1 + F_1(z)(z+1)} \{ d^*_1 + F_2(z) \lfloor d^*_2 + \cdots + F_{2r}(z) \rfloor d^*_{2r}$$

$$+ F_{2r+1}(z) \lfloor (d^*_{2r+1} + d_{2r+1}(z-1)) + \cdots + F_{p+r}(z)(d^*_{p+r} + d_{p+r}(z-1)) \rfloor \}$$

$$(5\text{-}30)$$

where

$$\begin{array}{l} F_i(z) \text{ or } z^{-1}F_i(z) \\ \text{(i odd)} \quad \text{(i even)} \end{array} = \frac{1}{c^*_i/(1 - z^{-1})} + \cdots, \qquad i = 1, 2, \ldots, 2r$$

$$\begin{array}{l} F_i(z) \text{ or } z F_i^{-1}(z) \\ \text{(i odd)} \quad \text{(i even)} \end{array} = \frac{1}{c^*_i/(1 - z^{-1}) + c_i(z - 1)} + \cdots, \qquad i = 2r+1, \ldots, p+r$$

and the constants d_i and d^*_i are computed from the coefficients $f_n(z)$ and $g_n(z)$ and the intermediate symmetric and antisymmetric polynomials $A_k(z)$ and $S_k(z)$ in the expanding procedures of continued fractions of (5-12) and (5-13).

Computation of the coefficients d_i and d^*_i in (5-29) and (5-30) can be performed recursively in tabular form. Truncating the last several terms of canonical expansion of (5-29) and (5-30) yields its reduced-order model. That is, the $m(= 2q)$th-order reduced transfer function

$$\hat{H}_{n-m}(z) = \frac{\hat{g}_{n-m}(z)}{\hat{f}_{n-m}(z)} = \frac{\Sigma_{i=0}^{m}\hat{b}_i z^{n-i}}{\Sigma_{i=0}^{m}\hat{a}_i z^{n-i}} \tag{5-31}$$

is obtained by truncating the terms containing $F_{r+(p-q)}$ and $d_{r+(p-q)}$ $(p-q>r)$ and replacing $F_i(z)$ with $F_{i,\,r+p-q}(z)$, where for type DE1,

$$F_{i,\,R}(z) = \frac{1}{c_i(z-1)} + \cdots + \frac{1}{c_R(z-1) + c_R/(1-z^{-1})}, \qquad i = 1,\ldots,2r \quad (n = 2p)$$

$$\tag{5-32}$$

$$F_{i,\,R}(z) = \frac{1}{c_i(z-1) + c_i^*/(1-z^{-1})} + \cdots + \frac{1}{c_R(z-1) + c_R^*/(1-z^{-1})},$$

$$i = 2r+1,\ldots,p+r \quad (n = 2p)$$

and for type DE2,

$$\begin{array}{c}F_{i,\,R}(z) \text{ or } z^{-1}F_{i,\,R}(z) \\ (\text{i even}) \qquad\quad (\text{i odd})\end{array} = \frac{1}{c_i/(1-z)^{-1}} + \cdots + \frac{1}{c_R^*/(1-z^{-1}) + c_R(z-1)},$$

$$i = 1, 2, \ldots, 2r$$

$$\begin{array}{c}F_i(z) \text{ or } z^{-1}F_i(z) \\ (\text{i odd}) \qquad\; (\text{i even})\end{array} = \frac{1}{c_i^*/(1-z^{-1}) + c_i(z-1)} + \cdots + \frac{1}{c_R^*/(1-z^{-1}) + c_R(z-1)},$$

$$i = 2r+1, \ldots, p+r$$

Then the coefficients $\hat{a}_i$ and $\hat{b}_i$ of $H_{n-m}(z)$ are computed from c_i and d_i by applying the reverse algorithm for obtaining expansions (5-29) and (5-30).

Stability Preservation

Since the stability condition for $f(z)$ cannot be expressed by the positivity of expansion coefficients c_i and c_i^*, the stability of the reduced model derived by truncating the continued fractions (5-12) and (5-13) is not assured automatically. Thus we should check the latter condition of Theorems 5-2 and 5-3 [i.e., Eqs. (5-16) and (5-19)].

Partial Padé Approximation

Now let the expansions at $z = \infty$ and $z = 1$ be given by

$$H_n(z) = \begin{cases} \sum_{j=0}^{+\infty} h_j z^{-j} \\ \sum_{i=0}^{+\infty} h_i^*(z-1)^i \end{cases}$$

$$\hat{H}_{n-m}(z) = \begin{cases} \sum_{j=0}^{+\infty} \hat{h}_j z^{-j} \\ \sum_{i=0}^{+\infty} \hat{h}_i^*(z-1)^i \end{cases}$$

Then the following equations are satisfied.

Type DE1

$$\begin{cases} h_j = \hat{h}_j, & j = 0, \ldots, 2r+1+k \text{ for } n = 2m+1 \ (2r+k \text{ for } n = 2m) \\ h_i^* = \hat{h}_i^*, & i = 0, \ldots, k-1 \end{cases} \tag{5-33}$$

Type DE2

$$\begin{cases} h_j = \hat{h}_j, & j = 0, \ldots, k-1 \\ h_i^* = \hat{h}_i^*, & i = 0, \ldots, 2r+1+k \text{ for } n = 2m+1 \ (2r+k \text{ for } n = 2m) \end{cases} \tag{5-34}$$

The coincidence of h_j and $\hat{h}_j$ for some first j means that the transient response of $H_n(z)$ is probably well approximated by $\hat{H}_{n-m}(z)$. On the other hand, the coincidence of h_i^* and $\hat{h}_i^*$ for some first i means that the steady-state response of $H_n(z)$ is well approximated by $\hat{H}_{n-m}(z)$. Therefore, we should choose the type of expansion and also set the integer r to be an appropriate number depending on the requirements for a reduced-order system.

5-4 CONCLUSIONS

Extended Routh approximations as the model simplification techniques of linear continuous- and discrete-time scalar systems have been discussed. To preserve the stability of the original system, new continued-fraction expansion forms derived from denominator polynomials have been used. They yield some canonical representations of a given system, and according to the selected expansion forms, the dynamics of the approximant are altered. In the present formulations, trade-offs of the approximation results between steady-state response and transient response characteristics are taken into account.

In the first part of this chapter, stability criteria based on continued-fraction expansion on two points in the s-plane have been presented. Lossless digital integrator (LDI) transformation gives discrete-time continued fraction forms by which analogous reduction methods in the z-domain are derived.

REFERENCES

1. P. N. Paraskevopoulus, "Techniques in model reduction for large-scale systems," Control Dyn. Syst., vol. 23, pp. 165-193 (1986).

2. M. J. Bosley and F. P. Lees, "A survey of simple transfer function derivation from high-order state-variable models," Automatica, vol. 8, pp. 765-775 (1972).

3. D. D. Siljak, Large-Scale Dynamic Systems, North-Holland, Amsterdam, 1978.

4. Mohammad Jamshidi, Large-Scale Systems, North-Holland, Amsterdam, 1983.

5. E. J. Davidson, "A method for simplifying linear dynamic systems," IEEE Trans. Autom. Control, vol. AC-11, no. 1, pp. 93-101 (1966).

6. Y. Shamash, "Linear system reduction using Padé approximation to allow retention of dominant modes," Int. J. Control, vol. 21, no. 2, pp. 257-272 (1975).

7. S. C. Chang, "Homographic transformation for the simplification of continuous-time transfer functions by Padé approximation," Int. J. Control, vol. 23, no. 6, pp. 821-826 (1976).

8. Y. Bistritz and U. Shaked, "Discrete multivariable system approximations by minimal Padé-type stable models," IEEE Trans. Circuits Syst., vol. CAS-31, no. 4, pp. 382-390 (1984).

9. M. J. Bosley, H. W. Klopholler, and F. P. Lees, "On the relation between the continued fraction expansion and moment matching methods of model reduction," Int. J. Control, vol. 18, pp. 461-474 (1973).

10. D. J. Wright, "The continued fraction representation of transfer functions and model simplification," Int. J. Control, vol. 18, pp. 449-454 (1973).

11. Y. Shamash, "Continued fraction methods for the reduction of discrete-time dynamic systems," Int. J. Control, vol. 20, pp. 267-275 (1974).

12. Y. P. Shih and W. T. Wu, "Simplification of Z-transfer functions by continued fractions," Int. J. Control, vol. 17, pp. 1089-1094 (1973).

13. M. S. Mahmoud and M. G. Sigh, Discrete Systems Analysis, Control and Optimization, Springer-Verlag, Berlin, 1984.

14. M. F. Hutton and B. Friedland, "Routh approximation for reducing order of linear, time-invariant systems," IEEE Trans. Autom. Control, vol. AC-20, pp. 329-337 (1975).

15. Y. Bistritz, "Discrete stability equation theorem and method of stable model reduction," Syst. Control Lett., vol. 1-6, pp. 373-381 (1982).

16. S. Yokota and R. Tagawa, "Hermete approximation of transfer function," IEE of Jpn., vol. 98-C, no. 12 (1978) (in Japanese).

17. E. Badreddin and M. Mansour, "A multivariable normal-form for model reduction of discrete-time systems," Syst. Control Lett., vol. 2, pp. 271-285 (Feb. 1983).

18. B. D. O. Anderson, E. I. Jury, and M. Mansour, "On model reduction of discrete time systems," Automatica, vol. 22, no. 6, pp. 717-721 (1986).

19. S.-Y. Kung and D. W. Lin, "Optimal Hankel-norm model reductions: Multivariable systems," IEEE Trans. Autom. Control, vol. AC-26, no. 4, pp. 832-852 (1981).

20. L. Pernebo and L. M. Silverman, "Model reduction via balanced state space representations," IEEE Trans. Autom. Control, vol. AC-27, no. 2, pp. 382-387 (1982).

21. M. Aoki, "Some approximation methods for estimation and control of large scale systems," IEEE Trans. Autom. Control, vol. AC-23, no. 2, pp. 173-182 (1978).

22. K. Horiguchi and N. Hamada, "Model reduction for linear discrete-time systems via aggregation," Trans. Soc. Instrum. Control Eng. Jpn., vol. 20, no. 1, pp. 22-28 (1984) (in Japanese).

23. B. D. O. Anderson, E. I. Jury, and M. Mansour, "Schwarz matrix properties for continuous and discrete time systems," Int. J. Control, vol. 23, no. 1, pp. 1-16 (1976).

24. F. R. Gantmacher, The Theory of Matrices, Vol. 2, Chelsea, New York, 1971.

25. K. Horiguchi, S. Hayashi, N. Hamada, "New sets of stability criteria as a generalization of Markov's stability theorem," Int. J. Control, vol. 43, no. 5, pp. 1581-1591 (1986).

26. Y. Bistritz, "Zero location with respect to the unit circle of discrete-time linear system polynomials," Proc. IEEE, vol. 72, no. 9, pp. 1131-1142 (1984).

27. A. M. Davis, "A new Z domain continued fraction expansion," IEEE Trans. Circuits Syst., vol. CAS-29, no. 10, pp. 658-662 (1982).

28. Y. Bistritz, "Z-domain continued fraction for stable discrete systems polynomials," IEEE Trans. Circuits Syst., vol. CAS-32, no. 11, pp. 1162-1166 (1985).

29. M. Ismail, "New Z-domain continued fraction expansions," IEEE Trans. Circuits Syst., vol. CAS-32, no. 8, pp. 754-758 (1985).

30. M. Ismail, "Synthesis of discrete-time reactance functions," Proc. IEEE Int. Symp. Circuits and Systems, San Jose, Calif., pp. 460-463, 1986.

31. C. Hwang and Y.-P. Shih, "Routh approximation for reducing order of discrete systems," Trans. ASME J. Dyn. Syst. Meas. Control, vol. 104, pp. 107-109 (1982).

32. L. Weinberg, Network Analysis and Synthesis, McGraw-Hill, New York, 1962, Chap. 9.

33. Y. Bistritz, "A new unit circle stability criterion," Int. Symp. Mathematical Theory of Network and Systems, Beer Shiva, Israel, pp. 69-87, 1983.

34. L. T. Bruton, "Low-sensitivity digital ladder filters," IEEE Trans. Circuits Syst., vol. CAS-22, no. 3, pp. 168-176 (1975).

35. A. Eydgahi and H. Singh, "A modified procedure for realization of the transfer-function matrix from a mixture of Markov parameters and moments," IEEE Trans. Autom. Control, vol. AC-30, no. 3, pp. 299-301 (1985).

36. M. Ismail and H. K. Kim, "A simplified stability test for discrete systems using a new Z-domain continued fraction method," IEEE Trans. Circuits Syst., vol. CAS-30, no. 7, pp. 505-507 (1983).

37. M. Abe and N. Hamada, "Discrete-time two-point Routh approximation," (to be submitted).

38. K. Horiguchi, "Model reduction of linear systems and N-point Padé approximation," Ph.D. thesis, Keio University, 1985.

6

Linear Systems with Transfer Functions of Bounded Type: Factorization Approach

YUJIRO INOUYE

Osaka University, Toyonaka, Osaka, Japan

6-1 INTRODUCTION

This chapter presents a concise account of some recent results [1–4] on linear multivariable systems with transfer functions of bounded type that can be obtained using what is called the "factorization approach." The central idea of this approach is to "factor" the (matrix-valued) transfer function of a system as a quotient of two stable (matrix-valued) functions. One of the motivations for the factorization approach is to obtain a simple parameterization of all compensators that stabilize a given plant. Such parameterizations were presented first by Kučera [5] in 1965 for the lumped discrete-time case, and independently by Youla et al. [6] in 1966 for the lumped continuous-time case. Their results were generalized to the distributed continuous-time case by Callier and Desoer [7, 8] and Vidyasagar et al. [9], and to the distributed discrete-time case by Cheng and Desoer [10]. Desoer et al. [11, 12] generalized such results to a general algebraic setting to include continuous- and discrete-time systems, both lumped and distributed. The algebraic theory developed by Desoer et al. requires the key assumption that the original ring in the algebraic setting be a principal ideal domain [12] or at least a Bézout domain [9]. It is well known [13] that the set of stable rational functions is a principal ideal domain, and the reader is referred to Vidyasager's book [14] for a factorization approach to linear lumped systems.

In this chapter we deal with linear discrete-time systems with matrix-valued transfer functions each entry of which is represented as a quotient of two bounded functions analytic outside the unit disk (i.e., in the open domain $1 < |z| \leq \infty$). In the mathematical literature [15], such a function is called a function of bounded type. That is, a scalar function h(z) of bounded type is written in the form h(z) = n(z)/d(z), where both n(z) and d(z) $\neq$ 0 belong to the Hardy space H_∞ (i.e., the set of analytic functions bounded in $1 < |z| \leq \infty$). Such a system will be called a system of bounded type. The notion of systems of bounded type is so general that their class contains the distributed discrete-time systems treated by Cheng and Desoer [10] and Desoer et al. [11, 12] as special cases. The algebraic theory developed by Desoer et al. cannot be applied directly to a system of bounded type, because the Hardy space H_∞ is not a Bézout domain [2]. It is thus natural to ask whether one can get a parameterization of all compensators that stabilize a given system of bounded type. Answering this question is the final objective of this chapter.

This chapter is organized as follows. We formulate a discrete-time convolution system and in Sec. 6-2 show a necessary and sufficient condition for it to become stable. We provide a canonical factorization theorem for matrix-valued functions of the Nevanlinna class N and show fundamental properties of outer functions of class N in Sec. 6-3. Then we present co-prime representations of transfer functions of bounded type in Sec. 6-4. In Sec. 6-5 we propose a quasi-McMillan form of a transfer function of bounded type, and in Sec. 6-6 we present some invariants of such a transfer function. In Sec. 6-7 we present necessary and sufficient conditions for a plant with a transfer function of bounded type to be stabilizable by a compensator, and finally give a parameterization of all compensators that stabilize such a plant.

The following notation is used throughout this chapter. The symbols T and E denote, respectively, the unit circle and the complement of the closed unit disk in the extended complex plane (i.e., $T = \{z : |z| = 1\}$, $E = \{z : 1 < |z| \leq \infty\}$). The Hardy space H_∞ denotes the set of complex-valued analytic functions bounded in E. The set of functions of bounded type is denoted by B. Let $H_\infty^{r \times m}$ and $B^{r \times m}$ denote the sets of r × m matrices whose elements all belong to H_∞ and B, respectively. Let $L_\infty^{r \times m}$, $N^{r \times m}$, and $N_+^{r \times m}$ be, respectively, the sets of r × m matrices whose elements all belong to the Lebesgue space L_∞ (i.e., the space of essentially bounded measurable functions on T), the Nevanlinna class N over E, and the Smirnov class N_+ over E [1,20]. Since any function $f \in B$ has the boundary value $f(e^{j\omega})$ almost everywhere (a.e.) on T defined by $f(e^{j\omega}) = \lim_{r \to 1} f(re^{j\omega})$ a.e., we shall also use the letter f to denote the boundary function $f(e^{j\omega})$ as a function $f \in B$ and hope that no confusion results. Rank H, det H, adj H, H^T, and H* denote, respectively, the rank, the determinant, the adjoint matrix, the transpose, and the conjugate transpose of a matrix H. Diag$\{\alpha_i\}$ denotes the square diagonal matrix with diagonal entries α_i. For Hermitian matrices K and L, $K \geq L$ means that K - L is positive semidefinite. Let I_r denote the r × r identity matrix; its subscript r is omitted when it is clear from the context.

6-2 CONVOLUTION SYSTEMS AND STABILITY

We consider a discrete-time convolution system [10, 16, 17], which can be represented by the convolution sum

$$y_t = \sum_{k=-\infty}^{\infty} H_k u_{t-k} \qquad (6\text{-}1)$$

where $\{u_t\}$ is a sequence of m-dimensional vectors, real or complex, defined for every integer t, and $\{y_t\}$ is a sequence of r-dimensional vectors, real or complex, only defined for the value of integer t for which the sum (6-1) converges. The sequence $\{u_t\}$ is called the input, and the sequence $\{y_t\}$, the output. The r × m matrix-valued sequence $\{H_t\}$ is called the impulse response (or the unit-sample response). We note that the impulse response $\{H_t\}$ is not necessarily zero for t < 0. If $H_t = 0$ for any t < 0, the system (6-1) is called causal. If $H_t = 0$ for every $t \leq 0$, it is called strongly causal.

The z-transform X(z) of a sequence $\{x_t\}$ is defined by the series

$$X(z) = \sum_{k=-\infty}^{\infty} x_k z^{-k} \qquad (6\text{-}2)$$

which is defined only for the value of z, real or complex, for which the series (6-2) converges. We know from the theory of functions of complex variables that the region R of convergence of the Laurent series (6-2) is a ring $r_1 < |z| < r_2$, a disk $|z| < r_2$, or the exterior $|z| > r_1$ of a disk, whose inner and outer radii r_1 and r_2 depend on the behavior of x_t as $t \to +\infty$ and $-\infty$, respectively [16, 17]. The z-transform H(z) of the impulse response $\{H_t\}$ of system (6-1) is called the transfer function.

An r-vector sequence $\{x_t\}$ is called ℓ_2-bounded if the ℓ_2-norm $\|x\|_2$ is finite, where the ℓ_2-norm is defined by $\|x\|_2 = (\Sigma_{i=1}^{r} \Sigma_{k=-\infty}^{\infty} |x_k^i|^2)^{\frac{1}{2}}$, with x_k^i being the ith element of vector x_k. We define a <u>stable</u> (or ℓ_2-<u>stable</u>) system as one for which every ℓ_2-bounded input produces an ℓ_2-bounded output; that is, for any positive number $K > 0$ there exist a number M such that the output y corresponding to any input u such that $\|u\|_2 < K$ satisfies $\|y\|_2 < M < \infty$.

The following theorem is a basic result for the stability of system (6-1).

<u>Theorem 6-1.</u> Let H(z) be the r × m transfer function matrix of a causal convolution system. Then it is stable if and only if $H(z) \in H_\infty^{r \times m}$. ∎

Theorem 6-1 seems at first glance to be self-evident, but its proof is found in [4].

Remark 6-1

When H(z) becomes irrational, there are several notions of stability: ℓ_∞-stability (or BIBO stability) and ℓ_p-stability, where $1 \leq p < \infty$ [18]. However, when H(z) becomes rational, all the notions of stability above are the same except in case of p = 1.

Remark 6-2

Let S' be the set of causal ℓ_∞-stable scalar transfer functions. Then $H(z) \in S'$ if and only if the sum $\Sigma_{k=0}^{\infty} |H_k|$ is finite, where $H(z) = \Sigma_{k=0}^{\infty} H_k z^{-k}$ [16]. Hence S' is contained in H_∞. However, we cannot treat S' in this chapter, because it is not yet clear whether any two functions in S' have a greatest common divisor.

6-3 CANONICAL FACTORIZATION

In this section we provide the canonical factorization theorem for the Nevanlinna class $N^{r \times n}$ and present the outer-inner factorization for the Hardy space $H_\infty^{r \times n}$. The reader is referred to Rudin [19] and Duren [20] for scalar functions and to Rosenblum and Rovnyak [21] for matrix functions belonging to the Hardy and Nevanlinna classes.

<u>Definition 6-1.</u> A matrix $F \in H_\infty^{r \times m}$ is called <u>row</u> (or <u>right</u>) <u>inner</u> if it satisfies $F(e^{j\omega})F(e^{j\omega})^* = I$, a.e. on T. A matrix $F_0 \in H_\infty^{r \times m}$ is called <u>column</u> (or <u>left</u>) <u>outer</u> if rank $F_0(e^{j\omega}) = m$ a.e. on T and if

$$F_0(\infty)F_0(\infty)^* \geq F(\infty)F(\infty)^* \qquad (6\text{-}3a)$$

for any $F \in H_\infty^{r \times m}$ whose boundary value $F(e^{j\omega})$ satisfies

$$F(e^{j\omega})F(e^{j\omega})^* = F_o(e^{j\omega})F_o(e^{j\omega})^* \quad \text{a.e. on } T \tag{6-3b}$$

∎

A column inner matrix and a row outer matrix are defined, respectively, in the same way; $F(z)$ is called column (or left) inner if $F(z)^T$ is row inner, and $F(z)$ is called row (or right) outer if $F(z)^T$ is column outer.

Let $F(z)$ be a square matrix. Then it is clear from the definitions above that $F(z)$ is row inner (respectively, row outer) if and only if it is column inner (column outer). Thus $F(z)$ is called inner (respectively, outer) if it is row inner (row outer).

Remark 6-3

Inner and outer functions are generalizations of all-pass and minimum-phase functions, respectively, in the lumped-network theory.

Remark 6-4

In the case of matrix functions with rational entries, the condition of column outer is equivalent to

$$\text{rank } F(z) = m \quad \text{for all } z \in E$$

See [23, Chap. 1, Theorem 10.1].

The following theorem provides an outer-inner factorization of a non-zero function $F(z) \not\equiv 0$ in the Hardy space $H_\infty^{n \times p}$ [22], which plays a fundamental role in this chapter.

Theorem 6-2 (Krein-Rozanov-Shmul'yan). Every nonzero $F \in H_\infty^{n \times p}$ can be expressed in the form

$$F(z) = F_o(z) F_i(z) \tag{6-4}$$

where F_o is column outer and F_i is row inner. Moreover, the factorization (6-4) is unique up to multiplication by a constant unitary matrix; that is, if $\tilde{F} = \tilde{F}_o \tilde{F}_i$ is another outer-inner factorization, there exists a constant unitary matrix U such that

$$\tilde{F}_o = F_o U \quad \text{and} \quad \tilde{F}_i = U^* F_i \tag{6-5}$$

∎

By applying Theorem 6-2 to a matrix $F(z)^T$, the following theorem, which gives an inner-outer factorization of a nonzero function $F(z) \not\equiv 0$ in the Hardy space $H_\infty^{n \times p}$, is easily derived.

Theorem 6-2'. Every nonzero $F \in H_\infty^{n \times p}$ can be represented in the form

$$F(z) = F_i(z) F_o(z) \tag{6-4'}$$

where F_i is column inner and F_o is row outer. Moreover, the factorization (6-4') is unique up to multiplication by a constant unitary matrix. ∎

Remark 6-5

We note that the original theorem in [22] corresponding to Theorem 6-2 is presented to specify all the factors of factorizations of a power spectral density matrix square integrable on $[-\pi, \pi]$. Hence it can be applied to a power spectral density bounded on $[-\pi, \pi]$.

The definition of outer functions for the Hardy class $H_\infty^{r \times m}$ is extended to matrix functions of the Nevanlinna class $N^{r \times m}$ as follows:

Definition 6-1'. A matrix $F \in N^{r \times m}$ is column (respectively, row) outer if there is a scalar outer function $\phi \in H_\infty$ such that ϕF is a column (respectively, row) outer function belonging to $H_\infty^{r \times m}$. ∎

Based on Theorem 6-2, we obtain the following theorem, which is an extension of the canonical factorization theorem by Smirnov [20, p. 24] for scalar functions to matrix functions in the Nevanlinna class.

Theorem 6-3 (Canonical Factorization Theorem). Every nonzero $F \in N^{n \times p}$ can be expressed in the form

$$F(z) = \frac{F_o(z) F_i(z)}{d_s(z)} \tag{6-6}$$

where F_o is column outer, F_i is row inner, and d_s is a scalar singular inner function (i.e., $0 < |d_s(z)| \le 1$ for $z \in E$). Moreover, F belongs to class $N_+^{n \times p}$ if and only if it can be expressed in the form (6-6) with $d_s(z) \equiv 1$. In this case, the factorization (called an outer-inner factorization of F) is unique up to multiplication by a constant unitary matrix; that is, if $F(z) = \tilde{F}_o(z) \tilde{F}_i(z)$ is another outer-inner factorization, there is a constant unitary matrix U such that

$$\tilde{F}_o(z) = F_o(z) U \quad \text{and} \quad \tilde{F}_i(z) = U^* F_i(z) \tag{6-7}$$

∎

Proof. Since every (i, j) entry f_{ij} of F is a scalar function of class N, using the Smirnov canonical factorization theorem [20, p. 25, Theorem 2.9], F has a representation

$$F(z) = \frac{N(z)}{d_o(z) d_s(z)} \tag{6-8}$$

where $N \in H_{\infty}^{n \times p}$, $d_o \in H_{\infty}$ is scalar outer, and $d_s \in H_{\infty}$ is scalar singular inner. Applying Theorem 6-2 to matrix $N \in H_{\infty}^{n \times p}$, N can be represented in the form

$$N(z) = N_o(z) N_i(z) \qquad (6-9)$$

where N_o is column outer and N_i is row inner. Put

$$F_o(z) = \frac{N_o(z)}{d_o(z)} \qquad (6-10a)$$

$$F_i(z) = N_i(z) \qquad (6-10b)$$

Then F_o becomes column outer by Definition 6-1'. Using (6-8) through (6-10), we obtain the representation (6-6). Moreover, it is clear that $F \in N_+^{r \times p}$ if and only if it holds $d_s(z) \equiv 1$ in the representation (6-8). The uniqueness of the outer-inner factorization for class $N_+^{n \times p}$ comes from that for class $H_{\infty}^{r \times p}$. ∎

We know from Theorem 6-3 that every column (or row) outer function $F \in N^{n \times p}$ must belong to class $N_+^{n \times p}$. By means of Theorem 6-3, we get the following.

   <u>Theorem 6-4 (Maximality of Outer Function)</u>. Let $F_o \in N_+^{r \times m}$. Then F_o is column outer if and only if

$$F_o(\infty) F_o(\infty)^* \geq \tilde{F}(\infty) \tilde{F}(\infty)^* \qquad (6-11)$$

for every $\tilde{F} \in N_+^{r \times m}$ such that

$$\tilde{F}(e^{j\omega}) \tilde{F}(e^{j\omega})^* = F_o(e^{j\omega}) F_o(e^{j\omega})^* \quad \text{a.e. on } T \qquad (6-12)$$
∎

   Proof. Suppose that $F_o \in N_+^{r \times m}$ is column outer and that $\tilde{F} \in N^{r \times m}$ satisfies (6-12). Then, using Theorem 6-3, $F_o(z)$ and $\tilde{F}(z)$ have representations

$$F_o(z) = \frac{N_o(z)}{d_o(z)}, \quad N_o \in H_{\infty}^{r \times m}, \quad d_o \in H_{\infty} \qquad (6-13a)$$

$$\tilde{F}(z) = \frac{\tilde{N}(z)}{\tilde{d}_o(z)}, \quad \tilde{N} \in H_{\infty}^{r \times m}, \quad \tilde{d}_o \in H_{\infty} \qquad (6-13b)$$

where N_0 is column outer, and both d_0 and $\tilde{d}_0$ are scalar outer. We get from (6-12) and (6-13)

$$|\tilde{d}_o(e^{j\omega})|^2 N_o(e^{j\omega}) N_o(e^{j\omega})^* = |d_o(e^{j\omega})|^2 \tilde{N}(e^{j\omega})\tilde{N}(e^{j\omega})^* \tag{6-14}$$

Since both $\tilde{d}_0 N_0$ and $d_0 N$ belong to class $H_\infty^{r\times m}$ and $d_0 N_0$ is column outer, using Definition 6-1,

$$|\tilde{d}_o(\infty)|^2 N_o(\infty) N_o(\infty)^* \geq |d_o(\infty)|^2 \tilde{N}(\infty)\tilde{N}(\infty)^*$$

holds, which implies (6-11) from (6-13a) and (6-13b).

Conversely, suppose that $F_o \in N_+^{r\times m}$ satisfies the maximality condition (6-11). By means of Theorem 6-3, F_o has a representation

$$F_o(z) = \frac{N(z)}{d_o(z)}, \qquad N \in H_\infty^{r\times m}, \quad d_o \in H_\infty \tag{6-15}$$

where d_0 is outer. Consider any $\tilde{N} \in H_\infty^{r\times m}$ such that

$$\tilde{N}(e^{j\omega})\tilde{N}(e^{j\omega})^* = N(e^{j\omega})N(e^{j\omega})^* \quad \text{a.e. on } T \tag{6-16}$$

This implies that

$$\tilde{F}(e^{j\omega})\tilde{F}(e^{j\omega})^* = F(e^{j\omega})F(e^{j\omega})^* \quad \text{a.e. on } T \tag{6-17}$$

where $\tilde{F} = \tilde{N}/d_o \in N_+^{r\times m}$. Then from condition (6-11),

$$F(\infty)F(\infty)^* \geq \tilde{F}(\infty)\tilde{F}(\infty)^*$$

which means that

$$N(\infty)N(\infty)^* \geq \tilde{N}(\infty)\tilde{N}(\infty)^* \tag{6-18}$$

Therefore, N is column outer in class $H_\infty^{r\times m}$. Hence F_o is column outer in class $N^{r\times m}$. ∎

Remark 6-6

In the previous work [1], the definition of outer functions was made by using the maximality property mentioned above, and then it was shown that any outer function $F \in N^{r \times m}$ has a representation $F = N_0/d_0$, where N_0 is outer in $H_\infty^{r \times m}$ and d_0 is scalar outer.

By means of Theorems 6-3 and 6-4 and some properties of outer functions for class $H_\infty^{r \times m}$, we obtain the following basic properties of outer functions for class $N^{r \times m}$, which are presented without proofs (see [1] and [3] for the proof).

<u>Theorem 6-5.</u> Let $F \in N_+^{r \times m}$. Then the following properties hold:

   (i) Suppose that $F(z)$ is square and becomes inner and outer. Then $F(z)$ is a constant unitary matrix.
  (ii) Suppose that $F(z)$ is square. Then $F(z)$ is outer if and only if det $F(z)$ is outer.
 (iii) If $F(z)$ is square and outer, then its inverse $F(z)^{-1}$ is also outer.
  (iv) Suppose that $F(z)$ has a factorization

$$F(z) = F_1(z) F_2(z), \quad F_1 \in N_+^{r \times r}, \quad F_2 \in N_+^{r \times m} \qquad (6\text{-}19)$$

Then $F(z)$ is row outer if and only if both $F_1(z)$ and $F_2(z)$ are row outer.
   (v) $F(z)$ is row outer if and only if there is no nonconstant common inner divisor of all the minors of order r in $F(z)$.
  (vi) Suppose that rank $F(e^{j\omega}) = r$, a.e. on T, and let $F(z) = F_i(z) F_o(z)$ be an inner–outer factorization of $F(z)$, where F_i is inner and F_o is row outer. Then det $F_i(z)$ is a greatest common inner divisor of all the minors of order r in $F(z)$. ∎

<u>Theorem 6-6 (Minimum–Energy Delay Property).</u> Suppose that $F \in N_+^{r \times m}$ satisfies rank $F(e^{j\omega}) = m$, a.e. on T. Then the following three properties are equivalent.

  (i) $F(z)$ is column outer.
 (ii) $F(z)$ satisfies

$$\sum_{k=0}^{q} F_k F_k^* \geq \sum_{k=0}^{q} \tilde{F}_k \tilde{F}_k^* \qquad (6\text{-}20)$$

for any integer $q \geq 0$ and any $\tilde{F} \in N_+^{r \times m}$ such that

$$\tilde{F}(e^{j\omega}) \tilde{F}(e^{j\omega})^* = F(e^{j\omega}) F(e^{j\omega})^* \quad \text{a.e. on T} \qquad (6\text{-}21)$$

where F_k and $\tilde{F}_k$ are the kth Taylor coefficients in the Taylor expansions of $F(z)$ and $\tilde{F}(z)$, respectively, at the point $z = \infty$.

(iii) $F(z)$ satisfies

$$\mathrm{tr}\left(\sum_{k=0}^{q} F_k F_k^*\right) \geq \mathrm{tr}\left(\sum_{k=0}^{q} \tilde{F}_k \tilde{F}_k^*\right) \tag{6-22}$$

for any integer $q \geq 0$ and any $\tilde{F} \in N_+^{r \times m}$ such that

$$\tilde{F}(e^{j\omega})\tilde{F}(e^{j\omega})^* = F(e^{j\omega})F(e^{j\omega})^* \quad \text{a.e. on } T \tag{6-23}$$

where the symbol $\mathrm{tr}(\cdot)$ denotes the trace of a matrix. ∎

Remark 6-7

The properties (iv) to (vi) mentioned in Theorem 6-5 are stated for row outer functions for convenience of later use. There are obvious analogs corresponding to them for column outer functions. Theorem 6-6 also has an obvious analog for row outer functions.

6-4 COPRIME REPRESENTATIONS

We introduce two kinds of notions of coprime representations of transfer functions of bounded type. One is the notion of the inner coprime representations, and the other is that of the coprime representations. The former was considered deeply in [1,2], and the latter was introduced in [4] in order to study the feedback stabilization of a linear plant using compensators.

Let $A \in H_\infty^{r \times \ell}$ and $B \in H_\infty^{r \times m}$. We say that A and B are left-inner coprime (LIC) if matrix $[A, B]$ is row outer. Moreover, they are called left coprime (LC) if $[A, B]$ is row outer and satisfies the condition that

(i) There exists a positive number $\epsilon > 0$ such that

$$[A(e^{j\omega}), B(e^{j\omega})][A(e^{j\omega}), B(e^{j\omega})]^* \geq \epsilon I \quad \text{a.e. on } T \tag{6-24}$$

It can be proved (see [4]) that condition (i) is equivalent to the condition that

(i') There exist two matrices $X \in L_\infty^{\ell \times r}$ and $Y \in L_\infty^{m \times r}$ such that

$$A(e^{j\omega})X(e^{j\omega}) + B(e^{j\omega})Y(e^{j\omega}) = I \quad \text{a.e. on } T \tag{6-25}$$

Then we have the following definition for <u>coprime representations</u> (CRs) for matrix functions of bounded type.

<u>Definition 6-2.</u> Let $P \in B^{r \times m}$. Then we say that $D^{-1}N$ is a <u>left-inner-coprime representation</u> (LICR) of P if $D \in H_\infty^{r \times r}$ and $N \in H_\infty^{r \times m}$ such that D and N are LIC and $P = D^{-1}N$. Moreover, $D^{-1}N$ is called a <u>left-coprime representation</u> (LCR) of P if D and N are LC such that $P = D^{-1}N$. ∎

Then we have the following existence theorem of LICRs and LCRs.

<u>Theorem 6-7.</u> Let $P \in B^{r \times m}$. Then it has both a LICR and a LCR. ∎

<u>Proof.</u> Since each element of matrix P can be represented as a quotient of two scalar functions in H_∞, we can write

$$P = \frac{N}{d}, \quad d \in H_\infty, \ N \in H_\infty^{r \times m} \tag{6-26}$$

where d is the product of all denominators of the elements of P, and $N = dP$. In general, $(dI)^{-1}N$ is not a LICR of P. The matrix $[dI, N]$ has an inner-outer factorization

$$[dI, N] = U[\tilde{D}, \tilde{N}] \tag{6-27}$$

where U is inner and $[\tilde{D}, \tilde{N}]$ is row outer. Thus $\tilde{D}^{-1}\tilde{N}$ is a LICR of P, because $\tilde{D}^{-1}\tilde{N} = P$. Furthermore, matrix $[\tilde{D}, \tilde{N}]$ has an outer-inner factorization

$$[\tilde{D}, \tilde{N}] = V[D_L, N_L] \tag{6-28}$$

where V is outer and $[D_L, N_L]$ is row inner, that is,

$$[D_L(e^{j\omega}), N_L(e^{j\omega})][D_L(e^{j\omega}), N_L(e^{j\omega})]^* = I \ \text{a.e.}$$

Hence $[D_L, N_L]$ satisfies the condition (6-24) with $\epsilon = 1$. On the other hand, by Theorem 6-5(iv), $\lfloor D_L, N_L \rfloor$ is row outer, because $[\tilde{D}, \tilde{N}]$ is row outer. It is easily seen that $D_L^{-1}N_L = P$. Therefore, $D_L^{-1}N_L$ is a LCR of P. ∎

By means of the Smirnov canonical factorization theorem (see Theorem 6-3), we obtain the following useful lemma due to Smirnov, which will be required later.

<u>Lemma 6-1.</u> If $F \in N_+^{n \times p}$ and $F \in L_\infty^{n \times p}$, then $F \in H_\infty^{n \times p}$. ∎

We present some basic properties of LICRs and LCRs of $P \in B^{r \times m}$.

<u>Theorem 6-8.</u> Let $P \in B^{r \times m}$. Then the following properties hold:

(i) Suppose that $D^{-1}N$ is a LICR of P. Then $(FD)^{-1}FN$ is also a LICR of P for any outer $F \in H_\infty^{r \times r}$. Moreover, if $D_1^{-1} N_1$ is another LICR of P, then there exists an outer function $F \in N_+^{r \times r}$ such that $D_1 = FD$ and $N_1 = FN$. Furthermore, suppose that $D_2^{-1} N_2$ is another <u>left representation</u> (LR) of P. Then there exists a matrix $L \in N_+^{r \times r}$ such that $D_2 = LD$ and $N_2 = LN$.

(ii) Assume that $D^{-1} N$ is a LCR of P. Then $(FD)^{-1} FN$ is also a LCR of P for any outer $F \in H_\infty^{r \times r}$ such that its inverse F^{-1} belongs to $H_\infty^{r \times r}$. Moreover, if $D_1^{-1} N_1$ is another LCR of P, there exists an outer $F \in H_\infty^{r \times r}$ such that $F^{-1} \in H_\infty^{r \times r}$, $D_1 = FD$ and $N_1 = FN$. Finally, suppose that $D_2^{-1} N_2$ is another LR of P. Then there exists a matrix $L \in H_\infty^{r \times r}$ such that $D_2 = LD$ and $N_2 = LN$. ∎

<u>Proof of (i)</u>. By Definition 6-2 and Theorem 6-5(iv), the fact that [D, N] is row outer means that [FD, FN] is row outer for any outer $F \in H_\infty^{r \times r}$. Therefore, $(FD)^{-1} FN$ is a LICR of P.

Since

$$P = D^{-1} N = D_1^{-1} N_1 \tag{6-29}$$

we obtain

$$N_1 = D_1 D^{-1} N \tag{6-30}$$

Putting $F = D_1 D^{-1} \in B^{r \times r}$, we get

$$D_1 = FD \quad \text{and} \quad N_1 = FN \tag{6-31}$$

that is,

$$[D_1, N_1] = F[D, N] \tag{6-32}$$

Since F is of bounded type, it has a representation

$$F = P_1^{-1} P_2, \quad P_1 \in H_\infty^{r \times r}, \quad P_2 \in H_\infty^{r \times r} \tag{6-33}$$

Let

$$P_1(z) = P_{1i}(z) P_{1o}(z)$$

$$P_2(z) = P_{2i}(z) P_{2o}(z)$$

be inner-outer factorizations of $P_1(z)$ and $P_2(z)$, respectively, where P_{1i} and P_{2i} both are inner, and P_{1o} and P_{2o} both are outer. Then we have

$$P_{1i}P_{1o}[D_1, N_1] = P_{2i}P_{2o}[D, N] \qquad (6\text{-}34)$$

Since $[D_1, N_1]$ and $[D, N]$ are both row outer, (6-34) represents two inner-outer factorizations. Thus by Theorem 6-2', there is a constant unitary matrix U such that

$$P_{1i} = P_{2i}U$$

$$P_{1o}[D_1, N_1] = U^*P_{2o}[D, N]$$

Thus we get

$$F(z) = P_{1o}(z)^{-1}U^*P_{2o}(z) \qquad (6\text{-}35)$$

which is outer in $N_+^{r \times r}$ by Theorem 6-5(iii) and (iv).
Let

$$[D_2, N_2] = V[\bar{D}, \bar{N}] \qquad (6\text{-}36)$$

be an inner-outer factorization of $[D_2, N_2]$, where V is inner and $[\bar{D}, \bar{N}]$ is row outer. Since

$$P = \bar{D}^{-1}\bar{N} = D^{-1}N$$

it follows from the discussion above that there is an outer $F \in N_+^{r \times r}$ such that

$$\bar{D} = FD \quad \text{and} \quad \bar{N} = FN \qquad (6\text{-}37)$$

Thus putting $L = VF \in N_+^{r \times r}$, it follows from (6-36) and (6-37) that

$$D_2 = LD \quad \text{and} \quad N_2 = LN \qquad (6\text{-}38)$$

$\blacksquare$

Proof of (ii). Let $F \in H_\infty^{r \times r}$ such that $F^{-1} \in H_\infty^{r \times r}$. Then it is clear that $FD \in H_\infty^{r \times r}$, $FN \in H_\infty^{r \times m}$, and $P = (FD)^{-1}FN$. It is shown below that $[FD, FN]$ is row outer and satisfies condition (6-24). We have

$$[D, N] = F^{-1}[FD, FN] \qquad (6\text{-}39)$$

By Theorem 6-5(iv) and (6-39), the fact that $[D, N]$ is row outer means that $[FD, FN]$ is also row outer. Since $[D, N]$ satisfies condition (6-24), there exists an $\epsilon > 0$ such that

$$[D, N][D, N]^* \geq \epsilon I \quad \text{a.e. on T} \tag{6-40}$$

Because $F^{-1} \in H_\infty^{r \times r}$, there exists an $\alpha > 0$ such that

$$F^{-1}(F^{-1})^* \leq \frac{1}{\alpha} I$$

which implies that

$$FF^* \geq \alpha I \tag{6-41}$$

Thus it follows from (6-40) and (6-41) that

$$[FN, FN][FN, FN]^* \geq \alpha \epsilon I \quad \text{a.e. on T} \tag{6-42}$$

Hence $[FD, FN]$ satisfies condition (6-24). Therefore, $(FD)^{-1}(FN)$ becomes a LCR of P.

We shall now verify the second statement of (ii). Because both $D^{-1}N$ and $D_1^{-1}N_1$ become LICRs of P, using (i) of this theorem, there exists an outer matrix $F \in N_+^{r \times r}$ such that $D_1 = FD$ and $N_1 = FN$. Since $D_1^{-1}N_1$ is a LCR of P, matrix $[D_1, N_1]$ satisfies condition (6-24), that is,

$$[D_1, N_1][D_1, N_1]^* = F[D, N][D, N]^* F^* \geq \epsilon_1 I \quad \text{a.e. on T} \tag{6-43}$$

where $\epsilon_1 > 0$. Since $[D, N] \in H_\infty^{r \times (r+m)}$, there exists a $\beta > 0$ such that

$$[D, N][D, N]^* \leq \beta I \tag{6-44}$$

It follows from (6-43) and (6-44) that

$$FF^* \geq \frac{\epsilon_1}{\beta} I \quad \text{a.e. on T} \tag{6-45}$$

which implies that

$$F^{-1}(F^{-1})^* \leq \frac{\beta}{\epsilon_1} I \quad \text{a.e. on T} \tag{6-46}$$

This means that $F^{-1} \in L_\infty^{r \times r}$. Since F is outer in $N_+^{r \times r}$, F^{-1} also belongs to $N_+^{r \times r}$ by means of Theorem 6-5(iii). By Lemma 6-1, this and $F^{-1} \in L_\infty^{r \times r}$ mean that $F^{-1} \in H_\infty^{r \times r}$. It follows from (6-40) that

$$F^{-1}[D_1, N_1][D_1, N_1](F^{-1})^* = [D, N][D, N]^* \geq \epsilon I \quad \text{a.e. on T} \tag{6-47}$$

Because $[D_1, N_1] \in H_\infty^{r\times(r+m)}$, there exists a $\beta_1 > 0$ such that

$$[D_1, N_1][D_1, N_1]^* \le \beta_1 I \tag{6-48}$$

Thus it follows from (6-47) and (6-48) that

$$F^{-1}(F^{-1})^* \ge \frac{\epsilon}{\beta_1} I \quad \text{a.e. on } T$$

which implies that

$$FF^* \le \frac{\beta_1}{\epsilon} I \quad \text{a.e. on } T$$

This means that $F \in L_\infty^{r\times r}$. By Lemma 6-1, this and $F \in N_+^{r\times r}$ mean that $F \in H_\infty^{r\times r}$.

Now, the final statement of the theorem will be verified as follows. Since $D_2^{-1}N_2$ becomes a LICR of P, by (i) of this theorem, there exists a matrix $L \in N_+^{r\times r}$ such that $D_2 = LD$ and $N_2 = LN$. Therefore, it follows from (6-40) that

$$L^{-1}[D_2, N_2][D_2, N_2](L^{-1})^* = [D, N][D, N]^* \ge \epsilon I \quad \text{a.e. on } T \tag{6-49}$$

Because $[D_2, N_2] \in H_\infty^{r\times(r+m)}$, there exists a $\beta_2 > 0$ such that

$$[D_2, N_2][D_2, N_2]^* \le \beta_2 I \tag{6-50}$$

It follows from (6-49) and (6-50) that

$$L^{-1}(L^{-1})^* \ge \frac{\epsilon}{\beta_2} I \quad \text{a.e. on } T$$

which implies that

$$LL^* < \frac{\beta_2}{\epsilon} I \quad \text{a.e. on } T$$

This means that $L \in L_\infty^{r\times r}$. Since $L \in N_+^{r\times r}$, by Lemma 6-1, this means that $L \in H_\infty^{r\times r}$. ∎

So far, we have talked only of LRs of P, but clearly, the two theorems above have obvious analogs for <u>right representations</u> (RRs) of P.

6-5 QUASI-McMILLAN FORMS

In this section we obtain the quasi-McMillan form of a transfer function of
bounded type. To this end we require the quasi-Smith form of a matrix over
the Hardy space H_∞, which was established by Nordgren [24]. Nordgren
introduced a relation of quasi-equivalence for matrices over H_∞ that generalizes the relation of equivalence for matrices over principal ideal domains.
Following him, we state some necessary parts of the results in [24].

At the outset, we note the following fact as a special case of Theorem
6-2. Every nonzero scalar function f in H_∞ can be factored into an inner
function f_i and an outer function f_o which are determined up to constant inner
divisors. We will require in this section that the first nonvanishing Taylor
coefficient in the Taylor expansion of $f_i(z)$ at the point $z = \infty$ be positive,
and in this case the inner-outer factorization of f is exactly unique.

Two inner functions $f \in H_\infty$ and $g \in H_\infty$ are relatively prime or
simply coprime if every greatest inner common divisor of f and g is a con-
stant. Let $A \in H_\infty^{r\times m}$ and $B \in H_\infty^{r\times m}$. We say that A is quasi-equivalent to
B if for any inner function $\phi \in H_\infty$ there exist two square matrices
$X \in H_\infty^{r\times r}$ and $Y \in H_\infty^{m\times m}$ such that XA = BY and that (det X)$_i$ and (det Y$_i$)
are relatively prime to ϕ. It can be seen that the definition of quasi-
equivalence above is the same as the original definition given by Nordgren.
Two functions $f \in H_\infty$ and $g \in H_\infty$ are called inner associates if there are
two outer functions $x \in H_\infty$ and $y \in H_\infty$ such that fx = gy. It can be verified
that two functions $f \in H_\infty$ and $g \in H_\infty$ are inner associates if and only if f is
quasi-equivalent to g.

Invariant factors for matrices over H_∞ may be defined in the usual
way. If A is an r × m matrix over H_∞, let $D_0(A) = 1$ and $D_k(A)$ be the
greatest common inner divisor of all minors of order k in A for k = 1, 2,
..., ℓ, where ℓ = rank A. The invariant factor of kth order of A is then
defined by

$$\lambda_k(A) = \frac{D_k(A)}{D_{k-1}(A)}, \quad \text{where } k = 1, 2, \ldots, \ell \tag{6-51}$$

Then we have the following result.

Theorem 6-9 (Nordgren). Let $A \in H_\infty^{r\times m}$.

(i) Then A is quasi-equivalent to the quasi-Smith form S_A defined by

$$S_A = \begin{bmatrix} \Lambda_A & \vdots & 0 \\ \cdots & \cdots & \cdots \\ 0 & \vdots & 0 \end{bmatrix} \tag{6-52}$$

where $\Lambda_A = \text{diag}\{\lambda_k(A)\}$. Furthermore, $\lambda_k(A)$ divides $\lambda_{k+1}(A)$ for $k = 1, 2, \ldots, \ell - 1$, where $\ell = \text{rank } A$.

(ii) A is quasi-equivalent to $B \in H_\infty^{r \times m}$ if and only if A and B have the same invariant factors.   ∎

Now we consider a transfer function $H \in B^{r \times m}$. Since each entry of $H(z)$ is a quotient of two H_∞ functions, it can be represented in the form

$$H(z) = \frac{N(z)}{d(z)} \tag{6-53}$$

where $N = dH \in H_\infty^{r \times m}$ and $d \in H_\infty$. By applying the Nordgren theorem to matrix $N(z)$, for any scalar inner function ϕ there exist two square matrices $X \in H_\infty^{r \times r}$ and $Y \in H_\infty^{m \times m}$ such that

$$X(z) N(z) = S_N(z) Y(z) \tag{6-54}$$

and that $(\det X)_i$ and $(\det Y)_i$ are relatively prime to ϕ, where $S_N(z)$ is the quasi-Smith form of $N(z)$ given by (6-52) with N substituted for A. The denominator $d(z)$ has the inner-outer factorization

$$d(z) = d_i(z) d_o(z) \tag{6-55}$$

Then define

$$M_H(z) = \frac{S_N(Z)}{d_i(z)} = \left[\begin{array}{c:c} \Lambda_N(z)/d_i(z) & 0 \\ \hdashline 0 & 0 \end{array} \right] \tag{6-56}$$

where $\Lambda_N(z)/d_i(z) = \text{diag}\{\lambda_k(z)/d_i(z)\}$, and $\lambda_k(z)$ is the invariant factor of kth order of $N(z)$. Cancel all common inner factors between the numerator and denominator of every entry on the leading diagonal of $M_H(z)$; that is, let

$$\frac{\lambda_k(z)}{d_i(z)} = \frac{\delta_k(z)}{\psi_k(z)} \quad \text{for } k = 1, 2, \ldots, \ell \tag{6-57}$$

where $\{\delta_k, \psi_k(z)\}$ are relatively prime and $\ell = \text{rank } N$. Then we can write

$$M_H(z) = \left[\begin{array}{c:c} \text{diag}\{\delta_k(z)/\psi_k(z)\} & 0 \\ \hdashline 0 & 0 \end{array} \right] \tag{6-58}$$

It follows from (6-53) through (6-56) that

$$d_0(z) X(z) H(z) = M_H(z) Y(z) \tag{6-59}$$

where $(\det d_0 X)_i = (\det X)_i$.

We should note that the matrix $M_H(z)$ is unique, while there are many representations in the form (6-53). In fact, suppose that there are two representations of $H(z)$:

$$H(z) = \frac{N_1(z)}{d_1(z)} = \frac{N_2(z)}{d_2(z)} \tag{6-60}$$

Then we have

$$d_2(z) N_1(z) = d_1(z) N_2(z) \tag{6-61}$$

Since $d_1(z)$ and $d_2(z)$ are scalar, the quasi-Smith forms of $d_2(z) N_1(z)$ and $d_1(z) N_2(z)$ become $d_{2i}(z) S_{N_1}(z)$ and $d_{1i}(z) S_{N_2}(z)$, respectively, where $d_{1i}(z)$ and $d_{2i}(z)$ are the inner functions of $d_1(z)$ and $d_2(z)$, respectively, and S_{N_1} and $S_{N_2}(z)$ are the quasi-Smith forms of $N_1(z)$ and $N_2(z)$, respectively. Because the quasi-Smith form is unique for a matrix, they are identical; that is,

$$d_{2i}(z) S_{N_1}(z) = d_{1i}(z) S_{N_2}(z) \tag{6-62}$$

which is equivalent to

$$\frac{S_{N_1}(z)}{d_{1i}(z)} = \frac{S_{N_2}(z)}{d_{2i}(z)} \tag{6-63}$$

Thus the matrix $M_H(z)$ is unique. $M_H(z)$ is called the quasi-McMillan form of $H(z)$, because it is constructed by the relation of quasi-equivalence. From (6-56) and (6-57) we can write $M_H(z)$ as a right or left representation

$$M_H(z) = \Delta(z) \Psi_R(z)^{-1} = \Psi_L(z)^{-1} \Delta(z) \tag{6-64a}$$

where

$$\Delta(z) = \begin{bmatrix} \text{diag}\{\delta_k(z)\} & \vdots & 0 \\ \cdots\cdots\cdots\cdots\cdots & & \\ 0 & \vdots & 0 \end{bmatrix}, \quad r \times m \tag{6-64b}$$

$$\Psi_R(z) = \begin{bmatrix} \text{diag}\{\psi_k(z)\} & \vdots & 0 \\ \cdots\cdots\cdots\cdots\cdots & \vdots & \cdots\cdots \\ 0 & \vdots & I_{m-\ell} \end{bmatrix}, \qquad m \times m \qquad (6\text{-}64c)$$

and

$$\Psi_L(z) = \begin{bmatrix} \text{diag}\{\psi_k(z)\} & \vdots & 0 \\ \cdots\cdots\cdots\cdots\cdots & \vdots & \cdots\cdots \\ 0 & \vdots & I_{r-\ell} \end{bmatrix}, \qquad r \times r \qquad (6\text{-}64d)$$

Thus from the discussions above we have established the following result.

Theorem 6-10. Let $H \in B^{r \times m}$. Then for any scalar inner function ϕ there exist two square matrices $X \in H_\infty^{r \times r}$ and $Y \in H_\infty^{m \times m}$ such that

$$X(z)H(z) = M_H(z)Y(z) \qquad (6\text{-}65)$$

and that $(\det X)_i$ and $(\det Y)_i$ are relatively prime to ϕ, where $M_H(z)$ is the quasi-McMillan form of $H(z)$, which can be represented as the right representation and left representation defined by (6-64). ∎

Some remarks on Theorem 6-10 follow.

Remark 6-8

When $d(z)$ and all the entries of $N(z)$ are relatively inner prime in (6-53), we have $\psi_1(z) = d_i(z)$, because the contrary would imply that every entry of $N(z)$ was divisible by an inner factor of $d(z)$.

Remark 6-9

Since each invariant factor in the quasi-Smith form $S_N(z)$ divides each succeeding invariant factor, it follows that if a common inner factor is canceled when forming δ_k, the same factor is canceled when forming each succeeding δ_i. From this, ψ_k divides $\psi_1, \psi_2, \ldots, \psi_{k-1}$, $k = 2, 3, \ldots, \ell$. Conversely, no inner factor is canceled when forming δ_k, which was present but not canceled in forming any preceding δ_i. Therefore, δ_k divides δ_{k+1}, $\delta_{k+2}, \ldots, \delta_\ell$, $k = 1, 2, \ldots, \ell - 1$.

Remark 6-10

Since $\{\delta_k(z), \psi_k(z)\}$ are relatively prime, where $k = 1, 2, \ldots, \ell$, it follows from Remark 6-9 that $\delta_1\delta_2 \cdots \delta_\ell$ and $\psi_1\psi_2 \cdots \psi_\ell$ are relatively prime. This implies that the greatest common inner divisor of all minors of order r in the matrix $[\Psi_L(z), \Delta(z)]$ is unity, and that of all the minors of order m in the matrix $[\Psi_R(z)^T, \Delta(z)^T]^T$ is unity. Thus by Theorem 6-5(v), $\Psi_L(z)$ and $\Delta(z)$ are LIC, and $\Psi_R(z)$ and $\Delta(z)$ are RIC. Therefore, $\Delta(z)\Psi_R(z)^{-1}$ and $\Psi_L(z)^{-1}\Delta(z)$ are a LICR and a RICR of $M_H(z)$, respectively.

6-6 INVARIANTS OF COPRIME REPRESENTATIONS

In this section we present some invariant quantities of coprime representations of a transfer function H(z) of bounded type. We have the following main result.

Theorem 6-11. Let $H \in B^{r \times m}$, and

$$M_H(z) = \Delta(z)\Psi_R(z)^{-1} = \Psi_L(z)^{-1}\Delta(z) \qquad (6\text{-}66)$$

be the quasi-McMillan form of H(z), where $\Delta(z)$, $\Psi_R(z)$, and $\Psi_L(z)$ are defined by (6-64).

   (i)   Then all the numerator matrices of RICRs or LICRs of H(z) possess the same invariant factors $\{\delta_k(z)\}$, where $\delta_k(z)$, $k = 1, 2, \ldots, \ell$ are the invariant factors of $\Delta(z)$ and $\ell = \text{rank } \Delta$.
   (ii) All the denominator matrices of RICRs or LICRs of H(z) possess the same invariant factors $\{\psi_k\}$ except unity invariant factors, where $\psi_k(z)$, $k = 1, 2, \ldots, \ell$ are the invariant factors of $\Psi_R(z)$ or $\Psi_L(z)$ except unity invariant factors. ∎

Proof. The theorem will be proved for RICRs of H(z). It can be verified in the same way for LICRs of H(z). By Theorem 6-10, for any scalar inner function ϕ there exist two square matrices $X \in H_\infty^{r \times r}$ and $Y \in H_\infty^{m \times m}$ such that

$$X(z)H(z) = M_H(z)Y(z) \qquad (6\text{-}67)$$

and that (det X)$_i$ and (det Y)$_i$ are relatively prime to ϕ. Let

$$H(z) = N_R(z)D_R(z)^{-1} \qquad (6\text{-}68)$$

be a RICR of H(z). It follows from (6-66), (6-67), and (6-68) that

$$M_H(z) = \Delta(z)\Psi_R(z)^{-1} = X(z)N_R(z)\{Y(z)D_R(z)\}^{-1} \qquad (6\text{-}69)$$

By Remark 6-10, $\Delta(z)\Psi_R(z)^{-1}$ is a RICR of $M_H(z)$. Applying Theorem 6-8(i) to the matrix $M_H(z)$ in (6-69), there exists a matrix $L \in N_+^{m \times m}$ such that

$$X(z) N_R(z) = \Delta(z) L(z) \tag{6-70a}$$

$$Y(z) D_R(z) = \Psi_R(z) L(z) \tag{6-70b}$$

It also follows from (6-66), (6-67), and (6-68) that

$$H(z) = N_R(z) D_R(z)^{-1} = X(z)^{-1} \Delta(z) \Psi_R(z)^{-1} Y(z)$$

$$= \{ \det Y(z) \text{ adj } X(z) \Delta(z) \} \{ \det X(z) \text{ adj } Y(z) \Psi_R(z) \}^{-1} \tag{6-71}$$

Since $N_R(z) D_R(z)^{-1}$ is a RICR of $H(z)$, applying Theorem 6-8(i) to the matrix $H(z)$ in (6-71), there exists a matrix $\tilde{L} \in N_+^{m \times m}$ such that

$$\det Y(z) \text{ adj } X(z) \Delta(z) = N_R(z) \tilde{L}(z) \tag{6-72a}$$

$$\det X(z) \text{ adj } Y(z) \Psi_R(z) = D_R(z) \tilde{L}(z) \tag{6-72b}$$

Since any scalar function h in N_+ has the form $h = a/b$, where $a \in H_\infty$, $b \in H_\infty$, and b is outer (see Theorem 6-3), $\tilde{L}(z)$ and $L(z)$ are represented as

$$L(z) = \frac{T(z)}{\ell(z)} \tag{6-73a}$$

$$\tilde{L}(z) = \frac{\tilde{T}(z)}{\tilde{\ell}(z)} \tag{6-73b}$$

where $T \in H_\infty^{m \times m}$, $\tilde{T} \in H_\infty^{m \times m}$, $\ell \in H_\infty$, $\tilde{\ell} \in H_\infty$, and both $\ell(z)$ and $\tilde{\ell}(z)$ are scalar outer functions. Then (6-70a) and (6-70b) become

$$\ell(z) X(z) N_R(z) = \Delta(z) T(z) \tag{6-74a}$$

$$\ell(z) Y(z) D_R(z) = \Psi_R(z) T(z) \tag{6-74b}$$

and (6-72a) and (6-72b) become

$$\tilde{\ell}(z) \det Y(z) \text{ adj } X(z) \Delta(z) = N_R(z) \tilde{T}(z) \tag{6-75a}$$

$$\tilde{\ell}(z) \det X(z) \text{ adj } Y(z) \Psi_R(z) = D_R(z) \tilde{T}(z) \tag{6-75b}$$

We get from (6-74b) and (6-75b)

$$\ell(z) \tilde{\ell}(z) Y(z) \det X(z) \text{ adj } Y(z) \Psi_R(z) = \Psi_R(z) T(z) \tilde{T}(z) \tag{6-76}$$

which becomes

$$\tilde{\ell}(z)\,\ell(z)\,\det X(z)\,\det Y(z)\,\Psi_R(z) \;=\; \Psi_R(z)\,T(z)\,\tilde{T}(z) \qquad (6\text{-}77)$$

Since $\Psi_R(z)$ is nonsingular, this gives

$$\left\{\,\ell(z)\,\tilde{\ell}(z)\,\det X(z)\,\det Y(z)\,\right\}^m \;=\; \det T(z)\,\det \tilde{T}(z) \qquad (6\text{-}78)$$

Because $\ell(z)$ and $\tilde{\ell}(z)$ are outer and $(\det X)_i$ and $(\det Y)_i$ are relatively prime to ϕ, (6-78) implies that $(\det T)_i$ and $(\det \tilde{T})_i$ are relatively prime to ϕ. Thus using (6-74), $N_R(z)$ is quasi-equivalent to $\Delta(z)$, and $D_R(z)$ is quasi-equivalent to $\Psi_R(z)$. By (ii) of Nordgren's theorem, $N_R(z)$ and $\Delta(z)$ have the same invariant factors $\delta_1(z)$, $\delta_2(z)$, $\cdots$, $\delta_\ell(z)$, and $D_R(z)$ and $\Psi_R(z)$ have the same invariant factors $\psi_1(z)$, $\psi(z)$, $\cdots$, $\psi_\ell(z)$ and $m - \ell$ unity invariant factors. ∎

We say that a matrix representation ND^{-1} is a <u>inner bicoprime (respectively, bicoprime) representation</u> of P if $P = ND^{-1}L$, N and D are RIC (respectively, RC), and D and L are LIC (respectively, LC). Theorem 6-11 gives the following property for such a representation, which is required in the next section.

Theorem 6-12. Let $ND^{-1}L$ be a inner bicoprime representation of $P \in B^{r \times m}$. Then $\det D(z)$ is outer if $P \in H_\infty^{r \times m}$. ∎

The proof of this theorem requires the following two lemmas.

Lemma 6-2. Two matrices $N \in H_\infty^{r \times \ell}$ and $D \in H_\infty^{\ell \times \ell}$ are RIC if and only if there exist two sequences of matrices $\{X_i\} \subset H^{\ell \times \ell}$ and $\{Y_i\} \subset H_\infty^{\ell \times r}$ such that

$$\lim_{i \to \infty} (X_i D + Y_i N) = I$$

where the limit stands for elementwise L_2-norm convergence. ∎

In the operator theory on Hilbert spaces [21], the property mentioned in Lemma 6-2 is used for the definition of outer functions, which can be seen to be equivalent to Definition 6-1. The proof is found in Dewilde [25, Theorem 3.1] and Helson and Lowdenslager [26, Part II, p. 203].

Lemma 6-3. Let $P = ND^{-1}L \in B^{r \times m}$, where $N \in H^{r \times \ell}$ and $D \in H^{\ell \times \ell}$ are RIC with $L \in H^{\ell \times m}$. Let $D^{-1}L = \tilde{N}\tilde{D}^{-1}$, where $\tilde{N} \in H_\infty^{\ell \times m}$ and $\tilde{D} \in H_\infty^{m \times m}$ are RIC. Then $(N\tilde{N})\tilde{D}^{-1}$ is a RICR of P. Moreover, if D and L are LIC, then $\det D$ and $\det \tilde{D}$ are inner associates. ∎

Proof. Since $\tilde{N}$ and $\tilde{D}$ are RIC, it follows from Lemma 6-2 that there exist sequences $\{X_i\} \subset H_\infty^{m \times m}$ and $\{Y_i\} \subset H_\infty^{m \times \ell}$ such that

$$\lim_{i \to \infty} (X_i \tilde{D} + Y_i \tilde{N}) = I \qquad (6\text{-}79)$$

Since N and D are also RIC, by Lemma 6-2, there exist sequences $\{A_i\} \subset H_\infty^{\ell \times \ell}$ and $\{B_i\} \subset H_\infty^{\ell \times r}$ such that

$$\lim_{i \to \infty} (A_i D + B_i N) = I \tag{6-80}$$

Premultiplying (6-80) with Y_k, we get

$$\lim_{i \to \infty} (Y_k A_i D + Y_k B_i N) = Y_k \tag{6-81}$$

Put

$$\Delta_{ki} = Y_k A_i D + Y_k B_i N - Y_k \tag{6-82}$$

Then $\lim_{i \to \infty} \Delta_{ki} = 0$ for every k. This implies that there exists a sub-sequence $\{i(k)\}$ of $\{i\}$ such that

$$\lim_{k \to \infty} \Delta_{ki(k)} = 0 \tag{6-83}$$

It follows from (6-82) that

$$(X_k + Y_k A_{i(k)} L) \tilde{D} + Y_k B_{i(k)} N \tilde{N} = (X_k + Y_k A_{i(k)} L) \tilde{D} + (Y_k - Y_k A_{i(k)} D + \Delta_{ki(k)}) \tilde{N}$$

$$= X_k \tilde{D} + Y_k \tilde{N} + Y_k A_{i(k)} (L\tilde{D} - D\tilde{N}) + \Delta_{ki(k)} \tilde{N}$$

$$= X_k \tilde{D} + Y_k \tilde{N} + \Delta_{ki(k)} \tilde{N}$$

Here we used $L\tilde{D} = D\tilde{N}$. This, together with (6-79) and (6-83), gives

$$\lim_{k \to \infty} \{(X_k + Y_k A_{i(k)} L) \tilde{D} + Y_k B_{i(k)} (N\tilde{N})\} = I \tag{6-84}$$

Thus $N\tilde{N}$ and $\tilde{D}$ are RIC by means of Lemma 6-2. Applying Theorem 6-11(ii) to $D^{-1}L = \tilde{N}\tilde{D}^{-1}$, we obtain that det D and det $\tilde{D}$ are inner associates. ∎

Proof of Theorem 6-12. Let $D^{-1}L = \tilde{N}\tilde{D}^{-1}$, where $\tilde{N} \in H_\infty^{\ell \times m}$ and $\tilde{D} \in H_\infty^{m \times m}$ are RIC. Then by Lemma 6-3, $(N\tilde{N})\tilde{D}^{-1}$ is a RICR of P. Suppose that $P \in H_\infty^{r \times m}$ and let $D_1 = I \in H_\infty^{r \times r}$ and $N_1 = P \in H_\infty^{r \times m}$. Then $D_1^{-1}N_1$ is a LICR of P. Applying Theorem 6-11(ii) to $D_1^{-1}N_1 = (N\tilde{N})\tilde{D}^{-1}$ implies that det $\tilde{D}$ and det D_1 are inner associates. By Lemma 6-2, det D and det $\tilde{D}$ are inner asso-ciates. Therefore, det D and det D_1 are inner associates, which implies that there are two outer functions $x \in H_\infty$ and $y \in H_\infty$ such that x det $D = y$ det $D_1 = y$. Since the inverse of an outer function is outer in N_+ (see

Theorem 6-5) and the product of two outer functions is outer in N_+ (see Theorem 6-5), det D is outer in N_+. This implies that det D is outer in H_∞, because det $D \in H_\infty$.   ∎

6-7 PARAMETERIZATION OF COMPENSATORS

A given plant P(z) (or a system to be compensated) and its compensator C(z) (or a system that stabilizes a given plant) are assumed to be taken from systems of bounded type. It will be seen that the stabilization problem for a given plant need not always have a solution. Thus necessary and sufficient conditions will be presented for a plant to be stabilizable by a compensator. Under the stabilizability condition, a parameterization of all compensators that stabilize a given plant is presented in this section.

Suppose that a compensator $C \in B^{m \times r}$ is connected to a plant $P \in B^{r \times m}$ in the feedback configuration shown in Fig. 6-1, where u_1 and u_2 denote the externally applied inputs, e_1 and e_2 denote the inputs to the compensator and plant, respectively, and y_1 and y_2 denote the outputs of the compensator and plant, respectively.

The system considered is completely described by

$$e = u - FGe \qquad\qquad (6\text{-}85a)$$

$$y = Ge \qquad\qquad (6\text{-}85b)$$

where

$$e = \begin{bmatrix} e_1 \\ e_2 \end{bmatrix}, \quad u = \begin{bmatrix} u_1 \\ u_2 \end{bmatrix}, \quad y = \begin{bmatrix} y_1 \\ y_2 \end{bmatrix}, \quad F = \begin{bmatrix} 0 & I \\ -I & 0 \end{bmatrix}, \quad G = \begin{bmatrix} C & 0 \\ 0 & P \end{bmatrix}$$

It is clear that $\det(I + FG) = \det(I + PC) = \det(I + CP)$. The system (6-85) is said to be <u>well-posed</u> if $\det(I + FG) \neq 0$. This condition is necessary and sufficient to ensure that (6-85) has a unique solution e in B^{r+m} corresponding

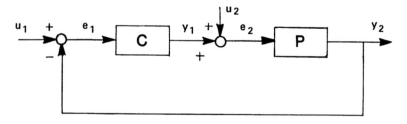

FIGURE 6-1 Feedback system with plant P and compensator C.

to every $u \in B^{r+m}$. Then under this condition, (6-85a) can be solved for e
as follows:

$$e = (I + FG)^{-1}u \qquad\qquad (6-86)$$

Let $H_e = (I + FG)^{-1}$. Then H_e is called the (external) input-(internal) input transfer function and can be expressed from the definitions of matrices
F and G as

$$H_e = \begin{bmatrix} (I + PC)^{-1} & -P(I + CP)^{-1} \\ C(I + PC)^{-1} & (I + CP)^{-1} \end{bmatrix} \qquad\qquad (6-87)$$

Equations (6-85a) and (6-85b) can be solved for y under the condition
of well-posedness as follows.

$$y = G(I + FG)^{-1}u \qquad\qquad (6-88)$$

Let $H_y = G(I + FG)^{-1}$. Then H_y is called the input-output transfer function and can be represented from (6-85b), (6-86), and (6-87) as

$$H_y = \begin{bmatrix} C(I + PC)^{-1} & -CP(I + PC)^{-1} \\ PC(I + PC)^{-1} & P(I + PC)^{-1} \end{bmatrix} \qquad\qquad (6-89)$$

It is easy to see that

$$H_e = I - FH_y \quad\text{and}\quad H_y = F(H_e - I) \qquad\qquad (6-90)$$

The feedback system (6-85) is called stable if it is well-posed and its
input-output transfer function H_y is stable (ℓ_2-stable), that is, if
$\det(I + PC) \neq 0$ and $H_y \in H_\infty^{n \times n}$ with $n = r + m$.

Remark 6-11

From (6-90), we get as in [14] the following fact: Suppose that the sys-
tem (6-85) is well-posed. Then $H_y \in H_\infty^{n \times n}$ if and only if $H_e \in H_\infty^{n \times n}$.
The plant P is said to be stabilizable if there exists a compensator
$C \in B^{m \times r}$ such that the feedback system (6-85) is stable. In this case, the
compensator C is said to stabilize the plant P.
The following theorem specifies the stabilizability condition of plant P.

Theorem 6-13. Let ND^{-1} and $\tilde{D}^{-1}\tilde{N}$ be a RCR and a LCR of $P \in B^{r \times m}$,
respectively. Then the following three statements are equivalent.

(i) P is stabilizable.

(ii) $[D^T, N^T]^T$ has a left inverse in $H_\infty^{m \times (m+r)}$; that is, there exist two
matrices $X \in H_\infty^{m \times m}$ and $Y \in H_\infty^{m \times r}$ such that

$$XD + YN = I \tag{6-91}$$

(iii) $[\tilde{D}, \tilde{N}]$ has a right inverse in $H_\infty^{(r+m) \times r}$; that is, there exist two
matrices $\tilde{X} \in H_\infty^{r \times r}$ and $\tilde{Y} \in H_\infty^{m \times r}$ such that

$$\tilde{D}\tilde{X} + \tilde{N}\tilde{Y} = I \tag{6-92}$$

∎

To prove Theorem 6-13, we need the following lemma, whose proof
can be found in [14, p. 96, Lemma 21].

Lemma 6-4. Let $P \in H_\infty^{m \times m}$ and $Q \in H_\infty^{r \times m}$. Suppose that $[P^T, Q^T]^T$
has full column rank. Then there exists $R \in H_\infty^{m \times r}$ such that $\det(P + RQ) \neq 0$.

∎

Proof of Theorem 6-6. We shall prove the equivalence between (i)
and (ii). The equivalence between (i) and (iii) can be proved in the same way.
We shall first prove that (i) implies (ii). By the definition of stability, there
exists a compensator $C \in B^{m \times r}$ for which the feedback system (6-85) is
well-posed and stable. Let $\tilde{A}^{-1}\tilde{B}$ be a LCR of C. Put

$$\Delta = \tilde{A}D + \tilde{B}N \in H_\infty^{m \times m} \tag{6-93}$$

Then, by $C = \tilde{A}^{-1}\tilde{B}$ and $P = ND^{-1}$, it holds that

$$\Delta = \tilde{A}(I + CP)D \tag{6-94}$$

Therefore, $\det \Delta \neq 0$, because $\det A \neq 0$, $\det D \neq 0$, and $\det(I + CP) \neq 0$ (the
last inequality comes from the definition of well-posedness). It follows from
(6-87) that

$$H_e = \begin{bmatrix} I - P(I + CP)^{-1}C & -P(I + CP)^{-1} \\ (I + CP)^{-1}C & (I + CP)^{-1} \end{bmatrix} \tag{6-95}$$

Here we used $C(I + PC)^{-1} = (I + CP)^{-1}C$ and $(I + PC)^{-1} = I - P(I + CP)^{-1}C$.
By (6-94) and (6-95) together with $C = \tilde{A}^{-1}\tilde{B}$ and $P = ND^{-1}$, we get

$$H_e = \begin{bmatrix} I & 0 \\ 0 & 0 \end{bmatrix} + \begin{bmatrix} -N \\ D \end{bmatrix} \Delta^{-1}[\tilde{B}, \tilde{A}] \tag{6-96}$$

Since N and D are RC, $[-N^T, D^T]^T$ and Δ are RC. Since $\tilde{A}$ and $\tilde{B}$ are LC,

Δ and $[\tilde{B}, \tilde{A}]$ are LC. Thus, by Theorem 6-12, $H_e \in H_\infty^{n \times n}$ means that $\det \Delta$ is outer, where $n = r + m$. Hence $(\det \Delta)^{-1}$ belongs to N_+. This implies that $\Delta^{-1}[\tilde{B}, \tilde{A}] \in N_+^{m \times n}$, because $\Delta^{-1} = \operatorname{adj} \Delta / \det \Delta$ and $\operatorname{adj} \Delta \cdot [\tilde{B}, \tilde{A}] \in H_\infty^{m \times n}$. Since N and D are RC, there exist two matrices $X_1 \in L_\infty^{m \times m}$ and $Y_1 \in L_\infty^{m \times r}$ such that

$$Y_1 N + X_1 D = I \quad \text{a.e. on } T \tag{6-97}$$

Premultiplying (6-96) by $[-Y_1, X_1]$, (6-96) gives

$$[-Y_1, 0] + \Delta^{-1}[\tilde{B}, \tilde{A}] = [-Y_1, X_1] H_e \in L_\infty^{m \times n} \tag{6-98}$$

which implies that $\Delta^{-1}[\tilde{B}, \tilde{A}] \in L_\infty^{m \times n}$. By means of Lemma 6-1, together with $\Delta^{-1}[\tilde{B}, \tilde{A}] \in N_+^{m \times n}$, this means that $\Delta^{-1}[\tilde{B}, \tilde{A}] \in H_\infty^{m \times n}$. Put

$$X = \Delta^{-1}\tilde{A} \quad \text{and} \quad Y = \Delta^{-1}\tilde{B} \tag{6-99}$$

Then $X \in H_\infty^{m \times m}$ and $Y \in H_\infty^{m \times r}$. Equation (6-91) comes from (6-93) and (6-99).

Next, we shall prove that (ii) implies (i). Suppose there exist two matrices $X \in H_\infty^{m \times m}$ and $Y \in H_\infty^{m \times r}$ such that (6-91) holds. Then we have

$$\begin{bmatrix} X & Y \\ -\tilde{N} & \tilde{D} \end{bmatrix} \begin{bmatrix} D & 0 \\ N & I \end{bmatrix} = \begin{bmatrix} I & Y \\ 0 & \tilde{D} \end{bmatrix} \tag{6-100}$$

Therefore, $[X^T, -\tilde{N}^T]^T$ has full column rank. By applying Lemma 6-4 to $[X^T, -\tilde{N}^T]^T$, there exists a matrix $R \in H^{m \times r}$ such that $\det(X - R\tilde{N}) \neq 0$. Premultiplying (6-100) by $[I, R]$, we get

$$(X - R\tilde{N}) D + (Y + R\tilde{D}) N = I \tag{6-101}$$

Put $\tilde{A} = X - R\tilde{N}$, $\tilde{B} = Y + R\tilde{D}$ and define the compensator $C(z)$ by $C = \tilde{A}^{-1}\tilde{B}$. Then we have, from (6-101),

$$\tilde{A}D + \tilde{B}N = I \tag{6-102}$$

which means that $\tilde{A}$ and $\tilde{B}$ are LC. Since

$$\det(I + CP) = \det(I + \tilde{A}^{-1}\tilde{B}ND^{-1}) = (\det \tilde{A})^{-1}(\det D)^{-1} \neq 0$$

the feedback system (6-85) with $P = ND^{-1}$ and $C = \tilde{A}^{-1}\tilde{B}$ is well-posed. In the same way as that of the derivation of (6-96) from (6-93), we get

$$H_e = \begin{bmatrix} I & 0 \\ 0 & 0 \end{bmatrix} + \begin{bmatrix} -N \\ D \end{bmatrix} [\tilde{B}, \tilde{A}] \in H_\infty^{n \times n} \qquad (6\text{-}103)$$

Hence the feedback system is stable, which implies statement (i). ∎

Remark 6-12

It can be seen from (6-93) and (6-99) that every compensator C that stabilizes a plant P has a LCR $X^{-1}Y$ such that

$$XD + YN = I \qquad (6\text{-}104)$$

where ND^{-1} is a RCR of P. It will be shown later that this LCR is unique for the given RCR $N^{-1}D$ of P.

It has been already shown in ⌊2⌋ that the Hardy space H_∞ is neither a Bézout domain nor a principal ideal domain. Thus we can find an unstabilizable plant, which is given in the following example. Put

$$a_n = 1 - \frac{1}{(2n)^2}$$
$$\qquad\qquad\qquad \text{for } n = 1, 2, \ldots$$
$$b_n = 1 - \frac{1}{(2n + 1)^2}$$

and let $f(z)$ and $g(z)$ be the Blaschke products with zeros at $\{a_n\}$ and zeros at $\{b_n\}$, respectively. Put $d(z) = f(z^{-1})$ and $n(z) = g(z^{-1})$. Then $d(z)$ and $n(z)$ have no common zero in E, and thus they are inner coprime. Let $P = n/d$. Then n and d are coprime, because they have no common zero and are inner functions [i.e., $|n(e^{j\omega})| = 1$ and $|d(e^{j\omega})| = 1$ a.e.]. On the other hand, it is shown in [2] that $xd + yn \neq 1$ for any $x, y \in H_\infty$. This implies that the plant P does not satisfy conditions (ii) and (iii) in Theorem 6-13. Therefore, the plant P is not stabilizable.

We shall now parameterize the set of all compensators that stabilize a given plant $P \in B^{r \times m}$. Let S(P) denote the set of all compensators $C \in B^{m \times r}$ that stabilize P.

Theorem 6-14. Suppose that $P \in B^{r \times m}$, and let ND^{-1} and $\tilde{D}^{-1}\tilde{N}$ be a RCR and a LCR of P, respectively. Under the stabilizability condition, select four matrices $X \in H_\infty^{m \times m}$, $Y \in H_\infty^{m \times r}$, $\tilde{X} \in H_\infty^{r \times r}$, and $\tilde{Y} \in H_\infty^{m \times r}$ such that $XD + YN = I$, $\tilde{D}\tilde{X} + \tilde{N}\tilde{Y} = I$. Then

$$S(P) = \{(X - R\tilde{N})^{-1}(Y + R\tilde{D}): R \in H_\infty^{m \times r}, \det(X - R\tilde{N}) \neq 0\}$$

$$= \{(\tilde{Y} + D\tilde{R})(\tilde{X} - N\tilde{R})^{-1}: \tilde{R} \in H_\infty^{m \times r}, \det(\tilde{X} - N\tilde{R}) \neq 0\}$$

∎

   Proof. Only the first representation is proved below. The second is proved in the same way.

   It can be seen from Remark 6-12 and the derivation of (6-96) from (6-93) that compensator C stabilizes P if and only if it has a LCR $\bar{X}^{-1}\bar{Y}$ such that

$$\bar{X}D + \bar{Y}N = I \tag{6-105}$$

Now, consider (6-105) in the unknowns $\bar{X}$ and $\bar{Y}$. Define the matrix U by

$$U = \begin{bmatrix} X & Y \\ -\tilde{N} & \tilde{D} \end{bmatrix} \in H_\infty^{n \times n} \tag{6-106}$$

where $n = r + m$. Since

$$\begin{bmatrix} X & Y \\ -\tilde{N} & \tilde{D} \end{bmatrix} \begin{bmatrix} D & -\tilde{Y} \\ N & \tilde{X} \end{bmatrix} = \begin{bmatrix} I & W \\ 0 & I \end{bmatrix} \tag{6-107}$$

where $W = -X\tilde{Y} + Y\tilde{X} \in H_\infty^{m \times r}$. Thus we get

$$\begin{bmatrix} X & Y \\ -\tilde{N} & \tilde{D} \end{bmatrix} \begin{bmatrix} D & -\tilde{Y} \\ N & \tilde{X} \end{bmatrix} \begin{bmatrix} I & -W \\ 0 & I \end{bmatrix} = \begin{bmatrix} I & 0 \\ 0 & I \end{bmatrix} \tag{6-108}$$

Therefore, U has the inverse $H_\infty^{n \times n}$, which is given by

$$U^{-1} = \begin{bmatrix} D & -\tilde{Y} \\ N & \tilde{X} \end{bmatrix} \begin{bmatrix} I & -W \\ 0 & I \end{bmatrix} = \begin{bmatrix} D & -\tilde{Y} - DW \\ N & \tilde{X} - NW \end{bmatrix} \in H_\infty^{n \times n}$$

By using (6-105), we have

$$[\bar{X}, \bar{Y}]U^{-1} = [I, R] \tag{6-109}$$

where $R = -\bar{X}(\tilde{Y} + DW) + \bar{Y}(\tilde{X} - NW) \in H_\infty^{m \times r}$. It follows from (6-106) and (6-109) that

$$[\bar{X}, \bar{Y}] = [I, R]U$$
$$= [X - R\tilde{N}, Y + R\tilde{D}] \tag{6-110}$$

Since all the compensators that stabilize P are represented in the form $C = \bar{X}^{-1}\bar{Y}$, the first representation is obtained by (6-110).   ∎

Remark 6-13

As in [14, p. 109], Theorem 6-14 characterizes the set of all compensators that stabilize a given plant in terms of a certain free parameter. The
correspondence between the parameter and the compensator is one-to-one;
that is, corresponding to each C stabilizing P, there is a unique $R \in H_\infty^{m \times r}$
such that $C = (X - R\tilde{N})^{-1}(Y + R\tilde{D})$, as well as a unique $\tilde{R} \in H_\infty^{m \times r}$ such that
$C = (\tilde{Y} + D\tilde{R})(\tilde{X} - N\tilde{R})^{-1}$.

As in [14], Theorem 6-14 leads to the following corollary, which can
be proved in the similar way in [14] with slight modifications.

Corollary. Let all the symbols be as in Theorem 6-14, and suppose
that the plant P is stabilizable. Then the following four statements hold.

(i) The set of $R \in H_\infty^{m \times r}$ such that $\det(X - R\tilde{N}) \neq 0$ is an open dense
set in $H_\infty^{m \times r}$, and the set of $\tilde{R} \in H_\infty^{m \times r}$ such that $\det(\tilde{X} - N\tilde{R}) \neq 0$
is an open dense set in $H_\infty^{m \times r}$.

(ii) If P is strongly causal, then every $C \in S(P)$ is causal.

(iii) If P is not strongly causal but causal, then there exists a strongly
causal $C \in S(P)$.

(iv) If P is stable and has a zero in E, then

$$S(P) = \left\{ (I - PR)^{-1} R : R \in H_\infty^{m \times r} \right\}$$

$$= \left\{ R(I - PR)^{-1} : R \in H_\infty^{m \times r} \right\}$$

In this case

$$H_e = \begin{bmatrix} I - PR & -P(I - RP) \\ R & I - RP \end{bmatrix}, \quad H_y = \begin{bmatrix} R & -PR \\ PR & P(I - RP) \end{bmatrix}$$
∎

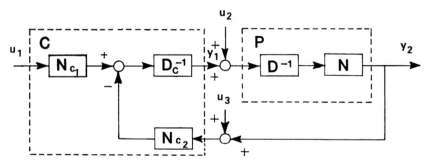

FIGURE 6-2 Feedback system with plant $P = ND^{-1}$ and two-parameter
compensator $C = D_c^{-1}[N_{c1}, N_{c2}]$.

The feedback system shown in Fig. 6-2 employs a compensator called a two-parameter compensator (or a compensator of two degrees of freedom). The reader is referred to [14] for the treatment of stabilization of feedback systems using two-parameter compensators in the lumped continuous case. We can use the same argument for the feedback system of bounded type above as in [14]. Following is the main result for the feedback system shown in Fig. 6-2.

Theorem 6-15. Suppose that $P \in B^{r \times m}$, and let ND^{-1} and $\tilde{D}^{-1}\tilde{N}$ be a RCR and a LCR of P, respectively. Under the stabilizability condition, select two matrices $X \in H_\infty^{m \times m}$ and $Y \in H_\infty^{m \times r}$ such that $XD + YN = I$. Then the set of all two-parameter compensators that stabilize P is given by

$$S_2(P) = \left\{ (X - R\tilde{N})^{-1}[Q, \ Y + R\tilde{D}] \right\}$$

where $Q \in H_\infty^{m \times \ell}$ is arbitrary (with ℓ being input dimension), and $R \in H_\infty^{m \times r}$ is arbitrary under the constraint $\det(X - R\tilde{N}) \neq 0$.   ∎

The proof of this theorem can be made in the same way as the proof of Theorem 5.6.15 in [14]. The reader is referred to [14, pp. 147-149] for several remarks concerning this theorem.

6-8 CONCLUSIONS

In this chapter we have explained some recent results on linear multivariable systems with transfer functions of bounded type obtained by means of the factorization approach. After all, under the stabilizability condition, it is possible to parameterize the set of all compensators that stabilize a given plant $P \in B^{r \times m}$, as in the lumped case.

We end by presenting the following two remarks. Suppose that $P \in B^{r \times m}$ is a symmetric function and that it is desired to parameterize all symmetric compensators $C \in B^{m \times r}$ that stabilize P [where a complex function f is called symmetric if $f(e^{j\omega}) = f(e^{-j\omega})$]. Then all the results in this chapter hold in this case, provided that H_∞ is replaced by the Hardy space consisting of bounded symmetric functions analytic in E. All the results in this chapter also hold for distributed continuous-time convolution systems, provided that H_∞ is replaced by the Hardy space consisting of bounded functions analytic in the open right half plane $\{s: \text{Re } s > 0\}$.

REFERENCES

1. Y. Inouye, "Linear systems with transfer functions of bounded type: canonical factorization," IEEE Trans. Circuits Syst., vol. CAS-33, pp. 581-589 (June 1986).

2. Y. Inouye, "Matrix-fraction descriptions and the quasi-McMillan form of a transfer function of bounded type," IEEE Trans. Circuits Syst., vol. CAS-34, pp. 127-132 (Feb. 1987).

3. Y. Inouye, "Notes on the minimum-energy delay property of impulse-response sequences of minimum-phase transfer functions," IEEE Trans. Circuits Syst., vol. CAS-34, pp. 188-190 (Feb. 1987).

4. Y. Inouye, "Parameterization of compensators for linear systems with transfer functions of bounded type," Technical report 88-01, Faculty of Engineering Science, Osaka University, Japan, Mar. 1988; also, Proc. of 27th IEEE Conf. Decision and Control, pp. 2083-2088 (1988).

5. V. Kučera, Discrete Linear Control, Wiley, New York, 1979, p. 109.

6. D. C. Youla, H. A. Jabr, and J. J. Bongiorno, Jr., "Modern Wiener-Hopf design of optimal controllers, Part II: The multivariable case," IEEE Trans. Autom. Control, vol. AC-21, pp. 319-338 (June 1976).

7. F. M. Callier and C. A. Desoer, "An algebra of transfer functions for distributed time-invariant systems," IEEE Trans. Circuit Syst., vol. CAS-25, pp. 651-662 (Sept. 1978); corrections in vol. CAS-26, p. 360 (May 1979).

8. F. M. Callier and C. A. Desoer, "Stabilization, tracking and disturbance rejection in multivariable convolution systems," Ann. Soc. Sci. Bruxelles, Sér. I, vol. 94, pp. 7-51 (1980).

9. M. Vidyasagar, H. Schneider, and B. A. Francis, "Algebra and topological aspects of feedback stabilization," IEEE Trans. Autom. Control, vol. AC-27, pp. 880-894 (Aug. 1982).

10. V. H. L. Cheng and C. A. Desoer, "Discrete time convolution control systems," Int. J. Control, vol. 36, no. 3, pp. 367-407 (1982).

11. C. A. Desoer, R. W. Liu, J. Murray, and R. Saeks, "Feedback system design: the fractional representation approach to analysis and synthesis," IEEE Trans. Autom. Control, vol. AC-25, pp. 399-412 (June 1980).

12. C. A. Desoer and C. L. Gustafson, "Algebraic theory of linear multivariable feedback system," IEEE Trans. Autom. Control, vol. AC-29, pp. 909-917 (Oct. 1984).

13. A. S. Morse, "System invariants under feedback and cascade control," in Mathematical System Theory, ed. G. Marchesini and S. K. Mitter, Springer-Verlag, New York, 1976.

14. M. Vidyasagar, Control System Synthesis: A Factorization Approach, The MIT Press, Cambridge, Mass., 1985.

15. R. Nevanlinna, Analytic Functions, Springer-Verlag, New York, 1970.

16. A. V. Oppenheim and R. W. Schafer, Digital Signal Processing, Prentice-Hall, Englewood Cliffs, N.J., 1975.

17. A. Papoulis, Signal Analysis, McGraw-Hill, New York, 1977.

18. C. A. Desoer and M. Vidyasagar, Feedback Systems: Input-Output Properties, Academic Press, New York, 1975.

19. W. Rudin, Real and Complex Analysis, McGraw-Hill, New York, 1966.

20. P. L. Duren, Theory of H^p Space, Academic Press, New York, 1970.

21. M. Rosenblum and J. Rovnyak, Hardy Classes and Operator Theory, Oxford University Press, Oxford, 1985.

22. Yu. L. Shmul'yan, "Optimal factorization of non-negative matrix functions," Theory of Probab. Its Appl. (USSR), vol. 9, pp. 346-349 (1964).

23. Yu. A. Rosanov, Stationary Random Processes (transl. A. Feinstein), Holden-Day, San Francisco, 1967.

24. E. A. Nordgren, "On quasi-equivalence of matrices over H^∞," Acta Sci. Math., vol. 34, pp. 311-316 (1973).

25. P. Dewilde, "Input-output description of roomy systems," SIAM J. Control Optim., vol. 14, pp. 712-736 (1976).

26. H. Helson and D. Lowdenslager, "Prediction theory and Fourier series in several variables," Acta Math., vol. 99, pp. 165-202 (1958); Part II, vol. 106, pp. 175-213 (1961).

7

Stability of Polynomial Matrices and Routh Approximation of MIMO Systems

HIROSHI NAGAOKA[*]

Tokyo Engineering University, Hachioji, Tokyo, Japan

7-1 INTRODUCTION

In this chapter we study the continuous-time stability of polynomial matrices (PMs) in the framework of inner products and orthogonal PMs, and as an application we construct the Routh approximation for multiple-input, multiple-output (MIMO) systems.

First, in Sec. 7-2 we define an inner product $\langle \cdot \rangle$, which takes a matrix value $\langle P(s), Q(s) \rangle$ for two PMs $\{P(s), Q(s)\}$ belonging to a certain class, from an arbitrarily given pair $(A(s), \Pi)$ of a nonsingular PM $A(s)$ and a positive-definite matrix Π. The inner product $\langle \cdot \rangle$ can be regarded as a representation of the solution of a Lyapunov equation, and has complete information about whether or not $A(s)$ is atable. Next, in Secs. 7-3 and 7-4 we introduce a family of PMs $\{R_j(s)\}$ that constitute an orthogonal system

[*]Present affiliation: Department of Information Engineering, Faculty of Engineering, Hokkaido University, Sapporo, Japan.

with respect to the inner product $\langle \cdot \rangle$. Each PM $R_j(s)$ is called an orthogonal PM associated with $(A(s), \Pi)$. Investigation of the algebraic structure of $\{R_j(s)\}$ yields a stability criterion for $A(s)$, which can be regarded as an extension of the Routh-Hurwitz test for scalar polynomials. This situation is analogous to that of the discrete-time case where an extension of the Schur-Cohn test is obtained via the notion of orthogonal PMs on the unit circle and the LWR (Levinson-Whittle-Wiggins-Robinson) algorithm (e.g., see [2], [3], [5], [8], etc.). Finally, in Sec. 7-5 we show that the Schwarz-form realization and the Routh approximation can be extended to MIMO systems by using the results in previous sections. For further results on the inner product and the orthogonal PMs, see [10].

7-2 INNER PRODUCTS OF POLYNOMIAL MATRICES

The purpose of this section is to define the (matrix-valued) inner product $\langle P(s), Q(s) \rangle$, where $P(s)$ and $Q(s)$ are PMs, associated with a given $p \times p$ positive definite matrix Π and a given $p \times p$ nonsingular PM $A(s)$. For a while we restrict ourselves to the case where $A(s)$ is <u>(continuous-time) stable</u> [i.e., all the zeros of det $A(s)$ have negative real parts]. Let $P(s) : k \times p$ and $Q(s) : \ell \times p$ be PMs such that $P(s)A^{-1}(s)$ and $Q(s)A^{-1}(s)$ are both strictly proper (i.e., tend to 0 as $s \to \infty$); for convenience we denote the condition by

$$P(s) \longrightarrow A(s) \quad \text{and} \quad Q(s) \longrightarrow A(s) \tag{7-1}$$

Then $P(s)A^{-1}(s)$ and $Q(s)A^{-1}(s)$ are both square integrable on the imaginary axis, and hence we can define a matrix-valued inner product as follows:

$$\langle P(s), Q(s) \rangle \overset{\Delta}{=} \frac{1}{2\pi} \int_{-\infty}^{\infty} P(j\omega) A^{-1}(j\omega) \Pi A_*^{-1}(j\omega) Q_*(j\omega) \, d\omega, \quad k \times \ell \tag{7-2}$$

where $A_*(s) \overset{\Delta}{=} A^T(-s)$ and $Q_*(s) \overset{\Delta}{=} Q^T(-s)$. It is evident from the definition that the inner product satisfies the following properties:

(i) $\langle P_1(s) + P_2(s), Q(x) \rangle = \langle P_1(s), Q(s) \rangle + \langle P_2(s), Q(s) \rangle$

(ii) $\langle KP(s), Q(s) \rangle = K \langle P(s), Q(s) \rangle$, where K is a constant matrix

(iii) $\langle P(s), Q(s) \rangle = \langle Q(s), P(s) \rangle^T$

(iv) $\begin{cases} \langle P(s), P(s) \rangle \geq 0 \\ \langle P(s), P(s) \rangle = 0 \quad \text{iff} \quad P(s) = 0 \\ \langle P(s), P(s) \rangle > 0 \quad \text{iff} \quad P(s) \text{ has full row rank} \end{cases}$

(v) $\langle sP(s), Q(s) \rangle + \langle P(s), sQ(s) \rangle = 0 \tag{7-3}$

        if

$$sP(s) \longrightarrow A(s) \quad \text{and} \quad sQ(s) \longrightarrow A(s) \tag{7-4}$$

We note that property (v) is a distinctive feature of inner products on the imaginary axis.

As the next step, we extend the domain of the definition of the inner product $\langle P(s), Q(s) \rangle$ from (7-1) to

$$P(s) \sim A(s) \quad \text{and} \quad Q(s) \longrightarrow A(s) \tag{7-5}$$

where $P(s) \sim A(s)$ means that $P(s)A^{-1}(s)$ is proper (i.e., tends to a finite value as $s \to \infty$). Suppose that $P(s) \sim A(s)$ and that $P(s) \nrightarrow A(s)$. Then $P(s)A^{-1}(s)$ is bounded on the imaginary axis, although it is not square integrable. Hence if $Q(s)A^{-1}(s)$ is absolutely integrable, the integrand $PA^{-1}\Pi A_*^{-1}Q_*$ in (7-2) is also absolutely integrable, and the inner product $\langle P(s), Q(s) \rangle$ can be defined by (7-2). We note that a necessary and sufficient condition for $Q(s)A^{-1}(s)$ to be absolutely integrable is that $sQ(s) \longrightarrow A(s)$. On the other hand, the situation becomes more delicate if $Q(s) \longrightarrow A(a)$ and $sQ(s) \nrightarrow A(s)$. In this case, $PA^{-1}\Pi A_*^{-1}Q_*$ is not absolutely integrable. Nevertheless, we can define $\langle P(s), Q(s) \rangle$ by taking the Cauchy principal value of the integral (7-2) on the basis of the following lemma, which is easily proved by Cauchy's integral theorem (e.g., see ⌊10⌋).

Lemma 7-1. If a rational function $f(s)$ is analytic in $\mathrm{Re}\lfloor s\rfloor \geq 0$ and is strictly proper, then

$$\lim_{r \to \infty} \int_{-r}^{r} f(j\omega) \, d\omega = \pi[f(s)]_{-1}$$

where

$$[f(s)]_{-1} \overset{\Delta}{=} \lim_{s \to \infty} sf(s)$$

∎

The following is immediate from Lemma 7-1.

(vi) If $Q(s) \longrightarrow A(s)$, then

$$\langle A(s), Q(s) \rangle = \frac{1}{2} \Pi [Q(s)A^{-1}(s)]_{-1}^{T} \tag{7-6}$$

It is concluded from (v) that $A(s)$ is orthogonal to every $Q(s)$ such that $sQ(s) \longrightarrow A(s)$.

Generally, an arbitrary PM $P(s)$ such that $P(s) \sim A(s)$ is written as

$$P(s) = KA(S) + R(s) \tag{7-7}$$

where K is a constant matrix and R(s) is a PM such that $R(s) \longrightarrow A(s)$. Hence we have

$$\langle P(s), Q(s) \rangle = \frac{1}{2} K\Pi \left[Q(s) A^{-1}(s) \right]_{-1}^{T} + \langle R(s), Q(s) \rangle \qquad (7-8)$$

Thus $\langle P(s), Q(s) \rangle$ has been defined for arbitrary P(s) and Q(s) satisfying (7-5). It is obvious that the fundamental properties (i) to (iv) are also valid for the extended inner product and that the condition (7-4) in property (v) can be replaced with the weakened condition

$$P(s) \longrightarrow A(s) \quad \text{and} \quad Q(s) \longrightarrow A(s) \qquad (7-9)$$

Now, let us consider how to calculate $\langle P(s), Q(s) \rangle$ for given P(s) and Q(s). We use the fact that there exist matrices F and G and a PM Z(s) such that

(F, G) is controllable   (7-10)

A(s) and Z(s) are right coprime   (7-11)

$(sI - F)^{-1}G = Z(s) A^{-1}(s)$ (7-12)

Given such a triplet (F, G, Z(s)), then

$$X \overset{\Delta}{=} \langle Z(s), Z(s) \rangle$$

is obtained as the unique solution of the Lyapunov equation

$$FX + XF^{T} + G\Pi G^{T} = 0 \qquad (7-13)$$

If P(s) and Q(s) satisfy (7-1), they are written as

$$P(s) = PZ(z) \quad \text{and} \quad Q(s) = QZ(s) \qquad (7-14)$$

where P and Q are constant matrices, and we have

$$\langle P(s), Q(s) \rangle = PXQ^{T} \qquad (7-15)$$

The case of (7-5) is reduced to the case of (7-1) by the use of (7-7) and (7-8).

So far, we have investigated the inner product defined by the integral on the imaginary axis (7-2) from given (A(s), Π) under the assumption that A(s) is stable. We now consider the problem of extending the definition of the inner product to the general case where A(s) is not necessarily stable, so that a stability criterion for A(s) will be derived from the defined inner product. The definition by the integral (7-2) is not suitable for this purpose, because the integral does not have information about whether or not A(s) is stable. Alternatively, we start from the fundamental properties (i) to (vi).

Let A(s) be a p × p nonsingular PM and Π be a p × p positive-definite matrix. Suppose that a matrix-valued inner product $\langle P(s), Q(s) \rangle$ is defined for arbitrary PMs P(s) and Q(s) satisfying (7-5) and the fundamental properties (i) to (iii), (v) [with the weakened condition (7-9)], and (vi) hold for the inner product. Then, for matrices F and G and a PM Z(s) satisfying (7-10) to (7-12), the matrix $X \overset{\Delta}{=} \langle Z(s), Z(s) \rangle$ becomes a solution of the Lyapunov equation (7-13). This fact can be proved by noting that

$$\langle sZ(s), Z(s) \rangle = FX + G \langle A(s), Z(s) \rangle \quad \text{[by (7-12), (i), (ii)]}$$

$$= FX + \frac{1}{2} G\Pi \, [Z(s)A^{-1}(s)]_{-1}^{T} \quad \text{[by property (vi)]}$$

$$= FX + \frac{1}{2} G\Pi G^{T} \quad \text{[by (7-12)]}$$

and by invoking (iii) and (v). The value of $\langle P(s), Q(s) \rangle$ for general P(s) and Q(s) is determined from X as in the case of stable A(s). Conversely, if we define an inner product $\langle P(s), Q(s) \rangle$ in this way from a solution of the Lyapunov equation (7-13), it is easy to verify that the inner product satisfies the fundamental properties (i) to (iii), (v), and (vi). Therefore, given (A(s), Π), there exists a unique inner product satisfying the fundamental properties if and only if the Lyapunov equation (7-13) has a unique solution. As is well known, this unique existence is equivalent to the condition that F and -F have no common eigenvalues (e.g., see [7]), or equivalently, that

$$\det A(s) \text{ and } \det A_{*}(s) \text{ have no common zeros} \tag{7-16}$$

From now on, we assume (7-16).

We should note that property (iv) (positivity) does not hold for the inner product in general. Actually, it is clear from the considerations above that

(iv) $\rightleftharpoons$ X is positive definite

$\rightleftharpoons$ F is a stable matrix $\tag{7-17}$

$\rightleftharpoons$ A(s) is a stable PM

7-3 STABILITY AND ORTHOGONAL POLYNOMIAL MATRICES: STRICTLY REGULAR CASE

Suppose that we are given a p × p positive definite matrix Π and a p × p nonsingular PM A(s) satisfying (7-16). Then, as in Sec. 7-2, we can define from

$(A(s), \Pi)$ the inner product $\langle P(s), Q(s) \rangle$ for PMs $P(s)$ and $Q(s)$ satisfying (7-5). In this section we assume further that $A(s)$ is strictly regular and monic in the sense that it is written as

$$A(s) = s^n I + s^{n-1} A_{n-1} + \cdots + A_0 \tag{7-18}$$

where A_j's are $p \times p$ matrices. In this situation, the condition (7-5) is equivalent to

$$\deg P(s) \le n \quad \text{and} \quad \deg Q(s) \le n - 1 \tag{7-19}$$

which means that $P(s)$ and $Q(s)$ are written as

$$P(s) = \sum_{j=0}^{n} s^j P_j$$

$$Q(s) = \sum_{j=0}^{n-1} s^j Q_j \tag{7-20}$$

The inner product $\langle P(s), Q(s) \rangle$ is defined for such $P(s)$ and $Q(s)$ and is written as

$$\langle P(s), Q(s) \rangle = \sum_{i=0}^{n} \sum_{j=0}^{n-1} P_i X_{ij} Q_j^T \tag{7-21}$$

where

$$X_{ij} \triangleq \langle s^i I, s^j I \rangle, \quad p \times p \tag{7-22}$$

In particular, if both $\deg P(s)$ and $\deg Q(s)$ are less than n, or equivalently if (7-1) holds, then (7-21) is written as

$$\langle P(s), Q(s) \rangle = [P_0, \ldots, P_{n-1}] X [Q_0, \ldots, Q_{n-1}]^T$$

where

$$X \triangleq \begin{bmatrix} X_{0,0} & \cdots & X_{0,n-1} \\ \vdots & & \vdots \\ X_{n-1,0} & \cdots & X_{n-1,n-1} \end{bmatrix} : np \times np$$

Hence the positivity (iv) in Sec. 7-2 is equivalent to the positive definiteness of X, which is also equivalent to the stability of A(s) [see (7-17)].

When a monic PM of degree j $(0 \leq j \leq n)$, say

$$R_j(s) = s^j I + s^{j-1} R_{j,j-1} + \cdots + R_{j,0}, \qquad p \times p \tag{7-24}$$

satisfies the orthogonal condition

$$\langle R_j(s), s^i I \rangle = 0 \quad \text{for } \forall i = 0, 1, \ldots, j-1 \tag{7-25}$$

we say that $R_j(s)$ is an __orthogonal PM of degree j__ defined from $(A(s), \Pi)$. Let X_j $(0 \leq j \leq n-1)$ be a submatrix of X defined as

$$X_j \triangleq \begin{bmatrix} X_{0,0} & \cdots & X_{0,j} \\ \vdots & & \vdots \\ X_{j,0} & \cdots & X_{j,j} \end{bmatrix} \tag{7-26}$$

Then (7-25) is written as

$$[R_{j,0}, \cdots, R_{j,j-1}] X_{j-1} + [X_{j,0}, \cdots, X_{j,j-1}] = 0 \tag{7-27}$$

Therefore, the condition that

$$X_j \text{ is nonsingular} \quad \forall j = 0, 1, \ldots, n-1 \tag{7-28}$$

ensures the existence and the uniqueness of $R_j(s)$ for every $j = 0, 1, \ldots, n$. From now on, we assume (7-28). Note that if A(s) is stable, then X is positive definite and the assumption is satisfied.

The orthogonal PMs $\{R_j(s); 0 \leq j \leq n\}$ constitute an orthogonal system; that is,

$$\langle R_i(s), R_j(s) \rangle = 0 \quad \text{if } i \neq j$$

Hence, defining a $p \times p$ symmetric matrix

$$\epsilon_j \triangleq \langle R_j(s), R_j(s) \rangle \tag{7-29}$$

for $j = 0, 1, \ldots, n-1$ and letting

$$
R \triangleq
\begin{bmatrix}
I & & & & 0 \\
R_{1,0} & I & & & \\
\vdots & & \ddots & & \\
\vdots & & & \ddots & \\
R_{n-1,0} & R_{n-1,1} & \cdots & & I
\end{bmatrix}, \quad np \times np
\tag{7-30}
$$

we have

$$
R X R^T = (E \triangleq) \text{ block diag}\{\epsilon_0, \epsilon_1, \ldots, \epsilon_{n-1}\}
\tag{7-31}
$$

where block diag$\{\cdots\}$ denotes the block diagonal matrix with diagonal elements $\{\cdots\}$. We can see from (7-31) that the assumption (7-28) implies the nonsingularity of ϵ_j, $\forall j = 0, 1, \ldots, n - 1$. Moreover, the previous argument on the stability condition of A(s) leads to the following theorem.

Theorem 7-1. A(s) is stable if and only if ϵ_j is positive definite for $\forall j = 0, 1, \ldots, n - 1$.   ∎

Theorem 7-1 itself might be a trivial result. Nevertheless, it will be shown later that a combination of the theorem and a recursive formula for $\{R_j(s)\}$ yields several important results.

The orthogonal PMs $\{R_j(s)\}$ can be used as a basis for representing PMs as well as $\{s^j I\}$. The representation of A(s) with respect to the basis $\{R_j(s)\}$ is as follows.

Theorem 7-2. A(s) is written as

$$
A(s) = R_n(s) + \frac{1}{2}\Pi\epsilon_{n-1}^{-1}R_{n-1}(s)
\tag{7-32}
$$

Proof. We note that property (vi) in Sec. 7-2 [Eq. (7-6)] is equivalent to

$$
\langle A(s), s^j I \rangle =
\begin{cases}
0 & \text{if } 0 \le j \le n - 2 \\[2mm]
\frac{1}{2}\Pi & \text{if } j = n - 1
\end{cases}
\tag{7-33}
$$

and that A(s) is characterized as a monic PM of degree n obeying (7-33). Hence (7-32) is readily proved by verifying that its right-hand side satisfies (7-33).   ∎

As the next step, we investigate the mutual relation among $\{R_j(s)\}$. For this purpose we introduce new quantities:

$$\theta_j \overset{\Delta}{=} \langle sR_j(s), R_j(s) \rangle \tag{7-34}$$

for $j = 0, 1, \ldots, n - 1$. It then turns out from property (v) in Sec. 7-2 [Eq. (7-3)] that θ_j is skew symmetric (i.e., $\theta_j^T = -\theta_j$).

<u>Theorem 7-3.</u> Starting from the initial condition

$$R_0(s) = I, \qquad R_1(s) = sI - \theta_0 \epsilon_0^{-1} \tag{7-35}$$

$\{R_j(s)\}$ can be produced recursively by

$$R_{j+1}(s) = (sI - \theta_j \epsilon_j^{-1}) R_j(s) + \epsilon_j \epsilon_{j-1}^{-1} R_{j-1}(s) \qquad (j = 1, 2, \ldots, n - 1) \tag{7-36}$$

■

 <u>Proof.</u> It is sufficient to verify that the right-hand side of (7-36) is orthogonal to $R_i(s)$, $\forall i = 0, 1, \ldots, j$. Owing to property (v) in Sec. 7-2, the inner product between the right-hand side and $R_i(s)$ is written as

$$- \langle R_j(s), sR_i(s) \rangle - \theta_j \epsilon_j^{-1} \langle R_j(s), R_i(s) \rangle + \epsilon_j \epsilon_{j-1}^{-1} \langle R_{j-1}(s), R_i(s) \rangle$$

For $i = 0, 1, \ldots, j - 2$, it is obvious from the orthogonality of $\{R_j(s)\}$ that the expression above is equal to 0. It is also clear from the definitions of $\{\epsilon_j\}$ and of $\{\theta_j\}$ that the expression above vanishes for $i = j - 1$ and for $i = j$.

■

 Example 7-1 (Scalar Case: Routh-Hurwitz Test)

Suppose that $A(s)$ is a monic scalar polynomial of degree n written as

$$A(s) = s^n + A_{n-1} s^{n-1} + \cdots + A_0 \tag{7-37}$$

Then, since the skew-symmetric matrices $\{\theta_j\}$ vanish, (7-35) and (7-36) are reduced to

$$R_0(s) = 1, \qquad R_1(s) = s \tag{7-38}$$

$$R_{j+1}(s) = sR_j(s) + e_j R_{j-1}(s) \tag{7-39}$$

where $e_j \overset{\Delta}{=} \epsilon_j/\epsilon_{j-1}$ $(1 \le j \le n - 1)$. It follows from the equations above that $R_j(s)$ is of the form

$$R_j(s) = s^j + R_{j,j-2}s^{j-2} + R_{j,j-4}s^{j-4} + \cdots \qquad (7\text{-}40)$$

On the other hand, Theorem 7-2 shows that

$$A(s) = R_n(s) + e_n R_{n-1}(s) \qquad (7\text{-}41)$$

where $e_n \triangleq \Pi/2\epsilon_{n-1}$. Invoking (7-40) and comparing (7-41) with (7-37), we can see that $R_n(s)$, $R_{n-1}(s)$, and e_n are obtained directly from $A(s)$ as

$$R_n(s) = s^n + A_{n-2}s^{n-2} + A_{n-4}s^{n-4} + \cdots$$

$$R_{n-1}(s) = s^{n-1} + \left(\frac{A_{n-3}}{A_{n-1}}\right)s^{n-3} + \left(\frac{A_{n-5}}{A_{n-1}}\right)s^{n-5} + \cdots \qquad (7\text{-}42)$$

$$e_n = A_{n-1}$$

Furthermore, Eq. (7-39) can be interpreted as the Euclidean algorithm generating $\{e_j, R_{j-1}(s)\}$ from $\{R_{j+1}(s), R_j(s)\}$. Thus it is seen that (7-38) to (7-41) are equivalent to the procedure of the Routh-Hurwitz stability test (e.g., see [4]) except for the order of computation. It is obvious that the stability criterion of the Routh-Hurwitz test

$$e_j > 0 \quad \forall j = 1, 2, \ldots, n$$

is equivalent to the criterion of Theorem 7-1.

7-4 STABILITY AND ORTHOGONAL POLYNOMIAL MATRICES: COLUMN-REDUCED CASE

An arbitrary rational matrix H(s) : $q \times p$ can be represented by a <u>right MFD</u> <u>(matrix fraction description)</u>

$$H(s) = B(s)A^{-1}(s) \qquad (7\text{-}43)$$

where A(s) is a $p \times p$ nonsingular PM and B(s) is a $q \times p$ PM. When the degree of the MFD (7-43) [i.e., the degree of the determinant of the denominator PM A(s)] is minimal among all the right MFDs of H(s), it is said that (7-43) is <u>irreducible</u>. An irreducible right MFD of H(s) is unique up to right unimodular factor; that is, if (7-43) is irreducible, another MFD H(s) = $\bar{B}(s)\bar{A}^{-1}(s)$ is also irreducible if and only if there exists a unimodular PM

U(s) [i.e., det U(s) is a nonzero constant] such that $\bar{A}(s) = A(s)U(s)$ and $\bar{B}(s) = B(s)U(s)$.

When (7-43) is irreducible, the necessary and sufficient condition for H(s) to be the transfer function of a continuous-time stable system is that A(s) is a stable PM. Hence, if A(s) is <u>strictly regular</u> in the sense that it is written as

$$A(s) = s^n A_n + s^{n-1} A_{n-1} + \cdots + A_0$$

with

$$\det A_n \neq 0$$

the results in Sec. 7-3 can be applied to the stability analysis of H(s). However, a rational matrix does not always have an irreducible MFD with a strictly regular denominator. So in the sequel we will extend the results of Sec. 7-3 to a wider class of PMs—column-reduced PMs.

Suppose that A(s) : p × p is a nonsingular PM, and let n_i $(1 \leq i \leq p)$ be the ith-column degree (i.e., the highest degree of the polynomials in the ith column) of A(s). Then it holds in general that

$$(N \overset{\Delta}{=}) \deg \det A(s) \leq \sum_{i=1}^{p} n_i \tag{7-44}$$

When the equality holds in the above, A(s) is said to be <u>column-reduced</u>. In the sequel we assume the column reducedness of A(s). This assumption brings no loss of generality because, for an arbitrary PM A(s), there exists a unimodular PM U(s) such that A(s)U(s) is column reduced [7], which means that an arbitrary rational matrix has an irreducible right MFD with a column-reduced denominator. Moreover, since the columns of A(s) can be rearranged arbitrarily by multiplication of a unimodular PM to the right, we can further assume that

$$(n \overset{\Delta}{=}) n_1 \geq n_2 \geq \cdots \geq n_p \tag{7-45}$$

We introduce a basis for representing PMs which is suitable for our purpose—generalization of the results of Sec. 7-3 to the column-reduced case. First, let

$$T(s) \overset{\Delta}{=} \text{diag}\{s^{n_1}, \ldots, s^{n_p}\}, \quad p \times p \tag{7-46}$$

Then A(s) is uniquely expressed as

$$A(s) = A_{high} T(s) + A_{low}(s) \tag{7-47}$$

by a $p \times p$ constant matrix A_{high} and a $p \times p$ PM $A_{low}(s)$ such that $A_{low}(s)T^{-1}(s)$ is a strictly proper, or equivalently that the ith-column degree of $A_{low}(s)$ is less than n_i $\forall i = 1, 2, \ldots, p$. It is noted [7] that the column reducedness of A(s) is equivalent to

$$\det A_{high} \neq 0 \tag{7-48}$$

Next, consider the following matrix for $j = 0, 1, \ldots, n$:

$$\frac{T(s)}{s^{n-j}} = \text{diag}\{s^{j-n+n_i}; i = 1, 2, \ldots, p\}$$

This is not generally a PM. Indeed, defining the integer $r(j)$ for each $j = 0, 1, \ldots, n$ as

$$r(j) \triangleq \max\{i \mid 1 \le i \le p \text{ and } j - n + n_i \ge 0\} \tag{7-49}$$

we see that the first $r(j)$ elements among the p diagonal elements of $T(s)/s^{n-j}$ are nonnegative powers of s, while the other diagonal elements are negative powers when $r(j) < p$. Hence picking up the first $r(j)$ rows of $T(s)/s^{n-j}$, we can define an $r(j) \times p$ PM $T_j(s)$ as follows:

$$T_j(s) \triangleq \begin{bmatrix} s^j & & & & \vdots & \\ & s^{j-n+n_2} & & & \vdots & \\ & & \ddots & & \vdots & 0 \\ & & & s^{j-n+n_{r(j)}} & \vdots & \end{bmatrix} \tag{7-50}$$

The following properties are immediate from the definition.

$$1 \le r(0) \le r(1) \le \cdots \le r(n) = p \tag{7-51}$$

$$\sum_{j=0}^{n-1} r(j) = \sum_{i=1}^{p} n_i = N \tag{7-52}$$

$$sT_j(s) = \Lambda_j T_{j+1}(s) \quad (0 \le j \le n-1) \tag{7-53}$$

where

$$\Lambda_j \triangleq [I_{r(j)} \mid O] : r(j) \times r(j+1) \tag{7-54}$$

We note that a PM with p columns P(s) can be written as

$$P(s) = \sum_{j=0}^{d} P_j T_j(s) \tag{7-55}$$

if and only if the ith-column degree of P(s) is not greater than that of $T_d(s)$ (i.e., $d - n + n_i$) $\forall$ i. [Here a negative column degree means that the corresponding column of P(s) is a zero vector.] In particular, A(s) can be written as

$$A(s) = \sum_{j=0}^{n} A_j T_j(s) \tag{7-56}$$

where A_j is a $p \times r(j)$ matrix. Evidently, A_n above is equal to A_{high} in (7-47) and therefore is nonsingular.

Now assume that A(s) satisfies (7-16), and consider the inner product $\langle \cdot \rangle$ defined from A(s) and a fixed positive definite matrix Π. When an $r(j) \times p$ PM $R_j(s)$ is written as

$$R_j(s) = T_j(s) + R_{j,j-1} T_{j-1}(s) + \cdots + R_{j,0} T_0(s) \tag{7-57}$$

and satisfies the orthogonal condition

$$\langle R_j(s), T_i(s) \rangle = 0 \quad \forall i = 0, 1, \ldots, j-1 \tag{7-58}$$

we say that $R_j(s)$ is an orthogonal PM of degree j defined from $(A(s), \Pi)$. The unique existence of such an $R_j(s)$ is ensured by a condition similar to (7-28).

We define the matrices $\{\epsilon_j; 0 \le j \le n-1\}$ and $\{\theta_j; 0 \le j \le n-1\}$ by (7-29) and (7-34). Note that the sizes of these matrices are

$$\epsilon_j : r(j) \times r(j), \quad \theta_j : r(j) \times r(j)$$

Then Theorems 7-1, 7-2, and 7-3 are immediately extended to the column-reduced case as follows:

Theorem 7-4. A(s) is stable if and only if ϵ_j is positive definite for $\forall j = 0, 1, \ldots, n-1$. ∎

Theorem 7-5. A(s) is written as

$$A(s) = A_n \left\{ R_n(s) + \frac{1}{2} \bar{\Pi} \Lambda_{n-1}^T \epsilon_{n-1}^{-1} R_{n-1}(s) \right\} \tag{7-59}$$

where

$$\bar{\Pi} \triangleq A_n^{-1} \Pi A_n^{T-1} \tag{7-60}$$

∎

Theorem 7-6. The orthogonal PMs $\{R_j(s)\}$ satisfy the following recurrence relationship:

$$R_0(s) = T_0(s), \quad \Lambda_0 R_1(s) = (sI - \theta_0 \epsilon_0^{-1}) R_0(s)$$

$$\Lambda_j R_{j+1}(s) = (sI - \theta_j \epsilon_j^{-1}) R_j(s) + \epsilon_j \Lambda_{j-1}^T \epsilon_{j-1}^{-1} R_{j-1}(s) \quad (1 \le j \le n-1) \tag{7-61}$$

∎

Remark 7-1

In the recursion in Theorem 7-6, the $r(j+1) - r(j)$ lowest rows of $R_{j+1}(s)$ are not determined from $\{R_j(s), R_{j-1}(s)\}$, and therefore we cannot use (7-61) for computing $\{R_j(s)\}$ except for the strictly regular case. Nevertheless, introducing some auxiliary PMs $\{Y_j(s); j = 0, 1, \ldots, n\}$, we can construct a recursive algorithm generating $\{R_j(s), Y_j(s)\}$ simultaneously in the general case [9, 10].

7-5 SCHWARZ MATRICES AND THE
       ROUTH APPROXIMATION

In Sec. 7-4 the quantities $\bar{\Pi}$, $\{\epsilon_j\}$, and $\{\theta_j\}$ were defined from given $(A(s), \Pi)$, where $A(s)$ is a column reduced PM and Π is a positive definite matrix. Using these quantities, let

$$e_j \triangleq \begin{cases} \epsilon_j \Lambda_{j-1}^T \epsilon_{j-1}^{-1} : r(j) \times r(j-1) & (1 \le j \le n-1) \\ \frac{1}{2} \bar{\Pi} \Lambda_{n-1}^T \epsilon_{n-1}^{-1} : r(n) \times r(n-1) & (j = n) \end{cases}$$

$$f_j \triangleq \theta_j \epsilon_j^{-1} : r(j) \times r(j) \quad (0 \le j \le n-1)$$

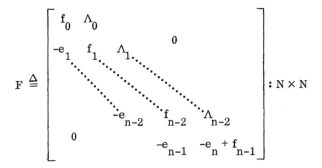

We call F the __block-Schwarz matrix__ defined from $(A(s), \Pi)$. Note that F is composed of a positive definite matrix $\bar{\Pi}$, nonsingular symmetric matrices $\{\epsilon_j\}$, and skew-symmetric matrices $\{\theta_j\}$.

In the case where $A(s)$ is a scalar polynomial, it turns out that $\Lambda_j = 1$ and that $f_j = 0$. Moreover, the quantities $\{e_j\}$ are obtained by the procedure of the Routh-Hurwitz stability test for $A(s)$ as shown in Example 7-1. Thus we can see that F is the Schwarz matrix defined from $A(s)$ in the usual definition (e.g., see [1]).

Based on Theorems 7-5 and 7-6, we have

$$(sI - F) R(s) = GA(s) \tag{7-62}$$

where

$$R(s) \triangleq [R_0^T(s), \ldots, R_{n-1}^T(s)]^T, \quad N \times p \tag{7-63}$$

$$G \triangleq [0 \mid A_n^{T-1} \Lambda_{n-1}^T]^T, \quad N \times p \tag{7-64}$$

It is noted that (7-62) characterizes F. This characterization leads to the following important property of F. Let

$$E \triangleq \text{block diag} \{\epsilon_0, \epsilon_1, \ldots, \epsilon_{n-1}\}, \quad N \times N$$

Then it follows that

$$\langle R(s), R(s) \rangle = E \tag{7-65}$$

Hence, as in the derivation of (7-13) from (7-12) and the fundamental properties (i) to (vi) at the end of Sec. 7-2, we can show from (7-62) and (7-65) that

$$FE + EF^T + G\Pi G^T = 0 \tag{7-66}$$

Suppose that $H(s)$ is a $q \times p$ strictly proper rational matrix such that

$$H(s) = B(s) A^{-1}(s) \qquad (7\text{-}67)$$

where $B(s)$ is a $q \times p$ PM. Then the ith-column degree of $B(s)$ is less than the corresponding column degree n_i of $A(s)$ for every i, and therefore $B(s)$ can be represented as

$$B(s) = KR(s) = \sum_{j=0}^{n-1} k_j R_j(s) \qquad (7\text{-}68)$$

where

$$k_j : q \times r(j)$$

$$K = [k_0, k_1, \ldots, k_{n-1}] : q \times N$$

Thus we have from (7-62) and (7-68)

$$H(s) = K(sI - F)^{-1}G$$

[i.e., (F, G, K) is a realization of $H(s)$]. We call it the <u>controllable Schwarz form realization</u> of $H(s)$ defined from the right MFD (7-67) and II. We note that the <u>observable Schwarz form realization</u> of a left MFD can also be defined in a similar way.

As an application of the results above, we construct the <u>Routh approximation</u> for MIMO (multiple-input, multiple-output) systems in the sequel. The Routh approximation method for reducing order of SISO (single-input, single-output) systems was introduced by Hutton and Friedland [6] and was shown to have some nice properties. This method consists of two steps: first, derive the Schwarz form realization of a given higher-order system, and next, construct the approximant by truncating a part of the state variables of the realization. The extension of the method to MIMO systems is quite an easy problem now.

Suppose that a $q \times p$ strictly proper rational matrix $H(s)$ is written in a right irreducible MFD (7-67), and let (F, G, K) be the controllable Schwarz form realization defined from the MFD and II. We further assume that $H(s)$ is stable. From Theorem 7-4 this condition is equivalent to

$$\epsilon_j > 0 \quad \forall j = 0, 1, \ldots, n - 1 \qquad (7\text{-}69)$$

For each $m = 1, 2, \ldots, n$, let

$$N_m \stackrel{\Delta}{=} \sum_{j=n-m}^{n-1} r(j)$$

$$F_m \stackrel{\Delta}{=} \begin{bmatrix} f_{n-m} & \Lambda_{n-m} & & & & & 0 \\ -e_{n-m+1} & f_{n-m+1} & \Lambda_{n-m+1} & & & & \\ & & \ddots & \ddots & \ddots & & \\ & & & & \ddots & \ddots & \\ 0 & & & -e_{n-2} & f_{n-2} & \Lambda_{n-2} \\ & & & & -e_{n-1} & -e_n + f_{n-1} \end{bmatrix}, \quad N_m \times N_m$$

$$G_m \stackrel{\Delta}{=} [0 \mid A_n^{T-1} \Lambda_{n-1}^T]^T \; : \; N_m \times p$$

$$K_m \stackrel{\Delta}{=} [k_{n-m}, \ldots, k_{n-1}] \; : \; q \times N_m$$

It should be noted that F_m is the block-Schwarz matrix composed of $\{\epsilon_j;$ $n - m \le j \le n - 1\}$, $\{\theta_j; n - m \le j \le n - 1\}$ and $\bar{\Pi}$. This fact proves the following:

$$F_m E_m + E_m F_m^T + G_m \Pi G_m^T = 0 \tag{7-70}$$

where

$$E_m \stackrel{\Delta}{=} \text{block diag}\{\epsilon_{n-m}, \ldots, \epsilon_{n-1}\}, \quad N_m \times N_m$$

Now we define the <u>mth-order Routh approximant</u> of H(s) by

$$H_m(s) \stackrel{\Delta}{=} K_m (sI - F_m)^{-1} G_m \tag{7-71}$$

In the SISO case, this definition coincides with that of [6].

Remark 7-2

Since the controllable Schwarz form realization (F, G, K) of $H(s)$ is defined from the right irreducible MFD (7-67) and Π, the definition of the Routh approximants seems to depend on a choice of a right irreducible MFD of $H(s)$. However, it can be shown [10] that the approximants are uniquely determined by $(H(s), \Pi)$. On the other hand, the definition depends actually on a choice of Π. Furthermore, the use of a left irreducible MFD and of the observable Schwarz form realization provides another definition. In the SISO case, all the ambiguity of definition vanishes.

The fundamental properties of the SISO approximation [6] are generalized to the MIMO case as follows. First, since the stability condition (7-69) implies the positive definiteness of E_m, it follows from (7-70) that F_m is a stable matrix. Thus we have:

(i)   The approximation preserves stability; that is, $H_m(s)$ is stable for all m.

The second property of the approximation is concerned with Markov parameters. Expand $H(s)$ and $H_m(s)$ as

$$H(s) = h_0 s^{-1} + h_1 s^{-2} + \cdots$$

$$H_m(s) = h_{m,0} s^{-1} + h_{m,1} s^{-2} + \cdots$$

The matrices $\{h_j\}$ and $\{h_{m,j}\}$ are called the <u>Markov parameters</u> of the systems and are represented as

$$h_j = KF^j G$$

$$h_{m,j} = K_m F_m^j G_m \qquad (j = 0, 1, \ldots)$$

Recalling that F_m, G_m, and K_m are right lower submatrices of F, G, and K, respectively, and noting the sparseness of the matrices, we can prove the following:

(ii)   The mth-order approximation preserves the first m Markov parameters; that is,

$$h_{m,j} = h_j \quad \text{for } \forall j = 0, 1, \ldots, m - 1$$

The last property shown below is quite peculiar to the Routh approximation. Let $h_m(t)$ $(t \geq 0)$ be the impulse response of $H_m(s)$, which is represented as

$$h_m(t) = K_m \exp(tF_m) G_m \tag{7-72}$$

We define the <u>impulse response energy</u> of $H_m(s)$ with respect to Π as

$$\Delta_m \triangleq \int_0^\infty h_m(t) \Pi h_m^T(t) \, dt, \quad q \times q$$

Noting that the solution E_m of the Lyapunov equation (7-70) is represented as

$$E_m = \int_0^\infty \exp(tF_m) G_m \Pi G_m^T \exp(tF_m^T) \, dt$$

we obtain from (7-72)

$$\Delta_m = K_m E_m K_m^T = \sum_{j=n-m}^{n-1} k_j \epsilon_j k_j^T$$

This leads to the following:

(iii)   The impulse response energy of the mth-order approximant decreases monotonically as m is lowered; that is,

$$\Delta_{m+1} \geq \Delta_m \quad (1 \leq \forall m \leq n - 1)$$

REFERENCES

1. B. D. O. Anderson, E. I. Jury, and M. Mansour, "Schwarz matrix properties for continuous and discrete time systems," Int. J. Control, vol. 23, no. 1, pp. 1-16 (1976).

2. B. D. O. Anderson and A. C. Tsoi, "Connecting forward and backward autoregressive models," IEEE Trans. Autom. Control, vol. AC-29, no. 10, pp. 917-926 (1984).

3. P. Delsarte, Y. V. Genin, and Y. G. Kamp, "Orthogonal polynomial matrices on the unit circle," IEEE Trans. Circuits Syst., vol. CAS-25, no. 3, pp. 149-160 (1978).

4. F. R. Gantmacher, Theory of Matrices, Chelsea, New York, 1959.

5. M. T. Hadidi, M. Morf, and B. Porat, "Efficient construction of canonical ladder forms for vector autoregressive processes," IEEE Trans. Autom. Control, vol. AC-27, no. 6, pp. 1222-1233 (1982).

6. M. F. Hutton and B. F. Friedland, "Routh approximations for reducing order of linear, time-invariant systems," IEEE Trans. Autom. Control, vol. AC-20, no. 3, pp. 329-337 (1975).

---

7. T. Kailath, Linear Systems, Prentice-Hall, Englewood Cliffs, N.J., 1980.

8. Y. Monden and S. Arimoto, "Generalized Rouche's theorem and its application to multivariable autoregressions," IEEE Trans. Acoust. Speech Signal Process., vol. ASSP-28, no. 6, pp. 733-738 (1980).

9. H. Nagaoka, Y. Monden, and S. Arimoto, "An approach to the continuous-time stability criterion of polynomial matrices via orthogonal polynomial matrices," Electron. Commun. Jpn., part 1, vol. 69, no. 8, pp. 1-9 (1986).

10. H. Nagaoka, "A study of stability and reduced order approximations of continuous-time linear systems via orthogonal polynomial matrices," Ph.D. dissertation, Osaka University, 1987.

Part III

Digital Lattice Filters as a Bridge Among Circuit
Theory, System Theory and Prediction Theory

8

Circular Lattice Filtering for Recursive Least-Squares and ARMA Modeling

HIDEAKI SAKAI

Kyoto University, Kyoto, Japan

8-1 INTRODUCTION

In this chapter we present the circular lattice (CL) method for multichannel signal processing, with applications to recursive least-squares (RLS) filtering and autoregressive-moving average (ARMA) modeling.

In Sec. 8-2 the basic circular Levinson algorithm for multichannel autoregressive (AR) time series is presented by using the notion of periodic autoregression, and its modification to singular time series is discussed. Also, the covariance characterization of general stationary multichannel time series is mentioned. Then, to estimate the parameters from data samples, we present the Yule-Walker and Burg algorithms, and an example of actual data analysis is given. The most salient feature of our algorithms is that they consist of calculations of scalar quantities, thus completely avoiding manipulations of matrices and vectors accompanying usual multichannel signal processing methods, and they are suitable to parallel processing.

In Sec. 8-3, first the least-squares circular lattice (LSCL) algorithm is presented for time recursive estimation of parameters. Then, as a special case of the LSCL algorithm, the triangular lattice (TL) is introduced for a general RLS filtering problem and is compared with other RLS methods, such as the Kalman, U-D, and escalator (ES) algorithms about their computational structures and numerical properties.

In Sec. 8-4, applications of the CL method to ARMA modeling are presented. First, the covariance characterization of an ARMA process is discussed by embedding an ARMA process into a degenerate two-channel AR process and a new ARMA lattice filter is proposed. The effect of coefficient quantization on the frequency response of the lattice filter is compared with those of other ARMA lattice filters. Finally, a parameter estimation problem of rational transfer functions based on the impulse response data is treated by using the Burg type CL algorithm in Sec. 8-2. Some numerical results are presented to show the effectiveness of the method proposed.

8-2 CIRCULAR LATTICE METHOD

8-2-1 Periodic and Multichannel Autoregressions

In this section we review some results of Pagano [1] needed for later discussions. Let us consider the following d-variates pth-order AR process:

$$\vec{X}(t) + \sum_{j=1}^{p} A(j)\vec{X}(t-j) = \vec{U}(t) \tag{8-1}$$

where the $\vec{U}(\cdot)$ are uncorrelated with zero mean vector and $\text{cov}(\vec{U}(t)) = W$.

We say that a process $Y(\cdot)$ is a periodic autoregression of period d and order
$(p_1, \ldots, p_d)$ if for all integer t,

$$Y(t) + \sum_{j=1}^{p_t} \alpha_t(j) Y(t - j) = \epsilon(t) \tag{8-2}$$

where the $\epsilon(\cdot)$ are uncorrelated with zero mean and $E[\epsilon^2(t)] = \sigma_t^2$, $p_t = p_{t+d}$,
$\sigma_t^2 = \sigma_{t+d}^2$, and $\alpha_t(j) = \alpha_{t+d}(j)$, $j = 1, \ldots, p_t$. We denote the autocovariance
of $Y(\cdot)$ by $R(s, t) = E[Y(s) Y(t)]$. From the definition of general periodically
correlated processes, we also note that

$$R(s, t) = R(s + d, t + d) \tag{8-3}$$

Then we have

Theorem 8-1 (Pagano [1]). If $\vec{X}(\cdot)$ and $Y(\cdot)$ are related by

$$X_j(t) = Y(j + d(t - 1)) \tag{8-4}$$

then $\vec{X}(\cdot)$ is an AR process of order p with positive definite W if, and only if,
$Y(\cdot)$ is a periodic autoregression of period d and order $(p_1, \ldots, p_d)$ with positive $\sigma_1, \ldots, \sigma_d$ and $p = \max_j[(p_j - j)/d] + 1$, where $X_j(t)$ is the jth component of $\vec{X}(t)$ and for integer j, $[x] = j$, for $j \le x < j + 1$. The relation between
various parameters in (8-1) and (8-2) are

$$A(j) = L^{-1} A'(j) \tag{8-5}$$

$$W = L^{-1} D (L^{-1})^T \tag{8-6}$$

with

$$L_{kj} = \alpha_k(k - j) \quad j < k \quad (\alpha_k(0) = 1, \ \alpha_k(j) = 0 \text{ for } j < 0)$$

$$A'_{kj}(v) = \alpha_k(dv + k - j), \quad v = 1, \ldots, p$$

$$D = \text{diag}(\sigma_1^2, \ldots, \sigma_d^2)$$

where T is the transpose operation and A_{kj} denotes the (k, j)th component
of A.   ∎

This theorem can easily be understood by noting the fact that if we
replace t in (8-2) by $k + d(t - 1)$, (8-2) is rewritten as

$$Y(k + d(t - 1)) + \sum_{j=1}^{p_k} \alpha_k(j) Y(k + d(t - 1) - j) = \epsilon(k + d(t - 1)) \tag{8-7}$$

Then, using this with (8-4), we have

$$\begin{bmatrix} 1 & & & 0 \\ \alpha_2(1) & 1 & & \\ \vdots & & \ddots & \\ & & & 1 \\ \alpha_d(d-1) & \cdots & \alpha_d(1) & 1 \end{bmatrix} \begin{bmatrix} Y(1 + d(t-1)) \\ Y(2 + d(t-1)) \\ \vdots \\ \\ Y(d + d(t-1)) \end{bmatrix}$$

$$+ \begin{bmatrix} \alpha_1(d) & \cdots & \alpha_1(1) \\ \alpha_2(d+1) & \cdots & \alpha_2(2) \\ \vdots & & \vdots \\ \alpha_d(2d-1) & \cdots & \alpha_d(d) \end{bmatrix} \begin{bmatrix} Y(1 + d(t-2)) \\ Y(2 + d(t-2)) \\ \vdots \\ Y(d + d(t-2)) \end{bmatrix} + \cdots = \begin{bmatrix} \epsilon(1 + d(t-1)) \\ \epsilon(2 + d(t-1)) \\ \vdots \\ \epsilon(d + d(t-1)) \end{bmatrix}$$

Thus, by estimating $\alpha_k(j)$ $(j + 1, \ldots, p_k, k = 1, \ldots, d)$ first, we can identify the multivariate AR model (8-1). Since L is a unit (ones on the diagonal) lower triangular matrix, inversion of L and calculation of (8-5) and (8-6) are quite easy. The Yule-Walker equations for (8-2) are obtained by multiplying both sides of (8-7) by $Y(k + d(t - 1) - v)$ and taking expectations as

$$R(k, k - v) + \sum_{j=1}^{p_k} \alpha_k(j) R(k - j, k - v) = \delta_{v,0} \sigma_k^2 \quad (v \geq 0, \ k = 1, \ldots, d) \tag{8-8}$$

8-2-2 Circular Levinson Algorithm

Here we derive a Levinson-type efficient algorithm for solving (8-8) with increasing $p_k = 0, 1, 2, \ldots$. Define

$$\vec{R}_k(j) = \begin{bmatrix} R(k,k) & R(k-1,k) & \cdots & R(k-j,k) \\ R(k,k-1) & R(k-1,k-1) & \cdots & R(k-j,k-1) \\ \vdots & \vdots & & \vdots \\ R(k,k-j) & R(k-1,k-j) & \cdots & R(k-j,k-j) \end{bmatrix} \tag{8-9}$$

$$\vec{a}_k(j) = [1 \quad \alpha_k(j,1) \quad \cdots \quad \alpha_k(j,j)]^T$$

Then (8-8) is written as

$$\vec{R}_k(j)\, \vec{a}_k(j) = [\sigma_k^2(j) \quad 0 \quad \cdots \quad 0]^T \tag{8-10}$$

with $p_k = j$. We also denote the explicit dependence of $\alpha_k(i)$ and σ_k^2 on p_k as $\alpha_k(p_k, i)$ and $\sigma_k^2(p_k)$, respectively. Obviously, (8-9) is segmented as

$$\vec{R}_k(j) = \left[\begin{array}{c} \vec{R}_k(j-1) \quad \rule{1cm}{0pt} \\ \rule{2cm}{0pt} \end{array} \right] = \left[\begin{array}{c} \rule{2cm}{0pt} \\ \vec{R}_{k-1}(j-1) \end{array} \right] \tag{8-11}$$

Also, from (8-3) we note the following cyclic property:

$$\vec{R}_0(j) = \vec{R}_d(j) \tag{8-12}$$

As is usual in the derivations of Levinson-type algorithms [2-5], we introduce an auxiliary vector defined by

$$\vec{b}_k(j) = [\beta_k(j,j) \quad \cdots \quad \beta_k(j,1) \quad 1]^T$$

$$\vec{R}_k(j)\, b_k(j) = [0 \quad \cdots \quad 0 \quad \tau_k^2(j)]^T \tag{8-13}$$

with $\beta_k(j,0) = 1$. Note that from positive definiteness of $\vec{R}_k(j)$, $\tau_k^2(j) > 0$. Then from (8-10), (8-11), and (8-13), we have

$$\vec{R}_k(j+1)\left[c_1 \left(\begin{array}{c} \vec{a}_k(j) \\ 0 \end{array} \right) + c_2 \left(\begin{array}{c} 0 \\ \vec{b}_{k-1}(j) \end{array} \right) \right] = c_1 \left(\begin{array}{c} \sigma_k^2(j) \\ 0 \\ \vdots \\ 0 \\ \Delta_k(j) \end{array} \right) + c_2 \left(\begin{array}{c} \Delta_k^*(j) \\ 0 \\ \vdots \\ 0 \\ \tau_{k-1}^2(j) \end{array} \right) \tag{8-14}$$

where

$$\Delta_k(j) = \sum_{m=0}^{j} R(k-m, k-j-1)\alpha_k(j,m) \tag{8-15}$$

$$\Delta_k^*(j) = \sum_{m=0}^{j} R(k-j-1+m, k)\beta_{k-1}(j,m) \tag{8-16}$$

for any constants c_1 and c_2. We can easily show the so-called Burg relation [3]

$$\Delta_k(j) = \Delta_k^*(j) \tag{8-17}$$

since from (8-15), (8-10), and (8-11),

$$\Delta_k(j) = [0 \quad \vec{b}_{k-1}^T(j)] \vec{R}_k(j+1)[\vec{a}_k^T(j) \quad 0]^T$$

$$= [\vec{a}_k^T(j) \quad 0]\vec{R}_k(j+1)[0 \quad \vec{b}_{k-1}^T(j)]^T = \Delta_k^*(j)$$

Taking $c_1 = 1$ and $c_2 = -\Delta_k(j)/\tau_{k-1}^2(j)$ yields

$$\vec{a}_k(j+1) = \begin{pmatrix} \vec{a}_k(j) \\ 0 \end{pmatrix} - \frac{\Delta_k(j)}{\tau_{k-1}^2(j)} \begin{pmatrix} 0 \\ \vec{b}_{k-1}(j) \end{pmatrix}$$

$$\tag{8-18}$$

$$\sigma_k^2(j+1) = \sigma_k^2(j) - \frac{\Delta_k^2(j)}{\tau_{k-1}^2(j)}$$

Similarly, taking $c_1 = -\Delta_k^*(j)/\sigma_k^2(j) = -\Delta_k(j)/\sigma_k^2(j)$ and $c_2 = 1$ yields

$$\vec{b}_k(j+1) = \begin{pmatrix} 0 \\ \vec{b}_{k-1}(j) \end{pmatrix} - \frac{\Delta_k(j)}{\sigma_k^2(j)} \begin{pmatrix} \vec{a}_k(j) \\ 0 \end{pmatrix}$$

$$\tag{8-19}$$

$$\tau_k^2(j+1) = \tau_{k-1}^2(j) - \frac{\Delta_k^2(j)}{\sigma_k^2(j)}$$

When the subscript $k - 1$ becomes 0 from the cyclic property (8-13), it is replaced by d. Hence we use the adjective "circular." Writing (8-18) and (8-19) componentwise, we have the following algorithm.

Circular Levinson Algorithm

(i) Initial conditions ($j = 0$)

$$\sigma_k^2(0) = \tau_k^2(0) = R(k,k), \quad \Delta_k(0) = R(k,k-1) \quad (k = 1, \ldots, d)$$

$$\tag{8-20}$$

(ii) Order update from j to j + 1

(a) Compute

$$\Delta_k(j) = \sum_{m=0}^{j} R(k - m, \ k - j - 1)\alpha_k(j, m) \qquad (8\text{-}15)$$

$$= \sum_{m=0}^{j} R(k - j - 1 + m, \ k)\beta_{k-1}(j, m) \qquad (8\text{-}16)$$

$$\alpha_k(j + 1, \ j + 1) = -\frac{\Delta_k(j)}{\tau_{k-1}^2(j)}$$

$$\beta_k(j + 1, \ j + 1) = -\frac{\Delta_k(j)}{\sigma_k^2(j)} \qquad (8\text{-}21)$$

(b) Update

$$\alpha_k(j+1, i) = \alpha_k(j, i) + \alpha_k(j+1, \ j+1)\beta_{k-1}(j, \ j+1-i)$$

$$\beta_k(j+1, i) = \beta_{k-1}(j, i) + \beta_k(j+1, \ j+1)\alpha_k(j, \ j+1-i) \qquad (8\text{-}22)$$

$$(i = 1, \dots, j)$$

$$\sigma_k^2(j+1) = \sigma_k^2(j)(1 - \alpha_k(j+1, \ j+1)\beta_k(j+1, \ j+1))$$

$$\tau_k^2(j+1) = \tau_{k-1}^2(j)(1 - \alpha_k(j+1, \ j+1)\beta_k(j+1, \ j+1)) \qquad (8\text{-}23)$$

where the subscript k - 1 = 0 is replaced by d in (8-21) to (8-23).

Note that $\alpha_k(j, j)$, $\beta_k(j, j)$ play the same role as the usual forward and backward reflection coefficient matrices [3]. In the scalar case, they become identical and their absolute magnitudes are always less than 1. But this property does not hold for scalar $\alpha_k(j, j)$, $\beta_k(j, j)$. That is, from (8-21) and (8-23),

$$0 \leq \alpha_k(j + 1, \ j + 1)\beta_k(j + 1, \ j + 1) < 1 \qquad (8\text{-}24)$$

but not necessarily $|\alpha_k(j + 1, \ j + 1)| < 1$ and $|\beta_k(j + 1, \ j + 1)| < 1$. However, if we define the normalized partial autocorrelation coefficient

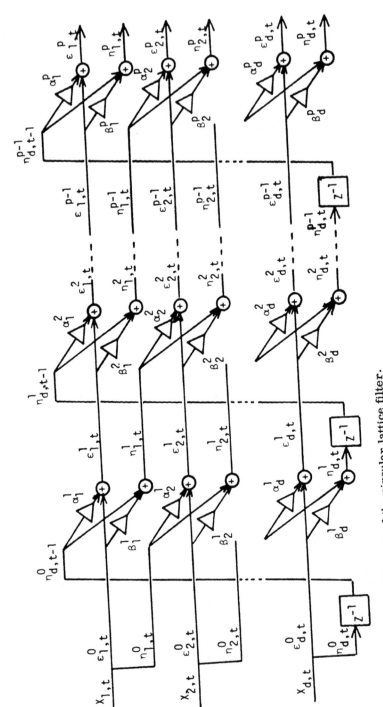

FIGURE 8-1 Block diagram of the circular lattice filter.

204

$$\rho_k(j+1) = -\frac{\Delta_k(j)}{\sigma_k(j)\,\tau_{k-1}(j)} \qquad (8\text{-}25)$$

as in [6], it follows that $|\rho_k(j+1)| < 1$. Later we treat the case where at some channel k, $|\rho_k(j)| = 1$ at some order j.

8-2-3 Circular Lattice Structure

Define the following jth-order kth channel forward and backward prediction errors

$$\epsilon(j, k+nd) = Y(k+nd) + \sum_{i=1}^{j} \alpha_k(j, i)\, Y(k+nd-i)$$

$$\qquad\qquad (8\text{-}26)$$

$$\eta(j, k+nd) = Y(k+nd-j) + \sum_{i=1}^{j} \beta_k(j, j+1-i)\, Y(k+nd-i+1)$$

for $k = 1, \ldots, d$ and an integer n, respectively. Then from (8-22), we readily obtain the following relations:

$$\epsilon(j+1, k+nd) = \epsilon(j, k+nd) + \alpha_k(j+1, j+1)\, \eta(j, k-1+nd)$$

$$\qquad\qquad (8\text{-}27)$$

$$\eta(j+1, k+nd) = \eta(j, k-1+nd) + \beta_k(j+1, j+1)\, \epsilon(j, k+nd)$$

with

$$\sigma_k^2(j) = E[\epsilon^2(j, k+nd)], \qquad \tau_{k-1}^2(j) = E[\eta^2(j, k-1+nd)]$$

$$\qquad\qquad (8\text{-}28)$$

$$\Delta_k(j) = E[\epsilon(j, k+nd)\, \eta(j, k-1+nd)]$$

Noting that $\eta(j, nd) = \eta(j, d + (n-1)d)$, (8-27) is illustrated in Fig. 8-1, where we write $\epsilon(j, k+td)$ as $\epsilon_{k,t}^{j}$, $\alpha_k(j,j)$ as α_k^{j}, and so on, and z^{-1} acts on t. This is a multivariate generalization of Itakura and Saito's lattice structure for scalar case [7]. Since signal flows between channels form a circle, we call (8-27) the circular lattice structure.

Next we consider the problem of estimating z(t), which is a scalar process correlated with $\{\vec{X}(t)\}$, by a linear combination of $\vec{X}(t)$, $\vec{X}(t-1)$, $\ldots$, $\vec{X}(t-m)$. This situation occurs in a noise canceling problem [8]. Griffiths [9] emphasized a necessity of making a scalar orthogonal basis of $\vec{X}(t)$, $\ldots$, $\vec{X}(t-m)$ and proposed a Gram-Schmidt (GS) type of method. In this respect, we have the following statement:

The signals $\{\eta(j, d + nd)\}$ $(j = 0, 1, \ldots, md + d - 1)$, that is, the backward prediction errors of the last channel form an orthogonal basis of $\vec{X}(t)$, $\ldots$, $\vec{X}(t - m)$.

Proof. The largest order that can be defined for the kth channel is $p_k^* = pd + k - 1$, since the time index of the oldest sample used for prediction at time n is $k + nd - p_k^* = (n - m)d + 1$. But $\{\vec{X}(t), \ldots, \vec{X}(t - m)\}$ has $(m + 1)d$ components, so that except for degenerate cases, any orthogonal basis must have $(m + 1)d$ components. Note also that $\eta(j, k + nd)$ is uncorrelated with $Y(k + nd - j + 1)$, $\ldots$, $Y(k + nd)$, so that $\{\eta(j, k + nd); k, \text{ fixed}; j = 0, 1, \ldots, p_k^*\}$ is an orthogonal system. Hence, only for $k = d$ can $\{\eta(j, k + nd)\}$ form a complete orthogonal basis. ∎

Thus the linear minimum mean square estimate $\hat{z}(t)$ of $z(t)$ is written as

$$\hat{z}(t) = \sum_{j=0}^{md+d-1} c_j \eta(j, td)$$

with $c_j = E[z(t)\eta(j, td)]/E[\eta^2(j, td)] = E[\nu(j - 1, t)\eta(j, td)]/E[\eta^2(j, td)]$, where we define

$$\nu(j, t) = z(t) - \sum_{i=1}^{j} c_i \eta(i, td)$$

$$= \nu(j - 1, t) - c_j \eta(j, td)$$

This CL joint process estimation system has

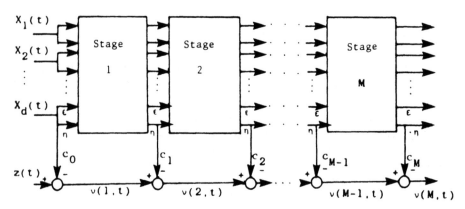

FIGURE 8-2 Joint process circular lattice structure for noise-canceling applications.

$$2 \sum_{k=1}^{d} p_k^* + (m + 1)d = 2md^2 + d(d - 1) + (m + 1)d$$

tap coefficients, while Griffiths' GS system has

$$\frac{1}{2}(m + 1)d[(m + 1)d - 1] + (m + 1)d$$

tap coefficients. For large m, the CL system significantly reduces the number of tap coefficients. (The reducing rate is about $4\ m^{-1}$.) In Fig. 8-2, a schematic diagram of the joint process circular lattice structure described above is presented where we assume that $\alpha_k(j,j) = \beta_k(j,j) = 0$ for $p_k^* + 1 \leq j \leq M = p_d^*$.

8-2-4 Covariance Matrices Characterization

It has been shown by Morf et al. [6] that the autocovariance matrices of a stationary multichannel time series can be uniquely characterized by a sequence of the normalized partial autocorrelation (PARCOR) matrices having singular values of less than 1. This is a nice generalization of the scalar case (Barndorff-Nielsen and Schou [10]) but it is not an easy task to parametrize the PARCOR matrices satisfying the constraint above. Here we give another characterization via the result in Sec. 8-2-1.

But before doing this, we discuss briefly how to treat the singular covariance case. This problem has been treated extensively by Inouye [11], but his algorithm requires computation of pseudoinverses. Since the circular Levinson algorithm in Sec. 8-2-1 calculates the prediction coefficients of each channel separately with scalar operations, we expect that the computation of pseudoinverses is avoided. This is actually so, since if $\sigma_k^2(j) = 0$ [similarly, $\tau_{k-1}^2(j) = 0$], then from (8-28), $\Delta_k(j) = 0$. This means that from (8-14) and (8-17) if $\tau_{k-1}^2(j) = 0$, $(\overrightarrow{a}_k^T(j)\ \ 0)^T$ is $\underline{one}$ solution to (8-10) with j replaced by $j + 1$, and similarly if $\sigma_k^2(j) = 0$, $(0\ \ \overrightarrow{b}_k^T(j))^T$ is $\underline{one}$ solution to (8-13) with j replaced by $j + 1$. Thus the modification becomes as follows.

If $\tau_{k-1}^2(j) = 0$,

$$\alpha_k(j + 1,\ i) = \alpha_k(j,i)\ (i + 1, \ldots, j),\ \alpha_k(j + 1,\ j + 1) = 0,$$

$$\sigma_k^2(j + 1) = \sigma_k^2(j) \tag{8-29}$$

and if $\sigma_k^2(j) = 0$,

$$\beta_k(j + 1,\ i) = \beta_{k-1}(j,i)\ (i = 1, \ldots, j),\ \beta_k(j + 1,\ j + 1) = 0,$$

$$\tau_k^2(j + 1) = \tau_{k-1}^2(j) \tag{8-30}$$

Using the results above, we have the following characterization. There is a one-to-one correspondence between a sequence of the autocovariance matrices $\{R_0, R_1, \ldots\}$ of d-channel stationary time series and d sequences of $\{R(k,k), \rho_k(j) \ (k = 1, \ldots, d; \ j = 0, 1, \ldots)\}$ satisfying

$$R(k,k) > 0, \quad |\rho_k(j)| \leq 1 \tag{8-31}$$

Proof. First we note from (8-4) that $(R_k)_{ij} = \text{cov}[X_i(t + k), \ X_j(t)]$ is expressed by the relation $(R_k)_{ij} = R(i, \ j - kd)$, that is

$$R_0 = \begin{bmatrix} R(1,1) & R(1,2) & \cdots & R(1,d) \\ R(2,1) & R(2,2) & \cdots & R(2,d) \\ \vdots & \vdots & & \vdots \\ R(d,1) & R(d,2) & \cdots & R(d,d) \end{bmatrix}, \ R_1 = \begin{bmatrix} R(1,1-d) & R(1,2-d) & \cdots & R(1,0) \\ R(2,1-d) & R(2,2-d) & \cdots & R(2,0) \\ \vdots & \vdots & & \vdots \\ R(d,1-d) & R(d,2-d) & & R(d,0) \end{bmatrix}, \ \cdots$$

$$\tag{8-32}$$

∎

Thus, given a sequence $\{R_0, R_1, \ldots\}$, the circular Levinson algorithm with the modification (8-29)–(8.30) yields d sequences of $R(k,k)$ and $\rho_k(j)$ defined by (8-25) satisfying $|\rho_k(j)| < 1$ and if $|\rho_k(j)| = 1$ at some j, we stop there for that channel. Conversely, given d sequences of $\{R(k,k), \rho_k(j)\}$ where $R(k,k) > 0$, $|\rho_k(j)| < 1$ and if $|\rho_k(j)| = 1$ at some j, we truncate the sequence there, then using the circular Levinson algorithm we can uniquely determine $R(k - j, k) \ (j = 0, 1, \ldots)$ as follows. First noting from (8-21) and (8-25) the relations

$$\alpha_k(j+1, j+1) = \frac{\sigma_k(j)}{\tau_{k-1}(j)}\rho_k(j+1), \quad \beta_k(j+1, j+1) = \frac{\tau_{k-1}(j)}{\sigma_k(j)}\rho_k(j+1)$$

we can generate $\alpha_k(j + 1, i)$ and $\beta_k(j + 1, i)$ by (8-22) and $\sigma_k(j + 1)$ and $\tau_k(j + 1)$ by (8-23). Then from (8-16) we can generate $R(k - 1, k)$, $R(k - 2, k)$, $\ldots$ by the equation

$$R(k-j-1, k) = - \sum_{m=1}^{j} R(k-j-1+m, k)\beta_{k-1}(j,m) - \rho_k(j+1)\sigma_k(j)\tau_{k-1}(j)$$

If $|\rho_k(j)| = 1$ at some j, the second term in the right-hand side of the equation above is dropped from that order with $\alpha_k(j+1, j+1) = \beta_k(j+1, j+1) = 0$.

8-2-5 Parameter Estimation

When a set of data $\{\vec{X}(1), \ldots, \vec{X}(N)\}$ is available, we first form a data set $\{Y(1), \ldots, Y(Nd)\}$ by (8-4), then estimate $R(k,v)$ by

$$R_N(k,v) = \frac{1}{N} \sum_{j=1}^{m} Y(k + dj) Y(v + dj) \qquad (8\text{-}33)$$

where $m = [N - \max(k,v)/d]$, substitute (8-33) into (8-8), and finally obtain the estimates $\hat{\vec{\alpha}}_k = (\hat{\alpha}_k(1), \ldots, \hat{\alpha}_k(p_k))^T$ for $\vec{\alpha}_k = (\alpha_k(1), \ldots, \alpha_k(p_k))^T$, $k = 1, \ldots, d$. Pagano also showed that the $\hat{\vec{\alpha}}_k$, $k = 1, \ldots, d$, are consistent and asymptotically efficient estimators for the $\vec{\alpha}_k$ and moreover, they are asymptotically independent, provided that $\vec{X}(\cdot)$ is Gaussian [1, Theorem 4]. This implies that the parametrization in terms of the $\vec{\alpha}_k$ allows us to perform statistical analysis for each channel separately. It is also shown in Sakai [12] that when $Y(t)$ obeys (8-2), the estimated normalized partial auto-correlation coefficients $\sqrt{N}\hat{\rho}_k(j + 1)$ $(j \geq p_k)$ are asymptotically distributed as Gaussian with zero mean and unit variance $\underline{N(0,1)}$ and are independent for each j and k.

If the orders $p_1, \ldots, p_d$ are unknown, we can obtain reasonable estimates by the well-known Akaike's method [13]. In [12] it is shown that the AIC is given by

$$AIC(p_1, \ldots, p_d) = \sum_{k=1}^{d} AIC(p_k) \qquad (8\text{-}34)$$

with

$$AIC(p_k) = N \log \hat{\sigma}_k^2(p_k) + 2p_k$$

where $\hat{\sigma}_k^2$ is the estimate of σ_k^2. Thus the total minimization of (8-34) can be accomplished by minimizing each $AIC(p_k)$ about p_k.

Since the Yule-Walker estimation method above, based on (8-33), uses the assumption that

$$Y(t) = 0 \quad \text{for } t \leq 0 \text{ or } Nd < t \qquad (8\text{-}35)$$

in some cases, especially when N is small, such an assumption degrades the quality of the estimate seriously. To avoid this, the Burg algorithm has been used extensively and Strand [14] derives its multichannel version. Here we present a Burg type CL parameter estimation algorithm. For simplicity, we use the same symbols as those used in (8-20) to (8-23) and (8-27), although the quantities below are estimated.

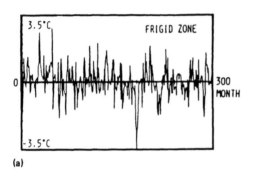

(a)

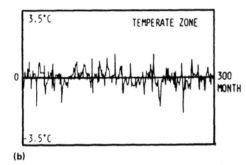

(b)

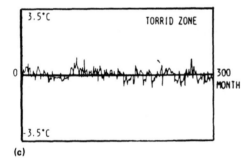

(c)

FIGURE 8-3 Monthly averaged temperatures of (a) frigid, (b) temperate, and (c) torrid zones.

TABLE 8-1 Minimum AICs of Periodic AR Models of Period d = 12 and Usual AR Models for Three Zones

| | AIC | | |
Zone	d = 1	d = 12	Differences
Frigid	2595.39	2508.14	87.25
Temperate	2221.63	2112.16	109.47
Torrid	1919.04	1910.44	8.6

As in the Burg algorithm, given jth-order forward and backward prediction errors, we first form the weighted sum of squares

$$\frac{1}{\sigma_k^2(j)} \sum_{t=L}^{N-1} \epsilon^2(j+1, k+td) + \frac{1}{\tau_{k-1}^2(j)} \sum_{t=L}^{N-1} \eta^2(j+1, k+td) \qquad (8\text{-}36)$$

where the upper and lower limits of the sum are determined by considering the following fact: that $\epsilon(j+1, k+td)$ and $\eta(j+1, k+td)$ are formed by the samples $Y(\cdot)$, where time indexes range from $k + td - j - 1$ to $k + td$. But the available samples are $Y(1), \ldots, Y(nd)$. Thus $k + (t_{min})d - j - 1 > 0$ and $k + (t_{max})d < Nd + 1$, so that $t_{min} = L = [(j + 1 - k)/d] + 1$, $t_{max} = N - 1$, where $[x]$ denotes the largest integer not exceeding x. Substituting (8-27) with (8-21) into (8-36) and minimizing the sum about $\Delta_k(j)$, we obtain the following CL version of the Strand algorithm.

Burg Type CL Algorithm

(i) Initialization

$$\sigma_k^2(0) = \tau_k^2(0) = \sum_{t=0}^{N-1} Y^2(k+td) \qquad (k = 1, \ldots, d)$$

$$\epsilon(0, k+td) = \eta(0, k+td) = Y(k+td) \qquad (t = 0, \ldots, N-1)$$

(ii) Order update from j to $j + 1$. Calculate

$$\Delta_k(j) = \frac{2 \sum_{t=L}^{N-1} \epsilon(j, k+td) \eta(j, k-1+td)}{[1/\sigma_k^2(j)] \sum_{t=L}^{N-1} \epsilon^2(j, k+td) + [1/\tau_{k-1}^2(j)] \sum_{t=L}^{N-1} \eta^2(j, k-1+td)}$$

$$(8\text{-}37)$$

then using (8-21), obtain $\alpha_k(j+1, j+1)$ and $\beta_k(j+1, j+1)$ and using (8-26), calculate $\epsilon(j+1, k+td)$ and $\eta(j+1, k+td)$ ($[(j + 2 - k)/d] + 1 \le t \le N - 1$). Also, use (8-23) to obtain $\sigma_k^2(j+1)$ and $\tau_k^2(j+1)$. It is obvious that the estimated $\rho_k(j+1)$ satisfies the required condition $|\rho_k(j+1)| \le 1$.

As an example of actual data analysis, we present some results obtained by applying the Yule-Walker estimation method above to the meteorological data in Fig. 8-3, where monthly averaged temperatures of the frigid, temperate, and torrid zones over 25 years (300 months) are shown. These were supplied by Professor H. Ogura (Kyoto University). To each datum point, periodic AR models (8-2) of various orders with period d = 12 were fitted and the minimum AIC (8-34) was calculated. For comparison, the minimum AIC of the stationary model (i.e., d = 1) was also calculated. Table 8-1 shows the results. In all zones, the periodic AR model is better. However, in the torrid zone, the difference between the AICs is very small and the variation in temperature in the torrid zone can be regarded as almost stationary. Also, the periodicity or seasonality is most eminent in the temperate zone.

8-2-6 Concluding Remarks

To conclude this section we discuss the computational complexity of the circular lattice method. For example, given a sequence of $d \times d$ auto-covariance matrices, the usual LWR algorithm [2] for obtaining p AR coefficient matrices requires $O(p^2 d^3)$ operations, while the circular Levinson algorithm requires $O(d \sum_{i=1}^{pd} i) = O(p^2 d^3)$ operations, since for the ith stage, $O(i)$ operations are required in each channel, and the number of the stages for each channel is about pd. This shows the methods are of equal complexity.

However, if we use a parallel processor, that is, there is a processor in each channel, the processing time reduces $O(p^2 d^2)$, provided that the time of communication between processors is negligible. The circular lattice algorithms are particularly suited to parallel processing since the complexity of interprocessor communications is minimal (i.e., we require only that the processors be combined to make a "ring").

8-3 RECURSIVE LEAST-SQUARES ALGORITHMS

8-3-1 Recursive Least-Squares Circular Lattice Algorithm

Lee et al. [15] derive the exact recursive least-squares (RLS) algorithms for lattice parameter estimation by using an elegant geometric method. Here we give the least-squares circular lattice (LSCL) algorithms by using their method [16].

We use the notations in [15]. First we define the ket vector by

$$|y\rangle_T = [y_0 \ y_1 \ \cdots \ y_T]^T$$

for scalar time series $\{y_t\}$ and the bra vector by $\langle y|_T = |y\rangle_T^T$. Then $|y\rangle_T$ lies in the space $H_T = \vec{R} \times \vec{R} \times \cdots \times \vec{R}$. We also define the inner product between $|y\rangle_T$ and $|z\rangle_T \in H_T$ by

$$\langle y \mid z \rangle_T = \sum_{i=0}^{T} y_i \lambda^{T-i} z_i$$

where λ is the forgetting factor and $0 < \lambda \leq 1$.

We assume d-variate time series data $\{\vec{X}(1) \ \vec{X}(2) \ \cdots \ \vec{X}(N)\}$. Writing the ith element of $\vec{X}(t)$ as $x_{i,t}$, we define the following vectors by

$$|x_i\rangle_T = [x_{i,0} \ x_{i,1} \ \cdots \ x_{i,T}]^T \qquad (i = 1, 2, \ldots, d) \tag{8-38}$$

To these vectors, an operator s^{-1} is defined as follows:

$$|s^{-1}x_1\rangle_T = [0 \ x_{d,0} \ x_{d,1} \ \cdots \ x_{d,T-1}]^T$$
$$\tag{8-39}$$
$$|s^{-1}x_i\rangle_T = |x_{i-1}\rangle_T \qquad (i = 2, 3, \ldots, d)$$

Let $Y_{i,1,p,T}$ be the supspace spanned by $|s^{-1}x_i\rangle_T$, $|s^{-2}x_i\rangle_T$, $\cdots$, $|s^{-p}x_i\rangle_T$ and $P_{i,1,p,T}$ be the projection operator on $Y_{i,1,p,T}$ defined by

$$P_{i,1,p,T} = |Y_{i,1,p}\rangle_T \langle Y_{i,1,p} | Y_{i,1,p}\rangle_T^{-1} \langle Y_{i,1,p} |_T$$

where $|Y_{i,1,p}\rangle_T = [|s^{-1}x_i\rangle_T \ |s^{-2}x_i\rangle_T \ \cdots \ |s^{-p}x_i\rangle_T]$. Also let $|\epsilon_i^p\rangle_T$, $|\eta_i^p\rangle_T$ be the projection errors of $|x_i\rangle_T$ and $|s^{-p}x_i\rangle_T$ on $Y_{i,1,p,T}$ and $Y_{i,0,p-1,T}$, respectively. That is

$$|\epsilon_i^p\rangle_T = P_{i,1,p,T}^{\perp} |x_i\rangle_T \tag{8-40}$$

$$|\eta_i^p\rangle_T = P_{i,0,p-1,T}^{\perp} |s^{-p}x_i\rangle_T \tag{8-41}$$

where $P^{\perp} = I - P$. From (8-40) it follows that

$$|\epsilon_i^p\rangle_T \in Y_{i,0,p,T}, \quad |\epsilon_i^p\rangle_T \perp Y_{i,1,p,T}$$

Also, from (8-41), we note the relation

$$|\eta_{i-1}^p\rangle_T = P_{i,1,p,T}^{\perp} |s^{-p-1}x_i\rangle_T \tag{8-42}$$

Then it follows that $| \eta_{i-1}^p \rangle_T \in Y_{i,1,p+1,T}$, $| \eta_{i-1}^p \rangle_T \perp Y_{i,1,p,T}$. Hence we have the representations $Y_{i,1,p+1,T} = Y_{i,1,p,T} \oplus | \eta_{i-1}^p \rangle_T$, $Y_{i,0,p,T} = Y_{i,1,p,T} \oplus | \epsilon_i^p \rangle_T$. From these we have

$$P_{i,1,p+1,T}^{\perp} = P_{i,1,p,T}^{\perp} - | \eta_{i-1}^p \rangle_T \langle \eta_{i-1}^p | \eta_{i-1}^p \rangle_T^{-1} \langle \eta_{i-1}^p |_T$$

$$P_{i,0,p,T}^{\perp} = P_{i,1,p,T}^{\perp} - | \epsilon_i^p \rangle_T \langle \epsilon_i^p | \epsilon_i^p \rangle_T^{-1} \langle \epsilon_i^p |_T \tag{8-43}$$

Hence, from (8-40) to (8-42), we have the order-updated recursions as

$$| \epsilon_i^{p+1} \rangle_T = | \epsilon_i^p \rangle_T - | \eta_{i-1}^p \rangle_T \langle \eta_{i-1}^p | \eta_{i-1}^p \rangle_T^{-1} \langle \eta_{i-1}^p | \epsilon_i^p \rangle_T$$

$$| \eta_i^{p+1} \rangle_T = | \eta_{i-1}^p \rangle_T - | \epsilon_i^p \rangle_T \langle \epsilon_i^p | \epsilon_i^p \rangle_T^{-1} \langle \epsilon_i^p | \eta_{i-1}^p \rangle_T \tag{8-44}$$

In (8-44) there appears $| \eta_0^p \rangle_T$ for $i = 1$. But from (8-39) and (8-41) we note the relation

$$| \eta_0^p \rangle_T = \begin{pmatrix} 0 \\ | \eta_d^p \rangle_{T-1} \end{pmatrix} \tag{8-45}$$

Writing down the last components of (8-44), we have

$$\epsilon_{i,T}^{p+1} = \epsilon_{i,T}^p + \alpha_{i,T}^{p+1} \eta_{i-1,T}^p$$

$$\eta_{i,T}^{p+1} = \eta_{i-1,T}^p + \beta_{i,T}^{p+1} \epsilon_{i,T}^p \tag{8-46}$$

where we define

$$\sigma_{i,T}^p = \langle \epsilon_i^p | \epsilon_i^p \rangle_T^{1/2}, \quad \tau_{i-1,T}^p = \langle \eta_{i-1}^p | \eta_{i-1}^p \rangle_T^{1/2}$$

$$\Delta_{i,T}^p = \langle \epsilon_i^p | \eta_{i-1}^p \rangle_T \tag{8-47}$$

$$\alpha_{i,T}^{p+1} = - \frac{\Delta_{i,T}^p}{\left(\tau_{i-1,T}^p \right)^2}, \quad \beta_{i,T}^{p+1} = - \frac{\Delta_{i,T}^p}{\left(\sigma_{i,T}^p \right)^2}$$

and from (8-45) we also note

$$\eta^p_{0,T} = \eta^p_{d,T-1} \tag{8-48}$$

The relations (8-46) to (8-48) satisfy the same CL structure in Fig. 8-1, with the time-dependent forward and backward PARCOR coefficients $\alpha^p_{i,T}$, $\beta^p_{i,T}$.

The time-updated recursions for $\sigma^p_{i,T}$, $\tau^p_{i-1,T}$, and $\Delta^p_{i,T}$ can be obtained from the general formula (51) in [15]. In our case, it becomes

$$\langle u \mid P^\perp_{i,1,p,T} \mid v \rangle_T = \lambda \langle u \mid P^\perp_{i,1,p,T} \mid v \rangle_{T-1}$$
$$+ \langle u \mid P^\perp_{i,1,p,T} \mid \pi \rangle_T \langle \pi \mid P^\perp_{i,1,p,T} \mid v \rangle_T \sec^2 \theta_{i,1,p,T} \tag{8-49}$$

for any $|u\rangle_T$, $|v\rangle_T \in H_T$ where $|\pi\rangle_T = [0 \cdots 0 \; 1]^T$ and

$$\cos^2 \theta_{i,1,p,T} = \langle \pi \mid P^\perp_{i,1,p,T} \mid \pi \rangle_T \tag{8-50}$$

From (8-40), (8-42), and (8-49), we have

$$(\sigma^p_{i,T})^2 = \lambda(\sigma^p_{i,T-1})^2 + (\epsilon^p_{i,T})^2 \sec^2 \theta_{i,1,p,T}$$
$$(\tau^p_{i-1,T})^2 = \lambda(\tau^p_{i-1,T-1})^2 + (\eta^p_{i-1,T})^2 \sec^2 \theta_{i,1,p,T} \tag{8-51}$$
$$\Delta^p_{i,T} = \lambda\Delta^p_{i,T-1} + \epsilon^p_{i,T}\eta^p_{i-1,T} \sec^2 \theta_{i,1,p,T}$$

The order-updated recursion for $\cos^2 \theta_{i,1,p,T}$ is obtained from (8-43), (8-50) as

$$\cos^2 \theta_{i,1,p+1,T} = \cos^2 \theta_{i,1,p,T} - \frac{(\eta^p_{i-1,T})^2}{(\tau^p_{i-1,T})^2} \tag{8-52}$$

The initial conditions are

$$\sigma^p_{i,0}, \ \tau^p_{i,0} = \text{some small positive values,} \quad \Delta^p_{i,0} = 0 \quad (T = 0)$$

$$\epsilon^0_{i,T} = \eta^0_{i,T} = x_{i,T}, \quad \cos^2\theta_{i,1,0,T} = 1 \quad (p = 0)$$

(8-53)

Next we present the parameter estimation algorithm of the joint process CL lattice treated in Sec. 8-2-2. We denote the least-squares estimator of z_T based on $Y_{d,0,p,T}$ and its error by z^{p+1}_T and u^{p+1}_T, respectively, where $0 \le p \le (m+1)d - 1$. That is,

$$|u^{p+1}\rangle_T = P^\perp_{d,0,p,T} |z\rangle_T \tag{8-54}$$

On the other hand, since $Y_{d,0,p,T} = Y_{d,0,p-1,T} \oplus |\eta^p_{d}\rangle_T$, we have

$$P^\perp_{d,0,p,T} = P^\perp_{d,0,p-1,T} - |\eta^p_{d}\rangle_T \langle\eta^p_{d}| \eta^p_{d}\rangle^{-1}_T \langle\eta^p_{d}|_T$$

so

$$u^{p+1}_T = u^p_T - \eta^p_{d,T}\delta^p_{d,T} \tag{8-55}$$

where $\delta^p_{d,T} = \langle\eta^p_{d}| \eta^p_{d}\rangle^{-1}_T \langle\eta^p_{d}| u^p\rangle_T$. The time-updated recursion for $\delta^p_{d,T}$ is obtained similarly as for (8-51).

8-3-2 Recursive Least-Squares Escalator Algorithm

Ahmed and Youn [17] derive an estimation algorithm for the (modified) Gram-Schmidt orthogonalization of n scalar stochastic processes $\{w_{1,t}, w_{2,t}, \ldots, w_{n,t}\}$. They call it the escalator (ES) algorithm, but it is not the exact least-squares algorithm. Here we present the recursive least-squares escalator algorithm [16]. Our orthogonalization starts from $w_{1,t}$ to $w_{n,t}$, so that the final orthogonalized process corresponds to the residual process when $w_{n,t}$ is "predicted" by $w_{1,t}, \ldots, w_{n-1,t}$.

First we define the vectors $|w_i\rangle_T$ $(i = 1, \ldots, n)$ as in (8-38). Let $W_{i,T}$ be the subspace spanned by $|w_1\rangle_T, |w_2\rangle_T, \ldots, |w_i\rangle_T$ and $P_{i,T}$ be the projection operator on $W_{i,T}$. That is,

$$P_{i,T} = |W_i\rangle_T \langle W_i| W_i\rangle^{-1}_T \langle W_i|_T$$

where $|W_i\rangle_T = [|w_1\rangle_T, \ldots, |w_i\rangle_T]$. We define the following vectors:

$$| u_i^k \rangle_T = P_{k,T}^\perp | w_i \rangle_T \quad (i = 1, 2, \ldots, k+1) \quad (| u_i^0 \rangle_T = | w_i \rangle_T)$$

Since $W_{k+1,T} = W_{k,T} \oplus | u_{k+1}^k \rangle_T$, we have

$$P_{k+1,T}^\perp = P_{k,T}^\perp - | u_{k+1}^k \rangle_T \langle u_{k+1}^k |_T^{-1} \langle u_{k+1}^k |_T$$

Operating $| w_i \rangle_T$ to both sides of this equation and writing down the last component, we have

$$u_{i,T}^{k+1} = u_{i,T}^k + a_{i,T}^{k+1} u_{k+1,T}^k \tag{8-56}$$

with

$$a_{i,T}^{k+1} = - \frac{\langle u_{k+1}^k | u_i^k \rangle_T}{\langle u_{k+1}^k | u_{k+1}^k \rangle_T} = - \frac{d_{i,T}^k}{s_{i,T}^2} \tag{8-57}$$

since $\langle u_{k+1}^k | w_i \rangle_T = \langle u_{k+1}^k | P_k^\perp w_i \rangle_T = \langle u_{k+1}^k | u_i^k \rangle_T$. The time-updated recursions for $d_{i,T}^k$ and $s_{k,T}^2$ are obtained by using the general formula (8-49) as

$$s_{k,T}^2 = \lambda s_{k,T-1}^2 + \frac{(u_{k+1,T}^k)^2}{\cos^2 \theta_{k,T}}$$

$$d_{i,T}^k = \lambda d_{i,T-1}^k + \frac{u_{k+1,T}^k u_{i,T}^k}{\cos^2 \theta_{k,T}} \tag{8-58}$$

with

$$\cos^2 \theta_{k+1,T} = \cos^2 \theta_{k,T} - \frac{(u_{k+1,T}^k)^2}{s_{k,T}^2} \tag{8-59}$$

The signal flow graph of the escalator algorithm is shown in Fig. 8-4. The adjective "escalator" comes from the structure of the algorithm.

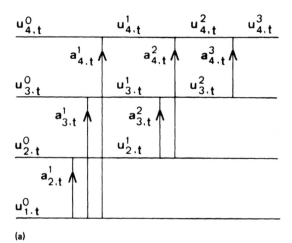

(a)

FIGURE 8-4 Descriptions of (a) the ES and (b) the TL algorithms for $d = 4$.

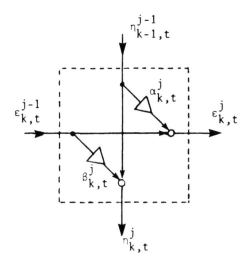

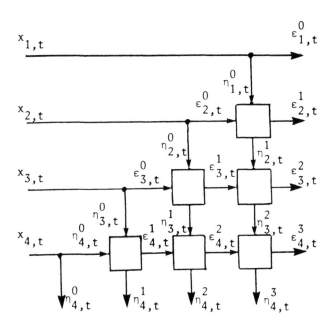

(b)

8-3-3 RLS Filtering Problem

Let $x_{i,t}$ $(1 \leq i \leq N)$ be scalar stationary input processes and y_t be another process to be estimated by a linear combination of $x_{1,t}, \ldots, x_{N,t}$. The general RLS filtering problem is formulated as

$$\min_{a_1, \ldots, a_N} \sum_{t=1}^{n} \lambda^{n-t} (y_t + a_1 x_{1,t} + \cdots + a_N x_{N,t})^2$$

$$= \sum_{t=1}^{n} \lambda^{n-t} (y_t + c_1(n) x_{1,t} + \cdots + c_N(n) x_{N,t})^2 \qquad (8\text{-}60)$$

where λ is a "forgetting" factor such that $1 - \lambda \ll 1$ to adapt in nonstationary environments. The error ϵ_n between y_n and its approximated value

$$\epsilon_n = y_n + c_1(n) x_{1,n} + \cdots + c_N(n) x_{N,n} \qquad (8\text{-}61)$$

is called the residual. In many practical applications such as noise and echo canceling and channel equalizations, the main interest is to obtain these residuals; the regression coefficients are of secondary concern.

Often, we consider the FIR (finite impulse response) filtering problem where $x_{i,t} = x_{t-i+1}$; that is, the ith signal is a time-shifted version of one common signal x_t. For this case, the so-called fast algorithms requiring only an amount $O(N)$ of calculations per each iteration have been developed. However, such an assumption does not hold for spatial signal processing. So we treat the general case here.

For (8-60) we have the well-known Kalman algorithm [18]

$$\underline{c}_n = \underline{c}_{n-1} + \underline{K}_n (y_n + \underline{x}_n^T \underline{c}_{n-1}) \qquad (8\text{-}62)$$

$$\underline{K}_n = -\frac{P_{n-1} \underline{x}_n}{\lambda + \underline{x}_n^T P_{n-1} \underline{x}_n} \qquad (8\text{-}63)$$

$$P_n = \frac{1}{\lambda} \left(P_{n-1} - \frac{P_{n-1} \underline{x}_n \underline{x}_n^T P_{n-1}}{\lambda + \underline{x}_n^T P_{n-1} \underline{x}_n} \right) \qquad (8\text{-}64)$$

where $\underline{c}_n = (c_1(n) \cdots c_N(n))^T$ and $\underline{x}_t = (x_{1,t} \cdots x_{N,t})^T$ with $\underline{c}_0 = \underline{0}$, $P_0 = w \cdot I$ ($w = 10 \sim 10^3$). This algorithm requires about $3N^2$ multiplications per each iteration. But performing this algorithm in single precision often produces unreliable results. This is primarily because the right-hand side of (8-64) is just a difference of two nonnegative definite matrices, and due to the canceling, the nonnegativeness of P_n is not kept in a long run. To remedy this drawback, Bierman [19] proposes an algorithm that updates the U-D factors U_n and D_n of the decomposition $P_n = U_n D_n U_n^T$ by using the Householder transformation technique. The nonnegativeness is now guaranteed as long as the elements of D_n are nonnegative. It is well known that the condition number of U_n is the square root of that of P_n, and this alleviates the numerical difficulties. For ease of reference, a summary of the algorithm is presented below.

Given the decomposition $P_{n-1} = \bar{U}\bar{D}\bar{U}^T$ where $\bar{U} = (\bar{v}_1 \cdots \bar{v}_N)$ and $\bar{D} = \text{diag}(\bar{d}_1 \cdots \bar{d}_N)$ and the vectors

$$\underline{f} = (f_1 \cdots f_N)^T = \bar{U}^T \underline{x}_n, \quad \underline{g} = (g_1 \cdots g_N)^T = \bar{D}\underline{f}$$

$P_n = \hat{U}\hat{D}\hat{U}^T$ is calculated as follows, where $\hat{U} = (\hat{v}_1 \cdots \hat{v}_N)$, $\hat{D} = \text{diag}(\hat{d}_1 \cdots \hat{d}_N)$.

(i) Initialization:

$$\alpha_1 = \lambda + f_1 g_1, \quad \underline{k}_1 = (g_1 \ 0 \cdots 0)^T$$

$$\hat{d}_1 = \frac{\hat{d}_1}{\alpha_1}, \quad \hat{\underline{v}}_1 = (1 \ 0 \cdots 0)^T$$

(ii) For $j = 2, \ldots, N$:

$$\alpha_j = \alpha_{j-1} + f_j g_j, \quad \hat{d}_j = \frac{\hat{d}_j \alpha_{j-1}}{\alpha_j \lambda}$$

$$\mu_j = \frac{f_j}{\alpha_{j-1}}, \quad \hat{\underline{v}}_j = \bar{\underline{v}}_j - \mu_j \underline{k}_{j-1}$$

$$\underline{k}_j = \underline{k}_{j-1} + g_j \bar{\underline{v}}_j$$

(iii) The Kalman gain $\underline{K}_n$ in (8-63) is given by $\underline{K}_n = -k_N/\alpha_N$.

This algorithm requires about $6N^2$ multiplications. The parallel computation architectures for this algorithm have been proposed by Jover and Kailath [20] and Hashimoto and Kimura [21], but the structures are not simple.

When we are interested in obtaining the residuals ϵ_n in (8-61), we can apply the escalator algorithm (8-56) to (8-59) by taking $w_{1,t} = x_{1,t}, \ldots,$ $w_{N,t} = x_{N,t}, w_{N+1,t} = y_t.$ The algorithm requires about $3.5N^2$ multiplications per each iteration. The LSCL algorithm in Sec. 8-3-1 can also be applied to this problem by taking $x_{k,t} = w_{k,t}$ (k = 1, ..., d) and the prediction orders as $p_k = k - 1$, where $d = N + 1$. The block diagram of this algorithm [(8-46) to (8-48), (8-51) to (8-53)] is shown in Fig. 8.4. We call this a triangular lattice (TL). A suboptimal TL algorithm was derived by Sharman and Durrani [22], and the exact LS algorithm was derived in [16] and Lev-Ari [23]. The TL algorithm is also equivalent to an orthogonalization method due to Cybenko [24] except for the time-update recursions (8-51). This orthogonalization method is based on the successive orthogonalization of adjacent vectors in both directions. Suppose that there are three vectors $\underline{X}_1, \underline{X}_2, \underline{X}_3$. In the first step, the correlated portions of $\underline{X}_1, \underline{X}_2, \underline{X}_2, \underline{X}_3$ are removed from $\underline{X}_2, \underline{X}_1, \underline{X}_3, \underline{X}_2$, respectively. In the second step, the correlated portions of the second vector in the first step is removed from the third one, and vice versa. Thus two orthogonal bases $\{\underline{u}_1 \ \underline{u}_2 \ \underline{u}_3\}$ and $\{\underline{v}_1 \ \underline{v}_2 \ \underline{v}_3\}$ are produced, where the former is the usual one in the ES algorithm and the latter is the one in the backward direction. The TL algorithm requires about $7.5N^2$ multiplications per each iteration. Thus the computational complexity of the TL algorithm is about twice that of the ES algorithm. This is because the TL algorithm performs two orthgonalizations. So one may wonder about the merit of the TL algorithm. We discuss this point next.

8-3-4 Systolic Array Implementations

The concept of systolic array was developed by Kung [25] and has attracted a great deal of attention in signal processing and computer architecture societies. A systolic array consists of a set of interconnected cells capable of performing simple operations in parallel and communicating the data only to neighboring cells. Using this parallel and pipelined structure, a higher computational throughput can be obtained, and this type of array is well suited to VLSI implementations.

Now the systolic array implementations of the TL and the ES algorithms are compared. The systolic array of the TL algorithm can be obtained by arranging processors on the lattice elements in Fig. 8.4. Using this array of $(d - 1)^2/2$ processors (d = N + 1), the processing time becomes independent of the size of d. The maximum parallelism is thus obtained. The latency is defined as the time between the arrival of the input signals $x_{i,t}$ and the output residual ϵ_t. This structure incurs d time units delay, where one time unit is defined as the input sampling time, and the operation of each processor element is assumed to be performed within this time unit. Similarly, for the ES algorithm, arranging processor elements on the nodes in Fig. 8-3 constructs a parallel and pipelined architecture that achieves maximum paral-

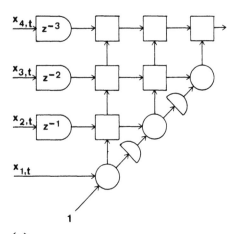

(a)

$$X_t \rightarrow \boxed{z^{-n}} \rightarrow X_{t-n}$$

$$X_t \rightarrow \!\!\!\!\!\!\triangleright \longrightarrow X_{t-1}$$

$$(u^p_{p+1,t} \ S^p_t \ \cos^2\theta^p_t)$$

$$u^p_{i,t} \rightarrow \boxed{D^p_{i,t}} \rightarrow u^{p+1}_{i,t}$$

$$(u^p_{p+1,t} \ S^p_t \ \cos^2\theta^p_t)$$

$$D^p_{i,t} = \lambda D^p_{i,t-1} + u^p_{p+1,t} u^p_{i,t}/\cos^2\theta^p_t$$

$$(u^p_{p+1,t} \ S^p_t \ \cos^2\theta^p_t)$$

$$u^p_{p+1,t} \longrightarrow \left(S^p_t\right) \nearrow \cos^2\theta^{p+1}_t$$

$$\cos^2\theta^p_t$$

$$(S^p_t)^2 = \lambda(S^p_{t-1})^2 + (u^p_{p+1,t})^2/\cos^2\theta^p_t$$

(b)

FIGURE 8-5 Block diagram of McWhirter's type systolic array: (a) entire structure; (b) cell descriptions.

TABLE 8-2 Comparison of Features of Each
Parallel Architecture

	Amount of computations	Latency	Locality
TL	$15d^2/2$	d	Yes
ES	$7d^2/2$	d	No
MSA	$7d^2/2$	2d	Yes

lelism and the least latency d. However, the elements or cells are not
locally connected. For example, the inputs from the first and the fourth
channels are required to compute $a_{4,t}^1$. Thus it is no longer a systolic array. The local connection property is quite important in VLSI circuit design,
but Kalson and Yao [26] have pointed out that a systolic array similar to the
one by McWhirter [27] for the RLS problem using the method of Givens rotation shown in Fig. 8-5 can be obtained by introducing the delays in the inputs
in the ES algorithm. The dotted lines in Fig. 8-5 denote the computational
wavefronts and the latency now becomes 2d. The features of these implementations are summarized in Table 8-2, where MSA stands for the McWhirter's

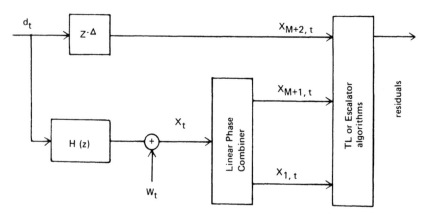

FIGURE 8-6 Block diagram of linear-phase channel equalization.

TABLE 8-3 Empirical Variances of the Residuals
of the Four RLS Algorithms with the Mantissa Bits
(a) B = 24, (b) B = 12, and (c) B = 9

	σ^2	Kalman	U-D	TL	ES
(a)	0.1	0.1405	0.1405	0.1256	0.1256
	0.01	0.0171	0.0171	0.0152	0.0152
	0.001	0.0018	0.0018	0.0016	0.0016
(b)	0.1	0.9493	0.1388	0.1236	0.1222
	0.01	10.010	0.0494	0.0158	0.0150
	0.001	1408.1	0.0318	0.0024	0.0017
(c)	0.1	—	0.1563	0.1354	0.1082
	0.01	260.45	0.0498	0.0407	0.0279
	0.001	—	0.0326	0.0334	0.0217

systolic array. Thus the TL algorithm realizes the local connection and the
least latency at the price of doubling the amount of computations introduced
by an extra orthogonalization.

8-3-5 Numerical Example

To see the finite-word-length effects of floating-point arithmetic on the four
RLS algorithms above, some simulation results are presented in Youn and
Prakash for a linear-phase channel equalization problem [28]. Figure 8-6
depicts the entire structure of the problem. The random training signal d_t
is bipolar ± 1 with equal probabilities and is delayed Δ time units at the re-
ceiver. Also, d_t is transmitted through a channel with a linear-phase transfer function $H(z)$ and the output is disturbed by white noise W_t with zero mean
and variance σ^2. The purpose is to approximate $x_{M+2,t} = d_{t-\Delta}$ with a linear
combination of $x_{1,t}, \cdots, x_{M+1,t}$, where

$$x_{i,t} = x_{t-M-1+i} + x_{t+M+1-i} \qquad (1 \leq i \leq M)$$

$$x_{M+1,t} = x_t$$

with $x_t = H(z)d_t + W_t$.

In the simulations, we set $M = 5$, $\Delta = 2$, $\lambda = 2028/2^{11}$, and the impulse response h_i of $H(z)$ is given by

$$h_i = \begin{cases} 0.5 + 0.5 \cos \dfrac{2\pi(i-3)}{3.1} & (1 \leq i \leq 5) \\ 0 & (6 \leq i) \end{cases}$$

In a situation such as $M = \infty$, the variance of the residuals is to be very close to σ^2. The steady-state empirical variances of the residuals of the four algorithms are obtained by calculating their ensemble and time averages over 50 data sets and $t = 500$ to 1000 for three different values of $\sigma^2 = 0.1$, 0.01, 0.001 and for three different mantissa bits of floating-point arithmetic $B = 24$, 12, 9, respectively. Assessment of numerical stability can be made by comparing these variances. Table 8-3 presents the results, where a dash indicates that the algorithm diverges. From these tables it may be concluded that the four algorithms perform almost equally well in the case of $B = 24$, but for $B = 12$ and 9, the Kalman algorithm is more sensitive than the other three. As for the comparison of the three, from Table 8-3(b) it is seen that for $B = 12$, $\sigma^2 = 0.001$, the U-D gives up, and from Table 8-3(c), for $B = 9$, $\sigma^2 = 0.001$, both the TL and ES now give up.

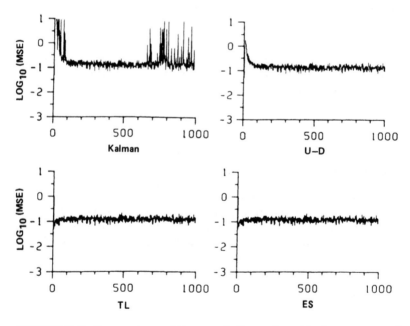

FIGURE 8-7 Comparison of time evolutions of residual variances of each algorithm.

Thus the TL and ES are very robust to rounding errors. Also, it is noted that the ES is slightly more insensitive than the TL. Finally, Fig. 8-7 shows the time evolution of the empirical variance of the residuals of each algorithm for $\sigma^2 = 0.1$, $B = 12$. It is shown that the ES and TL have a very rapid convergence property.

8-4 APPLICATIONS TO ARMA MODELING

8-4-1 Parameter Characterization of Scalar ARMA Processes

Here we apply the CL method for multichannel stationary time series in Sec. 8-2 to characterize the parameters of <u>scalar</u> ARMA processes. Consider the following ARMA(n, n) process:

$$x(t) + a_1 x(t-1) + \cdots + a_n x(t-n) = b_0 u(t) + b_1 u(t-1) + \cdots + b_n u(t-n) \tag{8-65}$$

where $\{u(t)\}$ is a sequence of white noise with zero mean and variance σ^2. Usually, the minimum phase property of the MA part is assumed, but here we <u>do not</u> assume it. As in Lee et al. [29], consider the following two-channel process:

$$\vec{X}(t) = \begin{pmatrix} u(t) \\ x(t) \end{pmatrix} \tag{8-66}$$

We apply the CL method in Sec. 8-2 to this process with $d = 2$. Thus, from (8-4),

$$Y(2t-1) = u(t), \quad Y(2t) = x(t)$$

Since u(t) is uncorrelated with $x(t-1)$, $u(t-1)$, ..., we have

$$\epsilon(j, 2t-1) = u(t) \quad (j \geq 0)$$

So $\alpha_1(j, i) = 0$ $(i = 1, \ldots, j; j \geq 1)$. In particular, $\alpha_1(j, j) = 0$ means that $\Delta_1(j) = 0$. This implies from (8-21) and (8-23) that

$$\rho_1(j) = 0 \quad (j \geq 1), \quad \sigma_1^2(j) = \sigma^2 \tag{8-67}$$

Also, from (8-65), the (2n + 1)th-order forward prediction error

$$\epsilon(2n+1, 2t)$$

$$= x(t) + \alpha_2(2n+1, 1)u(t) + \alpha_2(2n+1, 2)x(t-1) + \cdots + \alpha_2(2n+1, 2n+1)u(t-n) \tag{2-68}$$

is zero. This means that

$$a_i = \alpha_2(2n + 1, 2i), \quad b_i = -\alpha_2(2n + 1, 2i + 1) \quad (i = 0, \ldots, n) \quad (8\text{-}69)$$

where $\alpha_2(2n + 1, 0) = 1$. Also, we have $\sigma_2^2(2n + 1) = 0$. But from (8-23) and (8-25) this means that $|\rho_2(2n + 1)| = 1$. Since $\alpha_2(2n + 1, 2n + 1)$ and $\rho_2(2n + 1)$ have the same sign, from (8-69) it follows that

$$\rho_2(2n + 1) = -\text{sgn}(b_n) \qquad (8\text{-}70)$$

Since $\epsilon(0, 2t) = x(t)$, $\eta(0, 2t - 1) = u(t)$, from (8-21) and (8-28) $\alpha_2(1, 1) = -E[x(t)u(t)]/E[u^2(t)] = -b_0$. Thus, from (8-21), $\Delta_2(0) = b_0 \tau_1^2(0)$. Note from (8-25) that $\rho_2(1) = -\Delta_2(0)/\sigma_2(0)\tau_1(0)$. Since $\sigma_1(0) = \tau_1(0) = \sigma > 0$, we note that

$$\rho_2(1) = -b_0 \frac{\sigma}{\sigma_2(0)} \qquad (8\text{-}71)$$

In summary, the parameter $\{a_1 \cdots a_n b_0 \cdots b_n \sigma^2\}$ of an ARMA process (8-65) produces a sequence of the "generalized" partial autocorrelation coefficients $\{\rho_2(1) \cdots \rho_2(2n + 1)\}$ with the relations (8-70) and (8-71) and the property $|\rho_2(i)| < 1$ for $i = 1, \ldots, 2n$. The calculation of these $\rho_2(i)$ can be done easily by noting that the autocovariance matrix of (8-66) is given by

$$R_k = \begin{pmatrix} \sigma^2 \delta_{k,0} & \sigma^2 h_0 \delta_{k,0} \\ \sigma^2 h_k & r_k \end{pmatrix} \quad (k \geq 0)$$

where $\{h_k\}$ is the sequence of unit impulse responses of the ARMA system (8-65) and $\{r_k\}$ is the covariance sequence of $x(t)$ (i.e., $r_k = E[x(t+k)x(t)]$). Using the relation (8-32) and the circular Levinson algorithm in Sec. 8-2-1, we can generate $\rho_2(i)$ $(i = 1, \ldots, n)$. More specifically, since $\Delta_1(j) = 0$, from (8-21) and (8-22) we have $\beta_1(j, i) = \beta_2(j - 1, i)$ with $\beta_1(j, j) = 0$, and $\tau_1^2(j) = \tau_2^2(j - 1)$. Thus, dropping the subscript 2, we have

$$\alpha(j + 1, j + 1) = -\frac{\Delta(j)}{\tau^2(j - 1)}$$

$$\beta(j + 1, j + 1) = -\frac{\Delta(j)}{\sigma^2(j)}$$

$$\alpha(j + 1, i) = \alpha(j, i) + \alpha(j + 1, j + 1)\beta(j - 1, j + 1 - i) \quad (i = 1, \ldots, j)$$

$$\beta(j + 1, i) = \beta(j - 1, i) + \beta(j + 1, j + 1)\alpha(j, j + 1 - i) \qquad (8\text{-}72)$$

$$\sigma^2(j+1) = \sigma^2(j)(1 - \rho^2(j+1)), \quad \tau^2(j+1) = \tau^2(j-1)(1 - \rho^2(j+1))$$

with $\rho(j+1) = -\Delta(j)/\sigma(j)\tau(j-1)$ and $\tau^2(-1) = \sigma^2$, $\sigma^2(0) = \tau^2(0) = r_0$. Also, $\Delta(j)$ is given by

$$\Delta(j) = R(1-j, 2) + R(2-j, 2)\beta(j-1, 1) + \cdots + R(0, 2)\beta(j-1, j-1)$$

with $R(1 - 2k, 2) = \sigma^2 h_k$ and $R(2 - 2k, 2) = r_k$. The generation of $\{h_k\}$ and $\{r_k\}$ from the ARMA parameters is well known, so we do not present it. Also, a lattice relation is obtained by taking $n = t - 1$ in (8-27) and noting that $\eta(j, 2t - 1) = \eta(j - 1, 2t - 2)$ as

$$\epsilon_{j+1}(t) = \epsilon_j(t) + \alpha_{j+1}\eta_{j-1}(t-1)$$

$$\eta_{j+1}(t) = \eta_{j-1}(t-1) + \beta_{j+1}\epsilon_j(t)$$

$$(8\text{-}73)$$

where we put $\epsilon_j(t) = \epsilon(j, 2t)$, $\eta_j(t) = \eta(j, 2t)$, and $\alpha_j = \alpha(j, j)$, $\beta_j = \beta(j, j)$. It It is interesting to note that the subscript of the backward prediction error $\eta(\cdot)$ is $\underline{j - 1}$, not j as in the usual AR lattice.

8-4-2 CL ARMA Lattice Filter

Before presenting the CL ARMA lattice filter, we mention briefly an application to model reduction of the foregoing ARMA parameter characterization. Actually, the method above is very close to the well-known procedure due to Mullis and Roberts [30] where the first- and second-order information is used to reconstruct the ARMA transfer function. Suppose that we are given a large-order ARMA(n, n) system with $\sigma^2 = 1$. Then we generate $h_0, \ldots, h_n$, $r_0, \ldots, r_n$. Using the algorithm in (8-72), we obtain $\rho_1, \ldots, \rho_{2n+1}$, where $\rho_j = \rho_2(j)$. Due to (8-70), h_n is determined by $h_0, \ldots, h_{n-1}, r_0, \ldots, r_n$. Now we want to have a reduced-order ARMA system which approximates the original one in some sense. One immediate way is to truncate the ρ_j above at $j = 2m + 1$. That is, we replace ρ_{2m+1} by ± 1, where $m < n$. Using $\alpha(2m, i)$ and $\beta(2m - 1, i)$ with $\sigma^2(2m)$ and $\tau^2(2m - 1)$, we obtain a new ARMA(m, m) system by

$$\tilde{\alpha}(2m + 1, 2m + 1) = \pm \frac{\sigma(2m)}{\tau(2m - 1)}$$

$$\tilde{\alpha}(2m + 1, i) = \alpha(2m, i) + \tilde{\alpha}(2m + 1, 2m + 1)\beta(2m - 1, 2m + 1 - i)$$

$$(i = 1, \ldots, 2m)$$

It is obvious that the resulting $\tilde{h}_0, \ldots, \tilde{h}_m, \tilde{r}_0, \ldots, \tilde{r}_m$ generated by this new ARMA parameter $\tilde{a}_1, \ldots, \tilde{a}_m, \tilde{b}_0, \ldots, \tilde{b}_m$ have the following matching property:

$$\tilde{h}_i = h_i \text{ for } i = 0, \ldots, m - 1, \quad \tilde{r}_i = r_i \text{ for } i = 0, \ldots, m$$

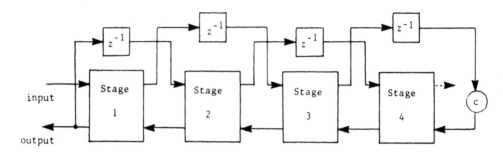

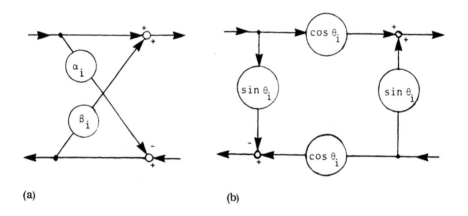

(a)   (b)

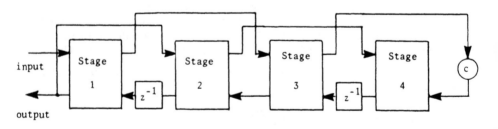

(c)

FIGURE 8-8 New ARMA lattice filters: (a) unnormalized structure with the block diagram of one lattice section and $c = -\alpha_5$; (b) normalized structure with the rotation matrix $\Sigma(\theta_i)$ for the operation of each lattice section and $c = \pm 1$; (c) another filter structure with minimal delays.

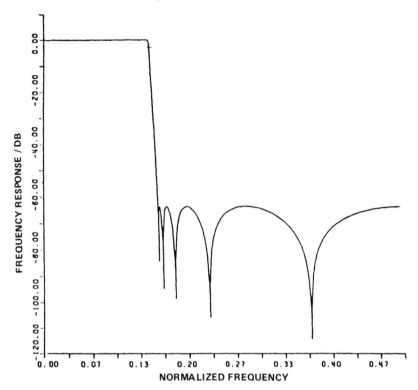

FIGURE 8-9 Magnitude of the frequency response of a low-pass filter in [32].

Next, we present the CL ARMA lattice filter with the input u(t) and the output x(t). First note that $\eta_{-1}(t - 1) = \eta(0, 1 + 2t) = u(t)$ and $\epsilon_0(t) = \epsilon(0, 2 + 2t) = x(t)$, and $\eta_0(t - 1) = x(t - 1)$ and $\epsilon_{2n+1}(t) = 0$. Using this with (8-73), we have the CL ARMA lattice filter whose block diagram is shown in Fig. 8-8, where n = 2. There are 2n sections of the fundamental lattice element and 2n unit delays. But it is easy to see that we can change the placement of the delays as in Fig. 8-8(c). This filter structure now has minimal n delays as those in [34] and [35] and improves that in [31]. Also, calculation of the lattice parameters from the ARMA parameters is very easy. To obtain the normalized CL ARMA lattice filter, we define the following normalized prediction errors by

$$\epsilon_j(t) = \sigma(j)\tilde{\epsilon}_j(t), \quad \eta_j(t) = \tau(j)\tilde{\eta}_j(t)$$

Then, from (8-72) and (8-73) we have

$$\begin{pmatrix} \tilde{\eta}_{j+1}(t) \\ \tilde{\epsilon}_j(t) \end{pmatrix} = \begin{pmatrix} \cos\theta_{j+1} & \sin\theta_{j+1} \\ -\sin\theta_{j+1} & \cos\theta_{j+1} \end{pmatrix} \begin{pmatrix} \tilde{\eta}_{j-1}(t - 1) \\ \tilde{\epsilon}_{j+1}(t) \end{pmatrix} \qquad (8\text{-}74)$$

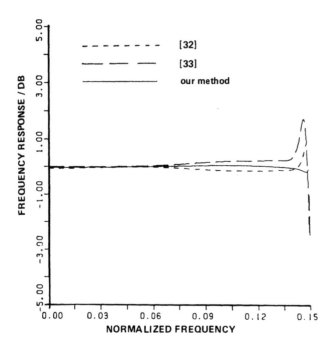

FIGURE 8-10 Comparison of the effects of coefficient quantization for
three ARMA lattice filters.

where $\rho_j = \sin \theta_j$ $(|\theta_j| < \pi/2)$. Note that the 2×2 matrix in the right-hand
side of (8-74) denoted by $\Sigma(\theta_{j+1})$ is an orthogonal rotation matrix.

We compare the sensitivities of coefficient quantization to the frequency response of our unnormalized filter and other lattice filters by Lim
and Parker [32] and Miyanaga et al. [33]. In Fig. 8-9 the magnitude of the
frequency response of a tenth-order low pass with the cutoff frequency 0.14
is shown. This example is taken from [32] and the ARMA coefficients are
given in [32]. Then the ARMA lattice parameters are generated for three
filters and each lattice coefficient is quantized with the word length 8 bits.
The magnitude of the frequency response of each perturbed lattice filter is
calculated and compared. Figure 8-10 shows the result where the ratios of
the magnitude to the ideal one of three filters are presented for the pass
and transient bands. At the transient band our filter is less sensitive than
the other two.

8-4-3 Parameter Estimation Algorithms for the CL ARMA Lattice Filter

Here we treat two kinds of parameter estimation problems for our ARMA
lattice filter. One is a common problem in time-series analysis where u(t)

TABLE 8-4 True and Estimated ARMA
Partial Autocorrelation Coefficients

j	ρ_j	$\hat{\rho}_j$
1	-0.267	-0.275
2	-0.911	-0.913
3	-0.783	-0.775
4	0.981	0.941
5	1.000	0.904
6		-0.480
7		-0.408
8		0.100
9		0.194
10		0.090

is white and a set of data $\{x(1) \cdots x(N)\}$ is given and we are asked to produce the estimates $\hat{\rho}_j$ of the generalized partial autocorrelation coefficients
ρ_j. In this case, we must assume that $b_0 = 1$ and the minimum phase property of the MA part if we use only the second-order information. The other
is to reconstruct the ARMA transfer function based on unit impulse response
data of finite length.

First, we treat the former one. The autocovariance r_k can easily be
estimated from the data, but to use (8-72) we need an estimate of the impulse
response h_k. One way frequently used in time series analysis field is first
to fit a high-order AR model to the data and obtain an estimate $\hat{h}_k$ by expanding the resulting AR transfer function. In Table 8-4 we give a simulation
result where a set of data of $N = 1000$ length from the ARMA(2, 2) process

$$x(t) - 1.5x(t - 1) + 0.7x(t - 2) = u(t) + 0.5u(t - 1) - 0.3u(t - 2)$$

with $\sigma^2 = 1$ is generated and we select the AR(14) model by the AIC method
[13] with the estimated white noise variance $\hat{\sigma}^2 = 0.979$. Using the procedure
stated above, we obtain the estimates $\hat{\rho}_j$ of ρ_j. From this table we see that
our method produces good results, although the error of the estimate of the
last coefficient ρ_5 is not as small. This may be due to the bias effect produced by using an AR model to obtain the estimates $\hat{h}_k$. We also present
$\hat{\rho}_6$ to $\hat{\rho}_{10}$, whose true values are not defined. Although all of these are not
close to zero, none of them are close to ±1. This indicates that it is possible

to estimate the order of the ARMA process by checking the pattern of $\hat{\rho}_j$. The procedure above is a batch processing method. A time-recursive estimation algorithm for the CL ARMA lattice filter has been proposed in Sakai et al. [36] by modifying the LSCL and joint process LSCL algorithms in Sec. 8-3-1 with a bootstrap method to supply the estimate of u(t). The number of multiplications per one sample of this algorithm is about 60% of that of Lee's algorithm [29].

Next, we treat the problem of estimating the ARMA parameter $\{a_1 \cdots a_n \; b_0 \cdots b_n\}$ based only on a set of unit impulse response data $\{h_0 \cdots h_{N-1}\}$. This problem has been well studied in realization theory and digital filter design. The important point is that an estimated transfer function must be stable whatever method we use. An obvious direct approach is to minimize the criterion

$$f_1(\vec{a}, \vec{b}) = \sum_{i=0}^{N-1} (h_i - g_i)^2$$

about $\vec{a}^T = (a_1 \cdots a_n)$, $\vec{b}^T = (b_0 \cdots b_n)$, where $\{g_i\}$ is the impulse response given by

$$H(z^{-1}) = \sum_{i=0}^{\infty} g_i z^{-i}$$

But this leads to a difficult constrained nonlinear optimization problem. Instead, Claerbout [37] and Morf et al. [38] modify the criterion as

$$f_2(\vec{a}, \vec{b}) = \sum_{i=0}^{\infty} \left(\sum_{j=0}^{n} a_j h_{i-j} - b_i \right)^2 \tag{8-75}$$

with $a_0 = 1$, $h_i = 0$ $(i < 0)$, and the assumption that

$$h_i = 0 \quad (i > N - 1) \tag{8-76}$$

Then the $\vec{a}$ and $\vec{b}$ that minimize (8-75) produce a stable transfer function. These $\vec{a}$ and $\vec{b}$ are expressed in a form which is more explicitly related to the multichannel linear prediction theory as follows. First define the two-variate vector

$$\vec{X}(j) = \begin{pmatrix} \delta_{j-1,0} \\ h_{j-1} \end{pmatrix} \quad (j = 1, \ldots, N) \tag{8-77}$$

Also, define the "autocovariance" matrices

$$R_i = \sum_{j=i}^{N-1} \vec{X}(j) \vec{X}^T(j - i)$$

and the coefficient matrices

$$F_{n,i} = \begin{pmatrix} 0 & 0 \\ -b_i & a_i \end{pmatrix}$$

Then, minimization of (8-75) leads to the two-channel Yule-Walker (Y-W) equation

$$R_i + \sum_{j=1}^{n} F_{n,j} R_{i-j} = \delta_{i,0} P_n \quad (i = 0, \ldots, n) \qquad (8\text{-}78)$$

where P_n is a certain matrix calculated from $h_0, \ldots, h_{N-1}$. So the stability of $A(z) = 1 + a_1 z^{-1} + \cdots + a_n z^{-n}$ follows from that of $I + F_{n,1} z^{-1} + \cdots + F_{n,n} z^{-n}$.

But as in spectral estimation problem, the assumption (8-76) is undesirable. The Burg algorithm has been used extensively for spectral estimation because it avoids assumptions such as (8-76). Pusey [39] applies the Strand algorithm [14], a multichannel version of the Burg algorithm, directly to the data (8-77). The estimated $\vec{a}$ and $\vec{b}$ also guarantee the stability. Pusey shows by numerical examples that his estimate is much better than the Y-W estimate in (8-78). But he also states that the method is rather sensitive to small amounts of additive noise in the impulse response data. To improve the Pusey method further, we examine the structure of the problem more closely by using the results in Sec. 8-4-2.

Applying the Burg type CL algorithm in Sec. 8-2-4 to the data (8-77), it is immediate to see by inductions that

$$\Delta_1(j) = 0 \quad (j \geq 0)$$

From (8-21) this means that $\alpha_1(j,i) = 0$ $(i = 1, \ldots, j)$. Thus we are in the same situation with that in Sec. 8-4-1. Hence we drop the subscript 2 and use the algorithm in (8-72) with $\Delta(j)$ corresponding to (8-37) given by

$$\Delta(j) = \frac{2 \sum_{t=L(j)}^{N-1} \epsilon_j(t + 1) \eta_{j-1}(t)}{\sum_{t=L(j)}^{N-1} \epsilon_j^2(t + 1)/\sigma^2(j) + \sum_{t=L(j)}^{N-1} \eta_{j-1}^2(t)/\tau^2(j - 1)}$$

where $L(j) = [(j + 1)/2]$ and $\epsilon_{j+1}(t + 1)$ and $\eta_{j+1}(t + 1)$ are calculated by (8-73) for $t = L(j + 1), \ldots, N - 1$. The initial conditions are

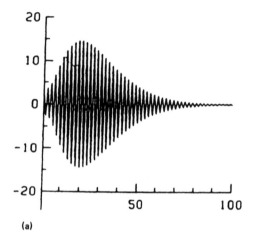

(a)

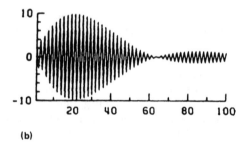

(b)

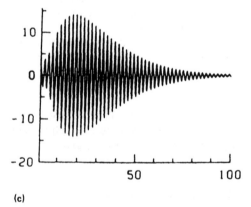

(c)

FIGURE 8-11 (a) True impulse response; (b) reconstructed by the algorithm without (8-79); (c) reconstructed by the algorithm with (8-79).

$$\epsilon_0(t + 1) = h_t, \quad \eta_{-1}(t) = \delta_{t,0} \quad (t = 0, \ldots, N - 1)$$

$$\sigma^2(0) = \tau^2(0) = \sum_{t=0}^{N-1} h_t^2, \quad \tau^2(-1) = \sum_{t=0}^{N-1} \delta_{t,0}^2 = 1$$

If we run the algorithm above from $j = 0$ to $2n + 1$, we obtain the ARMA parameter $a_1, \ldots, a_n, b_0, \ldots, b_n$ from (8-69). The resulting coefficients are almost identical to those obtained by Pusey's algorithm [39]. However, we know from (8-70) the special property of the last α_{2n+1} and β_{2n+1}. So it is expected to be beneficial to incorporate the property (8-70) into our estimation when the true transfer function is of ARMA(n, n) type. Thus we generate new ARMA parameters $\tilde{\alpha}(2n + 1, 1), \ldots, \tilde{\alpha}(2n + 1, 2n + 1) = \tilde{\alpha}_{2n+1}$ by

$$\tilde{\alpha}_{2n+1} = \frac{\sigma(2n)}{\tau(2n - 1)} \operatorname{sgn}(\rho_{2n+1}), \quad \tilde{\alpha}(2n+1, i) = \alpha(2n, i) + \tilde{\alpha}_{2n+1}\beta(2n - 1, 2n + 1 - i)$$

$$(i = 1, \ldots, 2n)$$

$$(8-79)$$

To see the effectiveness of the proposed method, a simulation result is presented here. We generate the impulse response data $\{h_0, \ldots, h_{99}\}$ from the ARMA$(4, 4)$ system with the coefficients

$$a_1 = 1.8, \quad a_2 = 0, \quad a_3 = -1.458, \quad a_4 = -0.6561$$
$$b_0 = 2, \quad b_1 = 1, \quad b_2 = -1, \quad b_3 = 0.5, \quad b_4 = 1.5$$

This is shown in Fig. 8-11(a). When there is no additive noise, the reconstructed impulse responses by the algorithms above without and with (8-79) are identical and the same with the true one. But if we add a small amount of white noise w_t with zero mean and variance σ_w^2 to each impulse response h_t and use the data $h_t^* = h_t + w_t$ $(t = 0, \ldots, 99)$, the results are fairly different, as shown in Fig. 8-11(b) and (c), where we set σ_w^2 so that the signal-to-noise ratio defined by $10 \log (\sum_{t=0}^{99} h_t^2)/(\sum_{t=0}^{99} w_t^2)$ becomes 40 dB. It is obvious that the algorithm with the modification (8-79) produces a better result. But it is to be noted that to obtain a good result by the algorithm with (8-79) we must know the true order n.

8-5 CONCLUSIONS

In this chapter we first presented the basic circular Levinson algorithm for calculating the predictor coefficients of multichannel time series for given covariance case. The algorithm is an alternative to the well-known LWR algorithm. Although the computational complexities of both algorithms are the same, our algorithm has several advantages. One of the most important points is that the circular lattice (CL) structure followed by our algorithm does not involve matrix-vector operations and is ideally suited to parallel processing where the processors are combined like a circle.

Then we presented a Burg algorithm for given data case and the recursive least-squares (RLS) estimation algorithm having the CL structure. As a special and important case, the triangular lattice (TL) algorithm has been treated and compared with other RLS algorithms in various aspects. An implication of the TL algorithm has been revealed by discussing several implementations in systolic arrays. Finally, we have applied the CL method to ARMA modeling where an ARMA process is embedded into a singular two-channel AR process. We have found some new properties of ARMA processes and devised ARMA lattice filters and a new algorithm for estimating the parameters based on the impulse response data.

From the developments above we conclude that the CL method has been applied successfully to the multichannel RLS estimation problem and ARMA modeling. We expect that the method may have applications in other multi-channel signal processing problems.

ACKNOWLEDGMENT

The author is grateful to Y. Iiguni, M. Makino, N. Tachibana, T. Yasumori, Y. Mekata, and Y. Nakaji for their help in preparing the materials in this chapter and to Assistant Professor Masakiyo Suzuki of Hokkaido University for a valuable suggestion concerning the filter structure in Fig. 8-8(c).

REFERENCES

1. M. Pagano, "Periodic and multiple autoregressions," Ann. Stat., vol. 6, pp. 1310-1317 (Nov. 1978).

2. R. A. Wiggins and E. A. Robinson, "Recursive solution to the multi-channel filtering problem," J. Geophys. Res., vol. 70, pp. 1885-1891 (1965).

3. J. P. Burg, "Maximum entropy spectral analysis," Ph.D. dissertation, Stanford University, 1975.

4. M. Morf, B. Dickinson, T. Kailath, and A. Vieira, "Efficient solution of covariance equations for linear prediction," IEEE Trans. Acoust. Speech Signal Process., vol. ASSP-25, pp. 429-433 (Oct. 1977).

5. M. Morf, A. Vieira, D. T. L. Lee, and T. Kailath, "Recursive multi-channel maximum entropy spectral estimation," IEEE Trans. Geosci. Electron., vol. GE-16, pp. 85-94 (Apr. 1978).

6. M. Morf, A. Vieira, and T. Kailath, "Covariance characterization by partial autocorrelation matrices," Ann. Stat., vol. 6, no. 3, pp. 643-648 (1978).

7. F. Itakura and S. Saito, "Digital filtering techniques for speech analysis and synthesis," Proc. 7th Int. Congr. Acoustics, Budapest, Paper 25-C-1, pp. 261-264, 1971.

8. L. J. Griffiths, "An adaptive lattice structure for noise cancelling applications," Proc. ICASSP, Tulsa, Okla., pp. 87-90, 1978.

9. L. J. Griffiths, "Adaptive structure for multi-input noise cancelling applications," Proc. ICASSP, Washington, D.C., pp. 925-928, 1979.

10. O. Barndorff-Nielsen and G. Schou, "On the parametrization of auto-regressive models by partial autocorrelation," J. Multivar. Anal., vol. 3, pp. 408-419 (1973).

11. Y. Inouye, "Autoregressive model fitting for multichannel time series of degenerate rank: limit properties," IEEE Trans. Circuits Syst., vol. CAS-32, pp. 252-259 (Mar. 1985).

12. H. Sakai, "Circular lattice filtering using Pagano's method," IEEE Trans. Acoust. Speech Signal Process., vol. ASSP-30, no. 2, pp. 279-287 (Apr. 1982).

13. H. Akaike, "A new look at the statistical model identification," IEEE Trans. Autom. Control, vol. AC-19, pp. 716-723 (Dec. 1974).

14. O. N. Strand, "Multichannel complex maximum entropy (autoregressive) spectral analysis," IEEE Trans. Autom. Control, vol. AC-22, pp. 634-640 (1977).

15. D. T. L. Lee, M. Morf, and B. Friedlander, "Recursive least squares ladder estimation algorithms," IEEE Trans. Acoust. Speech Signal Process., vol. ASSP-29, part III of no. 3, pp. 627-641 (June 1981).

16. T. Kawase, H. Sakai, and H. Tokumaru, "Recursive least squares circular lattice escalator estimation algorithms," IEEE Trans. Acoust. Speech Signal Process., vol. ASSP-31, pp. 228-231 (Feb. 1983).

17. N. Ahmed and D. E. Youn, "On a realization and related algorithm for adaptive prediction," IEEE Trans. Acoust. Speech Signal Process., vol. ASSP-28, no. 5, pp. 493-497 (Oct. 1980).

18. D. Godard, "Channel equalization using a Kalman filter for fast data transmission," IBM J. Res. Dev., pp. 267-273 (1974).

19. G. J. Bierman, Factorization Methods for Discrete Sequential Estimation, Academic Press, New York, 1977.

20. J. M. Jover and T. Kailath, "A parallel architecture for Kalman filter

measurement update and parameter estimation," Automatica, vol. 22, pp. 43-57 (1986).

21. K. Hashimoto and H. Kimura, "A parallel architecture for recursive least squares identification," Proc. ICASSP, Tokyo, pp. 1189-1192, 1986.

22. K. S. Sharman and T. S. Durrani, "A triangular adaptive lattice filter for spatial signal processing," Proc. ICASSP, Boston, pp. 348-351, 1983.

23. H. Lev-Ari, "Modular architectures for adaptive multichannel lattice algorithms," IEEE Trans. Acoust. Speech Signal Process., vol. ASSP-35, pp. 543-552 (Apr. 1987).

24. G. Cybenko, "A general orthogonalization technique with applications to time series analysis and signal processing," Math. Comput., vol. 40, pp. 323-336 (1983).

25. H. T. Kung, "Why systolic architecture?" IEEE Comput. Mag., vol. 15, pp. 37-46 (1982).

26. S. Kalson and K. Yao, "Systolic array processing for order and time recursive generalized least-squares estimation," Proc. SPIE, paper 564-05, vol. 564 (1985).

27. J. G. McWhirter, "Recursive least-squares minimization using a systolic array," Proc. SPIE, paper 431-15, vol. 431 (1983).

28. D. H. Youn and S. Prakash, "On realizations and related algorithm for adaptive linear phase filtering," Proc. ICASSP, San Diego, Calif., 3-11, 1984.

29. D. T. L. Lee, B. Friedlander, and M. Morf, "Recursive ladder algorithm for ARMA modeling," IEEE Trans. Autom. Control, vol. AC-27, pp. 753-764 (1982).

30. C. T. Mullis and R. A. Roberts, "The use of second-order information in the approximation of discrete-time linear systems," IEEE Trans. Acoust. Speech Signal Process., vol. ASSP-24, pp. 226-238 (June 1976).

31. D. Henrot and C. T. Mullis, "A modular and orthogonal digital filter structure for parallel processing," Proc. ICASSP, Boston, pp. 623-626, 1983.

32. Y. C. Lim and S. R. Parker, "On the synthesis of lattice parameter digital filters," IEEE Trans. Circuits Syst., vol. CAS-31, pp. 593-601 (1984).

33. Y. Miyanaga, N. Nagai, and N. Miki, "ARMA digital lattice filter with white Gaussian input," Trans. IECE (Jpn.), vol. J67-A, pp. 1270-1277 (1984).

34. S. K. Rao and T. Kailath, "Orthogonal digital filters for VLSI imple-
    mentation," IEEE Trans. Circuits Syst., vol. CAS-31, pp. 933-945
    (1984).

35. M. Suzuki, N. Nagai, and N. Miki, "Synthesis on digital lattice filters
    realizing stable rational transfer functions," Proc. ISCAS, Kyoto,
    pp. 499-502, 1985.

36. H. Sakai, H. Tokumaru, and A. Futamura, "ARMA modelling using
    circular lattice filter," Proc. IFAC 9th World Congress, Budapest,
    pp. 671-675, 1984.

37. J. F. Claerbout, Fundamentals of Geophysical Data Processing,
    McGraw-Hill, New York, 1976.

38. M. Morf, D. T. Lee, J. R. Nickolis, and A. Vieira, "A classification
    of algorithms for ARMA models and ladder realizations," Proc. ICASSP,
    Hartford, Conn., 1977.

39. L. C. Pusy, "Application of two-channel prediction filtering to the
    recursive filter design problem," IEEE Trans. Acoust. Speech Signal
    Process., vol. ASSP-31, pp. 1169-1177 (Oct. 1983).

9

Adaptive Identification of ARMA Model and Model Identification System

YOSHIKAZU MIYANAGA

Hokkaido University, Sapporo, Japan

9-1 INTRODUCTION

In the field of information processing, it is important to establish the characteristics of the signals observed. If the signals are deterministic, the characteristics are represented in the frequency domain by the Fourier transform [1, 2]. If the signals are stochastic, the characteristics are difficult to determine. The characteristics of stochastic signals are usually described as a stochastic reference model. To obtain the characteristics, a processing method estimates the parameters of the reference model. As the reference model, either an autoregressive (AR) model or an autoregressive-moving average (ARMA) model has been introduced [3, 4].

If the AR model is selected as a reference model, all resonances in the frequency domain can be represented [6, 9-11]. On the other hand, all resonances and antiresonances are described by using the ARMA model [5-8]. For example, if we try to analyze observed speech accurately, the speech production model chosen (i.e., a reference model) is generally an ARMA model [26-31], and thus ARMA analysis methods have been proposed in speech analysis. However, these methods are more complicated than AR analysis methods. Because of their low complexity, AR analysis methods are widely applied, whereas a speech production model is simple.

A stochastic model that changes its parameters over time is called a time-varying model [20, 21]. Identification methods of the time-varying model are becoming very important in many fields [22, 23, 29-31]. A method that estimates time-varying parameters is called an adaptive method [12-19]. Since the adaptive method can deal iteratively with new data each time, it fits a real-time estimation.

In this chapter we first discuss some difficulties of the adaptive method, which, optimally, estimates ARMA parameters only from observed signals. To overcome the difficulties, we introduce the basic theory which is applied to simultaneous estimation of model inputs and parameters. A new algorithm based on the theory is also proposed. We call this method the model identification system (MIS) [31]. From an analysis of results for real speech and synthesized waveforms, it is shown that the formants and antiformants are stably estimated by MIS.

9-2 STATEMENT OF PROBLEMS FOR ARMA PARAMETER ESTIMATION

In this section we show some difficulties in estimating ARMA parameters. The reference model is defined as

$$y(k) = -\sum_{i=1}^{n} a_i y(k-i) + u(k) + \sum_{j=1}^{m} b_j u(k-j) + v(k) \qquad (9-1)$$

where $y(k-i)$ $(i = 0, 1, \ldots, n)$ and $u(k-j)$ $(j = 0, 1, \ldots, m)$ are output signals and input signals, respectively. The signal $v(k)$ is a disturbance noise which is a zero-mean white Gaussian process with variance σ_v^2. The values a_i $(i = 1, \ldots, n)$ are AR parameters, and b_j $(j = 1, 2, \ldots, m)$ are MA parameters. The model in (9-1) is called an ARMA model.

If both the input signals and the output signals are observed and $\sigma_v^2 = 0$, then from (9-1) we get

$$
\begin{bmatrix}
y(k) & \cdots & y(k-n) & -u(k) & \cdots & -u(k-m) & 1 & 0 \\
y(k+1) & \cdots & y(k-n+1) & -u(k+1) & \cdots & -u(k-m+1) & \hat{a}_1^0(k) & 0 \\
y(k+2) & \cdots & y(k-n+2) & -u(k+2) & \cdots & -u(k-m+2) & \hat{a}_2^0(k) & 0 \\
\vdots & & \vdots & \vdots & & \vdots & \vdots & \vdots \\
y(k+n) & \cdots & y(k) & -u(k+n) & \cdots & -u(k-m+n) & \hat{a}_n^0(k) & = & 0 \\
y(k+n+1) & \cdots & y(k+1) & -u(k+n+1) & \cdots & -u(k+n-m+1) & \hat{b}_0^0(k) & 0 \\
\vdots & & \vdots & \vdots & & \vdots & \vdots & \vdots \\
y(k+n+m+1) & \cdots & y(k+m+1) & -u(k+n+m+1) & \cdots & -u(k+n+1) & \hat{b}_m^0(k) & 0
\end{bmatrix}
$$

$$(9\text{-}2)$$

where $k = \max(n, m) + 1$. The ARMA parameters can be estimated by solving (9-2). The values $\hat{a}_i^0(k)$ ($i = 1, \ldots, n$) and $\hat{b}_j^0(k)$ ($j = 0, 1, \ldots, m$) are the estimated AR and the estimated MA parameters, respectively. This method directly calculates the parameters from the observed waveform. But when the noise disturbs the waveform (i.e., $\sigma_v^2 \neq 0$), the right-hand side in (9-2) is not equal to zero vector. When $\sigma_v^2 \neq 0$, the following equation is derived from the reference model in (9-1):

$$
[1 \ \hat{a}_1(k) \ \cdots \ \hat{a}_n(k) \ \hat{b}_0(k) \ \cdots \ \hat{b}_m(k)]
\begin{bmatrix}
R_{-n+1}^y(k) & -R_{-n+1,\,m+1}^{yu}(k) \\
-(R_{-n+1,\,m+1}^{yu}(k))^T & R_{-m+1}^u(k)
\end{bmatrix}
$$

$$
= [r_0^{yv}(k) \ \cdots \ r_{-n}^{yv}(k) \ -r_0^{uv}(k) \ \cdots \ -r_{-m}^{uv}(k)] \tag{9-3}
$$

In (9-3), each undefined value is given as

$$
r_i^y(k) = \frac{1}{k} \sum_{s=1}^{k} y(s) y(s-i), \qquad r_i^u(k) = \frac{1}{k} \sum_{s=1}^{k} u(s) u(s-i),
$$

$$
r_i^{yu}(k) = \frac{1}{k} \sum_{s=1}^{k} y(s) u(s-i), \qquad r_i^{yv}(k) = \frac{1}{k} \sum_{s=1}^{k} y(s) v(s-i), \tag{9-4}
$$

$$
r_i^{uv}(k) = \frac{1}{k} \sum_{s=1}^{k} u(s) v(s-i)
$$

$$\underline{R}^y_{n+1} = \begin{bmatrix} r^y_0(k) & r^y_1(k) & \cdots & r^y_n(k) \\ r^y_{-1}(k) & r^y_0(k) & \cdots & r^y_{n-1}(k) \\ \vdots & \vdots & & \vdots \\ r^y_{-n}(k) & r^y_{1-n}(k) & \cdots & r^y_0(k) \end{bmatrix}$$

$$\underline{R}^{yu}_{n+1,m+1} = \begin{bmatrix} r^{yu}_0(k) & r^{yu}_1(k) & \cdots & r^{yu}_m(k) \\ r^{yu}_{-1}(k) & r^{yu}_0(k) & \cdots & r^{yu}_{m-1}(k) \\ \vdots & \vdots & & \vdots \\ r^{yu}_{-n}(k) & r^{yu}_{1-n}(k) & \cdots & r^{yu}_{m-n}(k) \end{bmatrix} \qquad (9\text{-}5)$$

$$\underline{R}^u_{m+1} = \begin{bmatrix} r^u_0(k) & r^u_1(k) & \cdots & r^u_m(k) \\ r^u_{-1}(k) & r^u_0(k) & \cdots & r^u_{m-1}(k) \\ \vdots & \vdots & & \vdots \\ r^u_{-m}(k) & r^u_{1-m}(k) & \cdots & r^u_0(k) \end{bmatrix}$$

where $y(s) = u(s) = v(s) = 0$ ($s \le 0$). A superscript T denotes transpose. In addition, the orders of matrices (i.e., $n + 1$ and $m + 1$) are represented as subscripts. When $k \to \infty$, each value in (9-4) can be replaced by

$$r^y_i(\infty) = E[y(k)y(k - i)], \qquad r^u_i(\infty) = E[u(k)u(k - i)]$$

$$r^{yu}_i(\infty) = E[y(k)u(k - i)], \qquad r^{uv}_i(\infty) = E[u(k)v(k - i)] \qquad (9\text{-}6)$$

$$r^{yv}_i(\infty) = E[y(k)v(k - i)]$$

An ergodic property is implicitly assumed in the calculation above. If the noise $v(k)$ is a zero-mean white Gaussian process with variance σ^2_v, we get

$$r^{uv}_i(\infty) = r^{yv}_i(\infty) = 0 \qquad (i \ne 0)$$

$$r^{uv}_0(\infty) = 0, \qquad r^{yv}_0(\infty) = \sigma^2_v \qquad (9\text{-}7)$$

Thus if a large number of data are observed in (9-3), and if (9-7) is approximately satisfied, then (9-3) is rewritten as

$$[1 \ \hat{a}_1(k) \ \cdots \ \hat{a}_n(k) \ \hat{b}_0(k) \ \cdots \ \hat{b}_m(k)] \begin{bmatrix} R^y_{n+1}(k) & -R^{yu}_{n+1,\,m+1}(k) \\ -(R^{yu}_{n+1,\,m+1}(k))^T & R^u_{m+1}(k) \end{bmatrix}$$

$$= [\hat{\sigma}^2_v(k) \ 0 \ \cdots \ 0] \qquad\qquad (9\text{-}8)$$

Thus from $r^y(k)$, $r^u(k)$, and $r^{yu}(k)$, the ARMA parameters without the influence of the disturbance noise can be estimated by using (9-8).

The ARMA parameters obtained by (9-2) are identical to the parameters of the reference model. But the estimated parameters in (9-8) are not exactly equal to those of the reference model since the approximation in (9-7) is used. Equation (9-8) means that the ARMA parameters which satisfy (9-8) minimize the following criterion:

$$V(k) = \frac{1}{k} \sum_{s=1}^{k} [y(s) - \hat{y}(s \mid k)]^2 \qquad\qquad (9\text{-}9)$$

where $\hat{y}(s \mid k)$ is defined as

$$\hat{y}(s \mid k) = - \sum_{i=1}^{n} \hat{a}_i(k)y(s - i) + \sum_{j=0}^{m} \hat{b}_j(k)u(s - j) \qquad\qquad (9\text{-}10)$$

The signal $\hat{y}(s \mid k)$ is an estimated output for $y(s)$ at time instant k. Equation (9-10) is an estimation model for (9-1). Thus the parameters calculated in (9-8) are regarded as the optimum parameters based on the least mean square estimation. Since $r_i(\infty) = E[r_i(k)]$, from (9-3) we get

$$E[\hat{a}_i(k)] = a_i \qquad (i = 1, \dots, n)$$
$$\qquad\qquad\qquad\qquad\qquad\qquad\qquad (9\text{-}11)$$
$$E[\hat{b}_i(k)] = b_j \qquad (j = 0, 1, \dots, m)$$

where $b_0 = 1$. The values of $E[(a_i - \hat{a}_i(k))^2]$ and $E[(b_j - \hat{b}_j(k))^2]$ depend on the value k in (9-9). These values decrease as time k increases. Equation (9-2) is the parameter estimation that uses the first information of deterministic signals. On the other hand, Eq. (9-8) estimates the parameters by using the second information of stochastic signals. The first information means a waveform (i.e., an impulse response). The second information means autocorrelation data.

Both input and output information is demanded whenever we apply either of them to signal processing. In signal processing, it is difficult to gain input information. Information about input signals should be evaluated from observed output signals. Thus in (9-8) it is necessary to estimate $r^{yu}(k)$ and $r^u(k)$ from output only. Instead of $u(k)$, we introduce a self-tuning input $x(k)$ which is calculated from the output signals. If the correlation data $r_s^{yx}(k)$ and $r_s^x(k)$ are defined, respectively, as

$$\frac{1}{k} \sum_{i=1}^{k} y(i)x(i-s) \quad \text{and} \quad \frac{1}{k} \sum_{i=1}^{k} x(i)x(i-s)$$

then (9-8) is rewritten as

$$[1 \ \hat{a}_1(k) \ \cdots \ \hat{a}_n(k) \ 1 \ \hat{b}_1(k) \ \cdots \ \hat{b}_m(k)] \begin{bmatrix} \underline{R}_{n+1}^y(k) & -\underline{R}_{n+1,m+1}^{yx}(k) \\ -(\underline{R}_{n+1,m+1}^{yx}(k))^T & \underline{R}_{m+1}^x(k) \end{bmatrix}$$

$$= [\hat{\sigma}_v^2(k) \ 0 \ \cdots \ 0] \tag{9-12}$$

where $\hat{b}_0(k) = 1$. Matrices $\underline{R}_{n+1,m+1}^{yx}(k)$ and $\underline{R}_{m+1}^x(k)$ are used in (9-12) instead of $\underline{R}_{n+1,m+1}^{yu}(k)$ and $\underline{R}_{m+1}^u(k)$. Equation (9-12) is obviously a nonlinear equation. It needs a large computational cost to estimate $\hat{a}_i(k)$ $(i = 1, \ldots, n)$, $\hat{b}_j(k)$ $(j = 1, \ldots, m)$, $r_s^{yx}(k)$ $(s = -n, \ldots, m)$, and $r_s^x(k)$ $(s = -m, \ldots, m)$ simultaneously. Generally speaking, if $u(k)$ is an arbitrary signal, it is impossible to solve (9-12).

Let us consider another difficulty with ARMA modeling, the order determination of an estimation model. The determination of its orders becomes a difficult problem when the orders of a reference model are unknown. In (9-10), let us change the orders of the estimation model on the assumption that the input is observed [i.e., $x(k) = u(k)$]. Since the following equations are derived from (9-1):

$$r_{-i}^y(\infty) + a_1 r_{1-i}^y(\infty) + \cdots + a_n r_{n-i}^y(\infty) = r_{-i}^{yu}(\infty) + b_1 r_{1-i}^{yu}(\infty) + \cdots + b_m r_{m-i}^{yu}(\infty)$$

$$\tag{9-13}$$

$$r_i^{yu}(\infty) + a_1 r_{i-1}^{yu}(\infty) + \cdots + b_m r_{i-n}^{yu}(\infty) = r_{-i}^u(\infty) + b_1 r_{1-i}^u(\infty) + \cdots + b_m r_{m-i}^u(\infty)$$

we get

$$\det \begin{bmatrix} \underline{R}_s^y(k) & -\underline{R}_{s,t}^{yu}(k) \\ -(\underline{R}_{s,t}^{yu}(k))^T & \underline{R}_t^u(k) \end{bmatrix} \neq 0 \quad (s < n+1 \text{ or } t < m+1) \qquad (9\text{-}14)$$

where $k \to \infty$ and the variables s and t denote the AR order and MA order of the estimation model, respectively. In addition, when the orders of the estimation model are $s \geq n+1$ and $t \geq m+1$, we get

$$\lim_{k \to \infty} \det \underline{E}(k)^{-1} = 0 \qquad (9\text{-}15)$$

where $\underline{E}(k)$ denotes the correlation matrix given in (9-12). Because of (9-15), the stable parameters cannot be estimated if $s \geq n+1$ and $t \geq m+1$.

As far as a reference model can be stable, an AR analysis method estimates AR parameters stably. The order of the AR estimation model does not influence its stability. But as to ARMA parameter estimation, these parameters are not stably estimated unless (9-14) is satisfied. Thus the orders of the estimation model have to be carefully determined. In addition, the accuracy of the parameter estimation method, which estimates both inputs and parameters, is greatly influenced by the order of the estimation model because the self-tuning estimated inputs are also influenced. Thus this is a critical problem for optimal ARMA parameter estimates.

9-3 ADAPTIVE ESTIMATION OF TIME-VARYING ARMA PARAMETERS

We define the reference model represented by time-varying parameters. An ARMA model with time-constant parameters is represented in (9-1). The parameters a_i and b_j ($i = 1, 2, \ldots, n; j = 1, 2, \ldots, m$) of the reference model are independent of time variations. On the other hand, the time-varying ARMA model is represented as

$$y(k) = -\sum_{i=1}^{n} a_i(k)y(k-i) + u(k) + \sum_{j=1}^{m} b_j(k)u(k-j) \qquad (9\text{-}16)$$

where $y(k-i)$ ($i = 0, 1, \ldots, n$) are the output signals of this model and can be observed. The variables $u(k-j)$ ($j = 0, 1, \ldots, m$) are white Gaussian input and cannot be observed. The mean and the variance of $u(k)$ are

$$E[u(k)] = 0$$
$$E[u(k)u(i)] = \delta(k, i)\sigma_u^2(k) \qquad (9\text{-}17)$$

where

$$\delta(k, i) = \begin{cases} 1, & k = i \\ 0, & k \neq i \end{cases}$$

In (9-16) AR parameters are given as $a_i(k)$ $(i = 1, 2, \ldots, n)$, which changes as the time index k increases, and the MA parameters are given as $b_j(k)$ $(j = 1, 2, \ldots, m)$. In addition, $\sigma_u^2(k)$ is a time-varying value. To represent the parameters and data in the vector form, we define the following vectors:

$$\underline{p}^0(k)^T = [a_1(k) \ a_2(k) \ \cdots \ a_n(k) \ b_1(k) \ b_2(k) \ \cdots \ b_m(k)]$$

$$\underline{h}^0(k)^T = [-y(k-1) \ -y(k-2) \ \cdots \ -y(k-n) \ u(k-1) \ u(k-2) \ \cdots \ u(k-m)]$$

(9-18)

From (9-18) the time-varying ARMA model of (9-16) is rewritten as

$$y(k) = \underline{h}^0(k)^T \underline{p}^0(k) + u(k) \qquad (9\text{-}19)$$

Since the ARMA parameters are time-varying values, it is necessary to represent the time variations of the parameters. When the ARMA parameters form the vector $\underline{p}^0(k)$ defined above, the time variations can be defined as

$$\underline{p}^0(k+1) = \underline{p}^0(k) + \underline{\gamma}(k) \qquad (9\text{-}20)$$

where $\underline{\gamma}(k)$ is the time-varying factor. In parameter estimation, it is necessary to estimate not only $\underline{p}^0(k)$ but also $\underline{\gamma}(k)$. The values in $\underline{\gamma}(k)$ depend on the properties of observed signals. In our algorithm it is assumed that the parameters of the reference model do not always change at every instant. In other words, the parameters change once in a certain interval. For example, a speech production model varies when the speech sound changes. This variation happens once in a certain time interval. Thus the assumption for the reference model above fits the behavior of the speech production model. Using a mathematical model, $\underline{\gamma}(k)$ is described as

$$\underline{\gamma}(k) = \begin{cases} 0, & k \neq T_i \\ \underline{q}(k), & k = T_i \end{cases} \qquad (9\text{-}21)$$

where T_i is the time instant when the parameters are changed. Note that there is no time variation between T_{i-1} and T_i [i.e., $\underline{\gamma}(k) = 0$]. In addition, it is assumed in (9-21) that there is enough of a time interval between T_{i-1} and T_i. The elements of $\underline{\gamma}(k)$ are independent of each other and are also independent of $\underline{p}^0(i)$ $(i = 1, 2, \ldots, k)$.

Let us consider an estimation model and a prediction model for (9-16).
The estimation model is defined as

$$\hat{y}(k) = -\sum_{i=1}^{n} \hat{a}_i(k)y(k-i) + \sum_{j=1}^{m} \hat{b}_j(k)x(k-j) \qquad (9\text{-}22)$$

where $\hat{a}_i(k)$ $(i = 1, 2, \ldots, n)$ and $\hat{b}_j(k)$ $(j = 1, 2, \ldots, m)$ are AR parameters
and MA parameters calculated at time k, respectively. The estimation
model at time k can be evaluated on the set of y(i) where $i = 1, 2, \ldots, k$.
Since only output y(k) can be observed, it is necessary to estimate the input
signals. Thus the estimation model in (9-22) should employ self-tuning input
signals x(k - i) $(i = 0, 1, \ldots, m)$.
    The prediction model that predicts the signal y(k) at time k - 1 is
defined by

$$y(k \mid k-1) = -\sum_{i=1}^{n} \hat{a}_i(k \mid k-1)y(k-i) + \sum_{j=1}^{m} \hat{b}_j(k \mid k-1)x(k-j) \qquad (9\text{-}23)$$

Since x(k) is an input signal estimated at time k, (9-23) cannot use it. Now
introduce the following vectors:

$$\underline{h}(k)^T = [-y(k-1) \ \ -y(k-2) \ \cdots \ -y(k-n) \ \ x(k-1) \ \cdots \ x(k-m)]$$

$$\underline{\hat{p}}(k)^T = [\hat{a}_1(k) \ \hat{a}_2(k) \ \cdots \ \hat{a}_n(k) \ \hat{b}_1(k) \ \cdots \ \hat{b}_m(k)] \qquad (9\text{-}24)$$

$$\underline{\hat{p}}(k \mid k-1)^T = [\hat{a}_1(k \mid k-1) \ \cdots \ \hat{a}_n(k \mid k-1) \ \hat{b}_1(k \mid k-1) \ \cdots \ \hat{b}_m(k \mid k-1)]$$

From (9-24) the estimated model in (9-22) and the prediction model in
(9-23) are rewritten as

$$\hat{y}(k) = \underline{h}(k)^T \underline{\hat{p}}(k)$$
$$\hat{y}(k \mid k-1) = \underline{h}(k)^T \underline{\hat{p}}(k \mid k-1) \qquad (9\text{-}25)$$

    We introduce a canonical representation of $\hat{p}(k \mid k-1)$ by using inno-
vation series. If the input signals are white Gaussian, the waveform synthesized by the input [i.e., y(k) in (9-16)] is also a Gaussian process. By
using the innovation sequence $\nu(i)$ $(i = 1, \ldots, k)$, $\hat{y}(k)$ can be represented as

$$\hat{y}(k) = \sum_{i=1}^{k} \xi(k, i)\nu(i) \qquad (9\text{-}26)$$

The innovation sequence $\nu(k)$ is given by

$$y(k) - E[y(k) \mid g(k - 1)] = \nu(k) \tag{9-27}$$

where $g(k - 1)$ is a space spanned by the observed data of $[y(i), \; i = 1, \ldots, k - 1]$. The mean operation $E[\; \mid g(k - 1)]$ denotes a conditional mean on $g(k - 1)$. Thus $\nu(k)$ represents a new random variable at time k. It is orthogonal to $g(k - 1)$.

Assume that the estimated parameter vector $\hat{\underline{p}}(k)$ at the time k is represented by the linear combination of the innovation sequence:

$$\hat{\underline{p}}(k) = \sum_{i=1}^{k} \underline{\gamma}(k, i) \nu(i) \tag{9-28}$$

Equation (9-28) is called a canonical representation for $\hat{\underline{p}}(k)$. The vector $\underline{\gamma}(k, i)$ is a $(n + m)$th vector.

From the property of (9-27) we get

$$E[\nu(k) \nu(i) \mid g(k)] = \sum_{s=1}^{k} \nu(s) \nu(s - k + j) = 0 \quad (j < k) \tag{9-29}$$

The optimum parameter vector $\hat{\underline{p}}(k)$ is defined as

$$E[(\underline{p}(k) - \hat{\underline{p}}(k)) \nu(i) \mid g(k)] = \underline{0} \quad (i \le k) \tag{9-30}$$

where the vector $\underline{p}(k)$ is a pseudoreference parameter vector. The pseudoreference parameter vector is defined as

$$y(k) = \underline{h}(k)^T \underline{p}(k) + x(k) \tag{9-31}$$

In addition, we also define the matrix

$$\underline{F}(k) = \sigma_x(k)^2 E[(\underline{p}(k) - \hat{\underline{p}}(k)) (\underline{p}(k) - \hat{\underline{p}}(k))^T \mid g(k)] \tag{9-32}$$

$$\underline{F}(k \mid k - 1) = \sigma_x(k - 1)^2 E[(\underline{p}(k) - \hat{\underline{p}}(k \mid k - 1)) (\underline{p}(k) - \hat{\underline{p}}(k \mid k - 1))^T \mid g(k - 1)]$$

The input of the pseudoreference model is assumed to be $x(k)$. Its mean and variance are

$$E[x(k)] = 0, \qquad E[x(k)x(t)] = \delta(k, t) \sigma_x(k)^2 \tag{9-33}$$

An optimum estimation for $y(k)$ is derived from (9-28) to (9-33)

$$\hat{\underline{p}}(k) = \hat{\underline{p}}(k \mid k-1) + \underline{F}(k \mid k-1)\underline{h}(k)[\lambda(k-1) + \underline{h}(k)^T\underline{F}(k \mid k-1)\underline{h}(k)]^{-1}\nu(k) \qquad (9-34)$$

$$\nu(k) = y(k) - \underline{h}^T(k)\hat{\underline{p}}(k \mid k-1) \qquad (9-35)$$

$$\underline{F}(k) = \lambda(k-1)^{-1}[\underline{F}(k \mid k-1)$$

$$- \underline{F}(k \mid k-1)\underline{h}(k)(\lambda(k-1) + \underline{h}(k)^T\underline{F}(k \mid k-1)\underline{h}(k))^{-1}\underline{h}(k)^T\underline{F}(k \mid k-1)]$$

$$\lambda(k-1) = \frac{\sigma_x(k)^2}{\sigma_x(k-1)^2} \qquad (9-36)$$

The estimated input signals are used in the vector $\underline{h}(k)$ in (9-33) and (9-35). From (9-32) it turns out that not $\underline{p}^0(k)$ in (9-19) but $\underline{p}(k)$ in (9-31) is calculated by the optimum estimation in (9-34) to (9-36).

Now we define all notations used in the proposed method. By using these notations, the adaptive algorithm for nonstationary stochastic signals will be derived. From (9-19) and (9-20) the reference model can be represented as

$$\underline{p}^0(k+1) = \underline{p}^0(k) + \gamma(k)$$

$$y(k) = \underline{h}^0(k)^T\underline{p}^0(k) + u(k) \qquad (9-37)$$

An optimal method to obtain $\underline{p}^0(k)$ is not derived from the optimal parameter estimation in (9-34) to (9-36) since the estimator of (9-33) uses $\underline{h}(k)$, which includes the self-tuning input signals. Thus the model to be estimated in (9-34) to (9-36) (i.e., the pseudoreference model) should be

$$\underline{p}(k+1) = \underline{p}(k) + \gamma(k)$$

$$y(k) = \underline{h}(k)^T\underline{p}(k) + x(k) \qquad (9-38)$$

where

$$\gamma(k) = \begin{cases} \underline{0}, & k \neq T_i \\ \underline{q}(k), & k = T_i \end{cases} \qquad (9-39)$$

We can only develop the optimal estimation of time-varying parameters for (9-38). But the vector which must be estimated is $\underline{p}^0(k)$ in (9-37). When the estimated inputs are used in $\underline{h}(k)$, its stochastic properties directly influence the parameter estimation. However, even if the self-tuning inputs are used, it can be shown that the optimal estimate for $\underline{p}^0(k)$ can be obtained.

The relationships among $\underline{p}^0(k)$, $\underline{p}(k)$, and $\hat{\underline{p}}(k)$ are discussed in the next section.

Now let us derive the relation between $\hat{\underline{p}}(k \mid k - 1)$ and $\hat{\underline{p}}(k - 1)$, and between $\underline{F}(k \mid k - 1)$ and $\underline{F}(k - 1)$. The parameter vectors $\hat{\underline{p}}(k \mid k - 1)$ and $\hat{\underline{p}}(k - 1)$ are predicted and estimated at time $k - 1$, respectively. Since $\underline{\gamma}$ is given in (9-21) and independent of signals observed until $k - 1$, the relation between $\hat{\underline{p}}(k \mid k - 1)$ and $\hat{\underline{p}}(k - 1)$ is

$$\hat{\underline{p}}(k \mid k - 1) = \hat{\underline{p}}(k - 1) \tag{9-40}$$

in other words, since $\underline{\gamma}(k)$ is independent of $y(i)$ ($i = 1, \ldots, k - 1$). The vector $\hat{\underline{p}}(k \mid k - 1)$ does not include the information of $\underline{\gamma}(k)$. From (9-40) it turns out that the optimal predicted parameter vector $\hat{\underline{p}}(k \mid k - 1)$ for $\underline{p}(k)$ is equal to $\underline{p}(k - 1)$.

Next let us derive the relation between $\underline{F}(k \mid k - 1)$ and $\underline{F}(k - 1)$:

$$\underline{F}(k \mid k - 1) = \sigma_x(k - 1)^2 E[(\underline{p}(k) - \hat{\underline{p}}(k \mid k - 1))\underline{p}(k) - \hat{\underline{p}}(k \mid k - 1))^T \mid g(k - 1)]$$

$$= \sigma_x(k - 1)^2 E[(\underline{p}(k - 1) + \underline{\gamma}(k - 1) - \hat{\underline{p}}(k \mid k - 1)) \tag{9-41}$$

$$\cdot (\underline{p}(k - 1) + \underline{\gamma}(k - 1) - \hat{\underline{p}}(k \mid k - 1))^T \mid g(k - 1)]$$

Since $\underline{\gamma}(k - 1)$ is independent of $\underline{p}(k - 1) - \hat{\underline{p}}(k - 1)$, we get

$$\underline{F}(k \mid k - 1) = \sigma_x(k)^{-2}[E[(\underline{p}(k - 1) - \hat{\underline{p}}(k - 1))(\underline{p}(k - 1)$$

$$- \hat{\underline{p}}(k - 1))^T \mid g(k)] + E[(\underline{\gamma}(k - 1)\underline{\gamma}(k - 1)^T) \mid g(k)] \tag{9-42}$$

$$= \underline{F}(k - 1) + \sigma_x(k - 1)^{-2}\delta(k - 1, T_i) E[\underline{q}(k - 1)\underline{q}(k - 1)^T \mid g(k)]$$

$$= \underline{F}(k - 1) + \sigma_x(k - 1)^{-2}\underline{Q}(k - 1)$$

where $\underline{Q}(k - 1) = \delta(k - 1, T_i) E[\underline{q}(k - 1)\underline{q}(k - 1)^T \mid g(k)]$. If the parameters are not varied, $\underline{Q}(k - 1) = \underline{0}$.

From the relation above, the estimation of time-varying ARMA parameters is obtained as follows:

$$\hat{\underline{p}}(k \mid k - 1) = \hat{\underline{p}}(k - 1)$$

$$\underline{F}(k \mid k - 1) = \underline{F}(k - 1) + \hat{\sigma}_x(k - 1)^{-2}\hat{\underline{Q}}(k - 1) \tag{9-43}$$

where the estimates of $\sigma_x(k - 1)^2$ and $\underline{Q}(k - 1)$ are used since these values cannot also be measured.

Let us evaluate $\hat{\sigma}_x(k - 1)^2$ and $\hat{Q}(k - 1)$. These can be estimated by using the likelihood function of $\nu(k)$. The likelihood function is given by

$$L[\nu(i) \mid \hat{Q}(i - 1), \hat{\sigma}_x(i)^2, \quad i = 1, 2, \ldots, k]$$

$$= -\frac{1}{2} \sum_{i=1}^{k} \log \sigma_\nu^2(i) - \frac{1}{2} \sum_{i=1}^{k} \frac{\nu(i)^2}{\sigma_\nu(i)^2} + L_0 \qquad (9\text{-}44)$$

where $\nu(i) = y(i) - \underline{h}(i)^T \hat{\underline{p}}(i \mid i - 1)$, L_0 is constant and $\sigma_\nu(i)^2 = E[\nu(i)^2 \mid g(k)]$ $(i = 1, 2, \ldots, k)$. The best likelihood values of $\hat{\sigma}_x(k)^2$ and $\hat{Q}(k - 1)$ maximize $(9\text{-}44)$. From $(9\text{-}38)$, $(9\text{-}32)$, and $(9\text{-}36)$, $\sigma_\nu^2(i)$ is written as

$$\sigma_\nu(i)^2 = E[\nu(i)^2 \mid g(i)] = E[(y(i) - \underline{h}(i)^T \hat{\underline{p}}(i \mid i - 1))^2 \mid g(i)]$$

$$= E[(\underline{h}(i)^T(\underline{p}(i) - \hat{\underline{p}}(i \mid i - 1)) + x(i))^2 \mid g(i)]$$

$$= \underline{h}(i)^T E[(\underline{p}(i) - \hat{\underline{p}}(i \mid i - 1))(\underline{p}(i)$$

$$\qquad - \hat{\underline{p}}(i \mid i - 1))^T \mid g(i)]\underline{h}(i) + E[x(i)x(i) \mid g(i)] \qquad (9\text{-}45)$$

$$= \hat{\sigma}_x(i - 1)^2 \underline{h}(i)^T \underline{F}(i \mid i - 1)\underline{h}(i) + \hat{\sigma}_x(i)^2$$

$$= \hat{\sigma}_x(i)^2 (\lambda(i - 1)^{-1} \underline{h}(i)^T \underline{F}(i \mid i - 1)\underline{h}(i) + 1)$$

Thus substituting $(9\text{-}45)$ into $(9\text{-}44)$, the likelihood function is rewritten as

$$L[\nu(i); i = 1, 2, \ldots, k \mid \hat{Q}(i - 1), \hat{\sigma}_x(i)^2; i = 1, 2, \ldots, k]$$

$$= -\frac{k}{2} \log \hat{\sigma}_x(k)^2 - \frac{1}{2} \sum_{i=1}^{k} \log(1 + \underline{h}(i)^T \underline{F}(i \mid i - 1)\underline{h}(i))$$

$$- \frac{1}{2\hat{\sigma}_x(k)^2} \sum_{i=1}^{k} \frac{\nu(i)^2}{1 + \underline{h}(i)^T \underline{F}(i \mid i - 1)\underline{h}(i)} + L_0 \qquad (9\text{-}46)$$

where the variance $\hat{\sigma}_x(i)^2$ of $x(i)$ is assumed to be constant in the certain limits (i.e., $i = 1, 2, \ldots, k$). Thus $\lambda(i - 1) = 1$ in $(9\text{-}46)$.

Since $\hat{\sigma}_x(k)^2$ is independent of $\nu(i)^2$ and $(1 + \underline{h}(i)^T \underline{F}(i \mid i - 1)\underline{h}(i))$ $(i = 1, 2, \ldots, k)$, $\hat{\sigma}_x(k)^2$, which maximizes the likelihood function, can be calculated as

$$\frac{d}{d\hat{\sigma}_x(k)^2} L[\nu(i); i = 1, 2, \ldots, k \mid \hat{Q}(i-1), \hat{\sigma}_x(i)^2; i = 1, 2, \ldots, k]$$

$$= -\frac{k}{2\hat{\sigma}_x(k)^2} + \frac{1}{2\hat{\sigma}_x(k)^4} \sum_{i=1}^{k} \frac{\nu(i)^2}{1 + \underline{h}(i)^T \underline{F}(i \mid i - 1)\underline{h}(i)}$$

$$= 0 \qquad\qquad\qquad\qquad\qquad\qquad\qquad\qquad\qquad\qquad (9\text{--}47)$$

Thus we get

$$\hat{\sigma}_x(k)^2 = \frac{1}{k} \sum_{i=1}^{k} \frac{\nu(i)^2}{1 + \underline{h}(i)^T \underline{F}(i \mid i - 1)\underline{h}(i)} \qquad\qquad (9\text{--}48)$$

Equation (9-48) can be represented as

$$\hat{\sigma}_x(k)^2 = \hat{\sigma}_x(k-1)^2 + \frac{1}{k}\left[\frac{\nu(k)^2}{1 + \underline{h}(k)^T \underline{F}(k \mid k - 1)\underline{h}(k)} - \hat{\sigma}_x(k-1)^2 \right] \quad (9\text{--}49)$$

Note that (9-46) applies only if every $\hat{\sigma}_x(k)^2$ ($i = 1, 2, \ldots, k$) is the same value. But if $\hat{\sigma}_x(k)^2$ is not constant beyond a certain limit ($i = k - \beta + 1$, $\ldots$, k), we should consider that $\hat{\sigma}_x(k)^2$ changes progressively. In this assumption we use the following equation instead of (9-49):

$$\hat{\sigma}_x(k)^2 = \hat{\sigma}_x(k-1)^2 + \frac{1}{\beta(k)}\left[\frac{\nu(k)^2}{1 + \underline{h}(k)^T \underline{F}(k \mid k - 1)\underline{h}(k)} - \hat{\sigma}_x(k-1)^2 \right],$$

$$(9\text{--}50)$$

Using (9-49), the variance of input white Gaussian process can be evaluated. Since $\hat{\sigma}_x(k)^2$ is estimated at time k, $\hat{\sigma}_x(k-1)^2$ used in (9-43) has already been obtained at time k.

Next let us calculate $\hat{Q}(k-1)$. To evaluate the time variation of the reference model, we use the momentary likelihood estimate calculated only from the innovation at time k. This estimate is not very appropriate because only an observed signal is used for the likelihood function despite a stochastic process. But if we use the mean value in the certain limit to evaluate the likelihood function, it is difficult to estimate accurately the time-varying value since $\hat{Q}(k)$ is independent of $\gamma(i)$ ($i < k$) and $\underline{p}^0(i)$ ($i = 1, 2, \ldots, k$). Thus we evaluate $\hat{Q}(k)$ from the likelihood function at time k. In addition, the time-varying location is detected from the change of the power of the prediction error. Using such kinds of processing techniques, we can evaluate the values and the locations of parameter variations at each instant. Substituting (9-43) and

$$\lambda(k - 1) = \frac{\hat{\sigma}_x(k)^2}{\hat{\sigma}_x(k - 1)^2} \tag{9-51}$$

into (9-45), we get

$$\sigma_\nu(i)^2 = \hat{\sigma}_x(k - 1)^2 (\lambda(k - 1) + \underline{h}(k)^T \underline{F}(k - 1)\underline{h}(k)) + \underline{h}(k)^T \underline{Q}(k - 1)\underline{h}(k) \tag{9-52}$$

If the diagonal elements of $\underline{Q}(k - 1)$ are equal to each other and the others are almost zero, we get approximately

$$\hat{\underline{Q}}(k - 1) = \hat{q}(k - 1)^2 \underline{I} \tag{9-53}$$

Generally speaking, there are many cases that do not satisfy (9-53). But since the estimation model can be constructed accurately by (9-34), (9-53) is applied only to the time-varying part. The momentary likelihood function is given by

$$L[\nu(k) \mid \hat{\underline{Q}}(k - 1)]$$

$$= -\frac{1}{2}\log \hat{\sigma}_x(k - 1)^2 - \frac{1}{2}\log[(\lambda(k - 1) + \underline{h}(k)^T \underline{F}(k - 1)\underline{h}(k))\hat{\sigma}_x(k - 1)^2 + \hat{q}(k - 1)^2 \underline{h}(k)^T \underline{h}(k)]$$

$$-\frac{\nu(k)^2}{2\hat{\sigma}_x(k - 1)^2(\lambda(k - 1) + \underline{h}(k)^T \underline{F}(k - 1)\underline{h}(k)) + 2\hat{q}(k - 1)^2 \underline{h}(k)^T \underline{h}(k)} \tag{9-54}$$

If the differentiation of (9-54) is set to be zero, we obtain $\hat{\underline{Q}}(k - 1)$, which maximizes (9-54). Thus we get

$$\frac{dL[\nu(k) \mid \hat{q}(k - 1)^2, \hat{\sigma}_x(k - 1)^2]}{d\hat{q}(k - 1)^2}$$

$$= \frac{\underline{h}(k)^T \underline{h}(k)}{2[\lambda(k - 1) + \underline{h}(k)^T \underline{F}(k - 1)\underline{h}(k)]\hat{\sigma}_x(k - 1)^2 + \hat{q}(k - 1)^2 \underline{h}(k)^T \underline{h}(k)}$$

$$+ \frac{\underline{h}(k)^T \underline{h}(k)\nu(k)^2}{[2\hat{\sigma}_x(k - 1)^2[\lambda(k - 1) + \underline{h}(k)^T \underline{F}(k - 1)\underline{h}(k)] + 2\hat{q}(k - 1)^2 \underline{h}(k)^T \underline{h}(k)]^2}$$

$$= 0 \tag{9-55}$$

From (9-52) we get

$$\hat{q}(k-1)^2 = \frac{\nu(k)^2 - \sigma_x(k-1)^2[\lambda(k-1) + \underline{h}(k)^T \underline{F}(k-1)\underline{h}(k)]}{\underline{h}(k)^T \underline{h}(k)} \tag{9-56}$$

Thus $\hat{Q}(k-1)$ is calculated from (9-53) and (9-56).

The location where (9-53) is applied is the time when the parameters are varied. Thus we need to examine whether the parameters are varied. Since the prediction error becomes large whenever the parameters are varied, let us use the following equation:

$$\nu(k)^2 > C\sigma_\nu(k)^2 \tag{9-57}$$

where we assume that $\hat{q}(k-1)^2 = 0$ during the calculation of $\sigma_\nu(k)^2$. If (9-57) is satisfied, (9-53) and (9-56) are used in MIS. Otherwise, $\hat{q}(k-1)^2$ must be zero. From (9-49), (9-54) is rewritten as

$$\nu(k)^2 > C\hat{\sigma}_x(k-1)^2[\lambda(k-1) + \underline{h}(k)^T \underline{F}(k-1)\underline{h}(k)] \tag{9-58}$$

where C is constant. If $\nu(k)$ is a white Gaussian process, the following criterion holds at the probability 0.999 when the parameters are not varied:

$$\nu(k)^2 < 9\hat{\sigma}_x(k-1)^2[\lambda(k-1) + \underline{h}(k)^T \underline{F}(k-1)\underline{h}(k)] \tag{9-59}$$

Thus if (9-57) holds where $C \geq 9$, the parameters of the reference model must be varied at that time.

From the equations above, the estimation method of time variation for a time-varying reference model is given by

$$\nu(k) = y(k) - \underline{h}(k)^T \hat{\underline{p}}(k \mid k-1):$$

(a) $\nu(k)^2 > C\hat{\sigma}_x(k-1)^2(\lambda(k-1) + \underline{h}(k)^T \underline{F}(k-1)\underline{h}(k))$

$$\hat{q}(k-1)^2 = \frac{\nu(k)^2 - \hat{\sigma}_x(k-1)^2[\lambda(k-1) + \underline{h}(k)^T \underline{F}(k-1)\underline{h}(k)]}{\underline{h}(k)^T \underline{h}(k)}$$

(b) $\nu(k)^2 \leq C\hat{\sigma}_x(k-1)^2[\lambda(k-1) + \underline{h}(k)^T \underline{F}(k-1)\underline{h}(k)]$ \hspace{1cm} (9-60)

$$\hat{q}(k)^2 = 0$$

$$\hat{\underline{Q}}(k - 1) = \hat{q}(k - 1)^2$$

$$\hat{\sigma}_x(k)^2 = \hat{\sigma}_x(k - 1)^2 + \frac{1}{\beta(k)} \left[\frac{\nu(k)^2}{1 + \underline{h}(k)^T \underline{F}(k \mid k - 1)\underline{h}(k)} - \hat{\sigma}_x(k - 1)^2 \right]$$

The block diagram of this method is shown in Fig. 9-1.

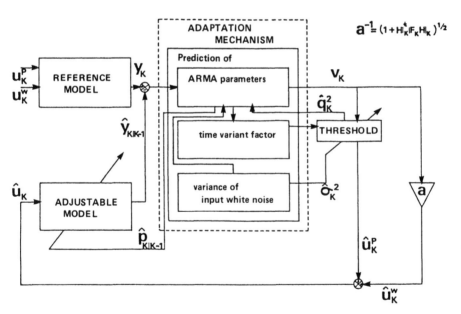

FIGURE 9-1 Block diagram of time-varying ARMA parameter estimation. Reference model, Eq. (9-16). Adjustable model, Eq. (9-22). Adaptation mechanism: prediction of ARMA parameters, Eq. (9-34); time-variant factor, Eq. (9-56); variance of input white noise, Eq. (9-50). Threshold, Eq. (9-58).

9-4 SELF-TUNING INPUT SYNTHESIZER

In (9-34) to (9-36) we have already employed a self-tuning input [i.e., $x(k)$].
In this section we show how to calculate the self-tuning input and also show
that the optimal estimation can be derived even by using the self-tuning input. Assume that the following input is a self-tuning input:

$$x(k) = y(k) - \underline{h}(k)^T \hat{\underline{p}}(k)$$

$$= \frac{\nu(k)}{1 + \underline{h}(k)^T \underline{F}(k \mid k - 1)\underline{h}(k)\lambda(k - 1)^{-1}} \tag{9-61}$$

Note that the self-tuning input is used only where the input of the reference
model is a zero-mean white Gaussian process. If we employ $x(k)$ in (9-61)
to the identification method of (9-34) to (9-36), it is possible to prove that
the estimated parameter $\hat{\underline{p}}(k)$ is optimal for $\underline{p}^0(k)$.

Let us assume an estimation model to be

$$\hat{y}(k) = -\sum_{i=1}^{n+t} \hat{a}_i(k)y(k - i) + \sum_{j=1}^{m+t} \hat{b}_j(k)x(k - j) \tag{9-62}$$

where each parameter is estimated on $g(\infty)$. They can be rewritten as

$$x(k - j) = y(k - j) - \hat{y}(k - j)$$

$$= \frac{\nu(k - j)}{1 + \underline{h}(k - j)^T \underline{F}(k - j \mid k - j - 1)\underline{h}(k - j)}$$

where $j = 0, 1, \ldots, m + t$, since we get

$$\hat{y}(k - j) = \underline{h}(k - j)^T \hat{\underline{p}}(k - j \mid k - j - 1) + \underline{h}(k - j)^T \underline{F}(k - j \mid k - j - 1)\underline{h}(k)$$

$$\bullet [1 + \underline{h}(k - j)^T \underline{F}(k - j \mid k - j - 1)\underline{h}(k - j)]^{-1} \nu(k - j)$$

$$= y(k) - \nu(k) + \underline{h}(k - j)^T \underline{F}(k - j \mid k - j - 1)\underline{h}(k - j) \tag{9-63}$$

$$\bullet [1 + \underline{h}(k - j)^T \underline{F}(k - j \mid k - j - 1)\underline{h}(k - j)]^{-1} \nu(k - j)$$

$$= y(k) - \frac{\nu(k)}{1 + \underline{h}(k - j)^T \underline{F}(k - j \mid k - j - 1)\underline{h}(k - j)}$$

On the other hand, the model estimated with a true input sequence is

$$\hat{y}^0(k) = -\sum_{i=1}^{n} \hat{a}_i^0(k)y(k-i) + \sum_{j=1}^{m} \hat{b}_j^0(k)u(k-j) \qquad (9\text{-}64)$$

The ARMA parameters that minimize $E[(y(k) - \hat{y}^0(k))^2 \mid g(\infty)]$ are optimum. Now, let us define

$$A(q^{-1}) = 1 + \sum_{i=1}^{n} a_i(k)q^{-1}, \qquad B(q^{-1}) = 1 + \sum_{j=1}^{m} b_j(k)q^{-j} \qquad (9\text{-}65)$$

$$\hat{A}(q^{-1}) = 1 + \sum_{i=1}^{n+t} \hat{a}_i(k)q^{-1}, \qquad \hat{B}(q^{-1}) = 1 + \sum_{j=1}^{m+t} \hat{b}_j(k)q^{-j} \qquad (9\text{-}66)$$

$$\hat{A}^0(q^{-1}) = 1 + \sum_{i=1}^{n} \hat{a}_i^0(k)q^{-1}, \qquad \hat{B}^0(q^{-1}) = 1 + \sum_{j=1}^{m} \hat{b}_j^0(k)q^{-j} \qquad (9\text{-}67)$$

where q^{-1} is shift operator. From (9-65) to (9-67) the reference models of (9-1), (9-62), and (9-64) are rewritten as

$$y(k) = \frac{B(q^{-1})}{A(q^{-1})}u(k) \qquad (9\text{-}68)$$

$$\hat{y}(k) = -\hat{A}(q^{-1})y(k) + y(k) + \hat{B}(q^{-1})x(k) - x(k) \qquad (9\text{-}69)$$

$$\hat{y}^0(k) = -\hat{A}^0(q^{-1})y(k) + y(k) + \hat{B}^0(q^{-1})u(k) - u(k) \qquad (9\text{-}70)$$

Thus from (9-62), (9-68), and (9-64), the estimation error is

$$y(k) - \hat{y}(k) = u(k) - \frac{\hat{B}(q^{-1})A(q^{-1}) - B(q^{-1})\hat{A}(q^{-1})}{A(q^{-1})\hat{B}(q^{-1})}u(k) \qquad (9\text{-}71)$$

From (9-64), (9-68), and (9-70), the other estimation error is

$$y(k) - \hat{y}^0(k) = u(k) - \frac{\hat{B}^0(q^{-1})A(q^{-1}) - \hat{A}^0(q^{-1})B(q^{-1})}{A(q^{-1})}u(k) \qquad (9\text{-}72)$$

From (9-71), $\hat{A}(q^{-1})$ and $\hat{B}(q^{-1})$, which minimize $E[(y(k) - \hat{y}(k))^2 \mid g(\infty)]$, are

$$\hat{B}(q^{-1})A(q^{-1}) = \hat{A}(q^{-1})B(q^{-1}) \qquad (9\text{-}73)$$

From (9-72), $\hat{A}^0(q^{-1})$ and $\hat{B}^0(q^{-1})$, which minimize $E[(y(k) - \hat{y}^0(k))^2 \mid g(\infty)]$, are

$$\hat{B}^0(q^{-1})A(q^{-1}) = \hat{A}^0(q^{-1})B(q^{-1}) \qquad (9\text{-}74)$$

where each parameter in $\hat{A}(q^{-1})$, $\hat{B}(q^{-1})$, $\hat{A}^0(q^{-1})$, and $\hat{B}^0(q^{-1})$ is estimated on $g(\infty)$. From (9-73) and (9-74), we get

$$\frac{\hat{B}(q^{-1})}{\hat{A}(q^{-1})} = \frac{\hat{B}^0(q^{-1})}{\hat{A}^0(q^{-1})} \tag{9-75}$$

Thus the ARMA parameters estimated by (9-34) to (9-36) and (9-60) are also optimum on $g(\infty)$ because $\hat{A}^0(q^{-1})$ and $\hat{B}^0(q^{-1})$ are optimal. In addition, from (9-75) we get

$$\hat{A}(q^{-1}) = \hat{A}^0(q^{-1})\hat{C}(q^{-1})$$
$$\hat{B}(q^{-1}) = \hat{B}^0(q^{-1})\hat{C}(q^{-1}) \tag{9-76}$$

Although we show that the parameter estimation with input estimates is optimal, an adaptive method has to evaluate the parameters with a small amount of sampled data. Thus the method shown above is required to calculate the optimal parameter within a short time.

The convergence speed is examined as follows: Since the convergence speed is measured by the power of an estimation error at each instant, we calculate the power of an estimation error at several ARMA orders. The estimation error power at the time k is given as $\hat{\sigma}_x(k)^2$. In addition, $\hat{\sigma}_x(k)^2$ is calculated by (9-48). Now assume that $\underline{h}_{s,t}(i)^T \underline{F}_{s,t}(i \mid i - 1)\underline{h}_{s,t}(i)$ and $\underline{h}_{n,m}(i)^T \underline{F}_{n,m}(i \mid i - 1)\underline{h}_{n,m}(i \mid i - 1)$ denotes $\underline{h}(i)^T \underline{F}(i \mid i - 1)\underline{h}(i)$ of ARMA order (s,t) and (n,m), respectively. If $\underline{F}_{s,t}(i \mid i - 1)$ and $\underline{F}_{n,m}(i \mid i - 1)$ are positive definite, $s > n$ and $t > m$, then the estimation error power of ARMA order (s,t) is less than that of order (n,m) at the same instant. Thus the convergence speed at order (s,t) is evidently faster than that at order (n,m). From (9-37) and (9-38), it is easy to show that $a_{n+1}(k)$ $(i = 1, \ldots, s - n)$ and $b_{m+j}(k)$ $(j = 1, \ldots, t - n)$ are not zero. Thus we can get

$$\underline{h}_{s,t}(i)^T \underline{F}_{s,t}(i \mid i - 1)\underline{h}_{s,t}(i) > \underline{h}_{n,m}(i)^T \underline{F}_{n,m}(i \mid i - 1)\underline{h}_{n,m}(i) \tag{9-77}$$

Finally, we obtain the fast convergence when $s > n$ and $t > m$.

9-5 ESTIMATION OF INPUT PULSE SERIES

In speech analysis, the input of speech production model is either white process or pseudoperiodical pulse train. Thus if an ARMA modeling method is applied to the speech analysis, we need the estimation of input pulse series. When the unexpected prediction error happens on the stochastic distribution of an innovation series, its time location can be regarded as the time when the pulse excites the reference model. Thus when the pulse excites the reference model, the following equation is satisfied:

$$\nu(k)^2 > C\sigma^2(k-1) = CE[\nu(i)^2 \mid g(k-1)] \tag{9-78}$$

Since $\nu(k)$ is a white Gaussian process, the value C is enough to be $C \geq 9$. Assuming that $e(k)$ is an error before the input pulse is estimated and $\hat{u}^p(k)$ is an estimated input pulse, the following input pulse estimation is obtained:

$$e(k) = y(k) - \underline{h}(k)^T \hat{\underline{p}}(k \mid k-1):$$

(a) $e(k)^2 > C\hat{\sigma}_u^2(k-1)(1+\underline{h}^T(k)\underline{F}(k-1)\underline{h}(k))$ and $k > T_i + T_0$

$$\hat{u}^p(k) = e(k), \quad \nu(k) = 0, \quad T_{i+1} = k \tag{9-79}$$

(b) $e(k)^2 \leq C\hat{\sigma}_u^2(k-1)(1+\underline{h}^T(k)\underline{F}(k-1)\underline{h}(k))$ or $k \leq T_i + T_0$

$$\hat{u}^p(k) = 0, \quad \nu(k) = e(k) \tag{9-80}$$

where C and T_0 are constant. In addition, the self-tuning input $x(k)$ in (9-61) is rewritten as

$$x(k) = \frac{\nu(k)}{1 + \underline{h}(k)^T \underline{F}(k \mid k-1)\underline{h}(k)\lambda(k-1)^{-1}} + \hat{u}^p(k) \tag{9-81}$$

Condition (a) is always selected when the parameters are varied and the input pulse excites the reference model. Thus the pulse (i.e., $\hat{u}^p(k)$) is misestimated when the parameters are varied. In addition, the time variation of the parameters (i.e., $\gamma(k)$) is misestimated when the input pulse excites the reference model. To alleviate this influence, we use

$$\underline{\hat{p}}^s(k) = \underline{\hat{p}}^s(k-1) + \frac{1}{T_s}[\underline{\hat{p}}(k) - \underline{\hat{p}}^s(k-1)] \tag{9-82}$$

Equation (9-82) can smooth the misestimation error of $\hat{\underline{p}}(k)$ by using the time average.

9-6 MODEL REDUCTION ALGORITHM

Let us show a model reduction algorithm that removes $\hat{C}(q^{-1})$ in (9-76) from an estimation model. Note that under ordinary circumstances, the common polynomial $\hat{C}(q^{-1})$ in $\hat{A}(q^{-1})$ and $\hat{B}(q^{-1})$ cannot be determined exactly. In other words, some roots of $\hat{A}(q^{-1})$ are not identical to any roots of $\hat{B}(q^{-1})$ but quite similar to some roots of $\hat{B}(q^{-1})$. Since the polynomials $\hat{A}(q^{-1})$ and $\hat{B}(q^{-1})$ are estimated from stochastic variables, these estimates may include calculation errors. Thus it should be considered that $\hat{A}(q^{-1})$ and $\hat{B}(q^{-1})$ have each other as approximate common roots.

Let us define the following matrix:

$$\underline{z}^T \underline{V} \underline{z} = [\hat{A}(q^{-1}) G_m(q^{-1}) - \hat{B}(q^{-1}) D_n(q^{-1})]^2 \qquad (9\text{-}83)$$

where

$$\underline{z}^T = [q^{-1} \quad q^{-2} \quad \cdots \quad q^{-n-m-t}]$$

$$G_m(q^{-1}) = 1 + \sum_{i=1}^{m} g_i q^{-i} \qquad (9\text{-}84)$$

$$D_n(q^{-1}) = 1 + \sum_{j=1}^{n} d_j q^{-j}$$

From (9-84), $G_m(q^{-1})$ and $D_n(q^{-1})$, which minimize $\underline{z}^T \underline{V} \underline{z}$, are

$$G_m(q^{-1}) = \hat{B}^0(q^{-1})$$

$$D_n(q^{-1}) = \hat{A}^0(q^{-1}) \qquad (9\text{-}85)$$

because

$$\min[\hat{A}(q^{-1}) G_m(q^{-1}) - \hat{B}(q^{-1}) D_n(q^{-1})]^2$$

$$\cong \min[\hat{A}^0(q^{-1}) G_m(q^{-1}) - \hat{B}^0(q^{-1}) D_n(q^{-1})]^2 \hat{C}(q^{-1})^2 \qquad (9\text{-}86)$$

Thus if we estimate g_i ($i = 1, 2, \ldots, m$) and d_j ($j = 1, 2, \ldots, n$) which minimize $\underline{z}^T \underline{V} \underline{z}$, the optimum parameters are calculated. Let us calculate $\underline{V}$ from (9-83). Since

$$\hat{A}(q^{-1})G_m(q^{-1}) = [1 \ \underline{z}^T] \begin{bmatrix} 1 & 0 & \cdots & 0 \\ \hat{a}_1(k) & 1 & \cdots & 0 \\ \hat{a}_2(k) & \hat{a}_1(k) & \cdots & 0 \\ \vdots & \vdots & & \vdots \\ \hat{a}_{n+t}(k) & \hat{a}_{n+t-1}(k) & \cdots & \\ 0 & \hat{a}_{n+t}(k) & \cdots & \\ 0 & 0 & \cdots & \\ 0 & 0 & \cdots & \hat{a}_{n+t}(k) \end{bmatrix} \begin{bmatrix} 1 \\ g_1 \\ g_2 \\ \vdots \\ \\ \\ g_m \end{bmatrix}$$

(9-87)

$$= [1 \ \underline{z}^T] \begin{bmatrix} 1 & \underline{0}^T \\ \underline{\hat{a}} & \hat{A} \end{bmatrix} \begin{bmatrix} 1 \\ \underline{g} \end{bmatrix}$$

$$= 1 + \underline{z}^T \underline{\hat{a}} + \underline{z}^T \hat{A}\underline{g}$$

$$\hat{B}(q^{-1})D_n(q^{-1}) = [1 \ \underline{z}^T] \begin{bmatrix} 1 & 0 & \cdots & 0 \\ \hat{b}_1(k) & 1 & \cdots & 0 \\ \hat{b}_2(k) & \hat{b}_1(k) & \cdots & 0 \\ \vdots & \vdots & & \vdots \\ \hat{b}_{m+t}(k) & \hat{b}_{m+t-1}(k) & \cdots & \\ 0 & \hat{b}_{m+t}(k) & \cdots & \\ 0 & 0 & \cdots & \\ \vdots & \vdots & & \vdots \\ 0 & 0 & \cdots & \hat{b}_{m+t}(k) \end{bmatrix} \begin{bmatrix} 1 \\ d_1 \\ \vdots \\ d_n \end{bmatrix}$$

(9-88)

$$= [1 \ \underline{z}^T] \begin{bmatrix} 1 & \underline{0} \\ \underline{\hat{b}} & \hat{B} \end{bmatrix} \begin{bmatrix} 1 \\ \underline{d} \end{bmatrix}$$

$$= 1 + \underline{z}^T \underline{\hat{b}} + \underline{z}^T \hat{B}\underline{d}$$

we get

$$\underline{V} = [\hat{\underline{a}} - \hat{\underline{b}} + \hat{\underline{A}}\underline{g} - \hat{\underline{B}}\underline{d}][\hat{\underline{a}} - \hat{\underline{b}} + \hat{\underline{A}}\underline{g} - \hat{\underline{B}}\underline{d}]^T \qquad (9\text{-}89)$$

Hence the minimization of $\underline{z}^T \underline{V} \underline{z}$ is exactly the same as the minimization of the following criterion:

$$V = \text{trace } \underline{V}$$

$$= \sum_{i=1}^{n+m+t} [\hat{a}_i(k) - \hat{b}_i(k) + \hat{\underline{a}}^T \underline{g} - \hat{\underline{b}}^T \underline{g}]^2 \qquad (9\text{-}90)$$

where

$$\hat{a}_i(k) = 0 \quad i < 0, \ i > n; \quad \hat{a}_0(k) = 1$$

$$\hat{b}_j(k) = 0 \quad j < 0, \ j > m; \quad \hat{b}_0(k) = 1$$

$$\hat{\underline{a}}_i^T = [\hat{a}_{i-1}(k) \ \hat{a}_{i-2}(k) \ \cdots \ \hat{a}_{i-m}(k)]$$

$$\hat{\underline{b}}_i^T = [\hat{b}_{i-1}(k) \ \hat{b}_{i-2}(k) \ \cdots \ \hat{b}_{i-n}(k)]$$

Thus if we assume that

$$\eta_i = \hat{a}_i(k) - \hat{b}_i(k)$$

$$\underline{\eta}_i^T = [\hat{\underline{b}}_i^T \ -\hat{\underline{a}}_i^T] \qquad (9\text{-}91)$$

$$\underline{\mu}^T = [\underline{h}^T \ \underline{d}^T]$$

then the criterion is rewritten as

$$V = \sum_{i=1}^{n+m+t} (\eta_i - \underline{\eta}_i^T \underline{\mu})^2 \qquad (9\text{-}92)$$

The vector $\underline{\mu}$ that minimizes the criterion V is given as

$$\frac{dV}{d\mu} = \underline{0} \qquad (9\text{-}93)$$

From (9-93) we get

$$\mu = \left(\sum_{i=1}^{n+m+t} \eta_i \eta_i^T \right)^{-1} \sum_{i=1}^{n+m+t} \eta_i \eta_i \tag{9-94}$$

and the minimum value of V is

$$\min V = \sum_{i=1}^{m+n+t} \eta_i (\eta_i - \eta_i^T \mu) \tag{9-95}$$

By using (9-94) the optimum parameters can be calculated from the redundant estimated parameters (i.e., the parameters given from the redundant model). In addition, the value min V shows the accuracy for the optimum parameters.

9-7 COMPARISONS

Let us show the parameter tracking ability on the MIS method. The RLS method and the extended least-squares (ELS) method are also examined in the experiments. The RLS method requires all input and output sequence information. Thus this algorithm cannot be applied to speech signal since no input information is obtained. Thus the ELS and the RLS are simply compared with MIS on the convergence property and the estimation accuracy for synthesized signals. Evidently, the RLS shows the fastest convergence property. The ELS method is one of the simultaneous estimation of input and parameters. In ELS, the estimation error is used as an input signal [24, 25]. Thus the basic concept for a self-tuning input is the same as MIS. However, the model reduction algorithm is not introduced in ELS.

Each parameter of the reference model is:

k = 1 - 400:

$a_1 = -1.03$, $a_2 = 1.29$, $a_3 = -0.86$, $a_4 = 1.2$, $a_5 = -0.91$, $a_6 = 0.83$

(1000/100), (2000/100), (3500/100)

[Formant frequency (Hz)/bandwidth (Hz)]

$b_1 = -1.12$, $b_2 = 1.82$. $b_3 = -1.02$, $b_4 = 0.828$

(1500/150), (2500/150)

[Antiformant frequency (Hz)/bandwidth (Hz)]

k = 500 - 800

$a_1 = 0.65$, $a_2 = 0.41$, $a_3 = -0.33$, $a_4 = 0.27$, $a_5 = 0.45$, $a_6 = 0.78$

(1000/200), (3000/100), (4000/100)

$b_1 = 0.51$, $b_2 = 1.16$, $b_3 = 0.51$, $b_4 = 0.83$

(2000/100), (3500/200)

During $k = 401$ to 499, every parameter is linearly varied. The input signals of the reference model are zero-mean white Gaussian with variance 4.3. The orders of the estimation model are

MIS: $n = 13$, $m = 11$ $(t = 7)$

RLS, ELS: $n = 6$, $m = 4$

Figure 9-2 shows the estimated parameters $\hat{a}_1(k)$ and $\hat{b}_2(k)$. Figure 9-3 shows the estimated spectra at $k = 300$, 450, and 700. According to Fig. 9-2, the convergence speeds on the MIS method are similar to those of the RLS method in $k \leq 400$. In addition, the estimated accuracy of the MIS

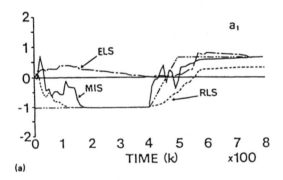

(a)

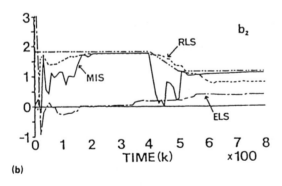

(b)

FIGURE 9-2 Comparisons of the true and the estimated parameters. (a) $a_1(k)$, $\hat{a}_1(k)$. (b) $b_2(k)$, $\hat{b}_2(k)$.

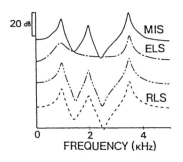
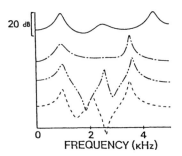

(a)   (b)

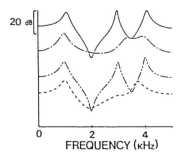

(c)

FIGURE 9-3 Real and estimated spectra. (a) k = 300, (b) k = 450, and
(c) k = 700.

method is better than that of the ELS method. In k > 500, the accuracy of
the MIS method is better than the others, because the MIS method can be
applied to a time-varying process while the RLS and the ELS method are
the estimation methods for a time-constant model. Figure 9-4 shows square
distances between the real parameters and the estimated parameters. In
Fig. 9-4(a), the vertical axis is defined as

$$\sum_{i=1}^{6} [a_i(k) - \hat{a}_i^0(k)]^2 \qquad\qquad (9\text{-}96)$$

In Fig. 9-4(b), the vertical axis is defined as

$$\sum_{i=1}^{4} [b_i(k) - \hat{b}_i^0(k)]^2 \qquad\qquad (9\text{-}97)$$

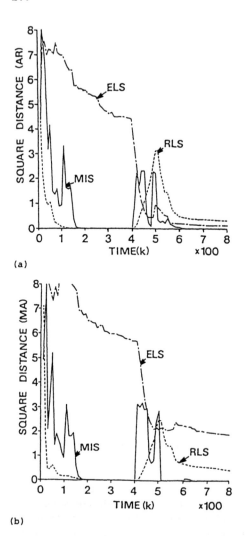

(a)

(b)

FIGURE 9-4 Square distances between the real and the estimated parameters.

Figure 9-4 shows that the MIS method can estimate the ARMA parameters
within a short time. In these experiments the initial conditions are
$\underline{F}(0 \mid -1) = 100\,\underline{I}$, $\hat{\underline{p}}(0 \mid -1) = \underline{0}$, $C_1 = 9$, $T_\beta = 50$, and $T_s = 1$.
    To observe the suitability of the MIS method for speech processing,
other experimental results for synthesized waveforms and real speech are
shown. The analysis results for synthesized waveforms are illustrated in
Fig. 9-5. The formants and the antiformants of a time-varying reference

model is shown in Fig. 9-5(a). The sampling frequency in these experiments is 10 kHz. Along the horizontal axis both formants (symbol X) and the antiformants (symbol O) of the reference model are represented at every 1 ms. A 3-dB bandwidth is represented as the symbol I for each formant and antiformant. These formants and antiformants are calculated from the roots of the denominator and the numerator polynomial of the estimated ARMA transfer function. Cross points of the formant and antiformant trajectories can be seen at three points in Fig. 9-5(a). Figure 9-5(b) illustrates the analysis results for the waveform synthesized by the reference

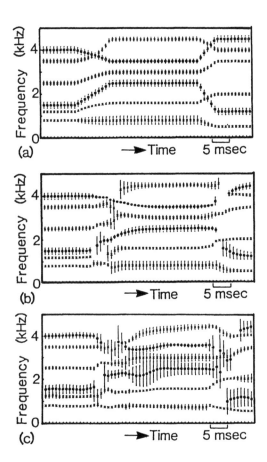

FIGURE 9-5 Experimental results: (a) reference model; (b) estimated formants and antiformants when the input of the reference model is a pulse series; (c) estimated formants and antiformants when the input signal of the reference model is a white Gaussian process.

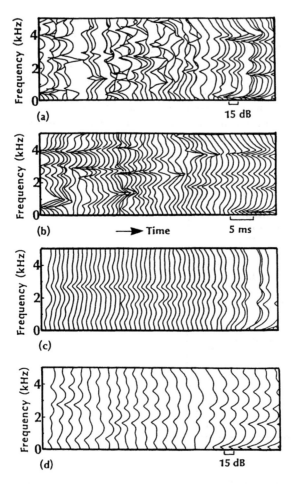

FIGURE 9-6 Analysis results for the real speech / tsu/: (a)-(c) MIS spectra; (d) LPC spectra; (e) waveform; (f) FFT spectra.

model when the input is a pseudoperiodical pulse train whose amplitude and periodicity are 100 and 5 ms, respectively. The analysis results are displayed in $k \geq 300$. The conditions in the MIS method are $\hat{\underline{p}}(0 \mid -1) = \underline{0}$, $\underline{F}(0 \mid -1) = \text{diag}[100 \cdots 100]$, $C = 9$, $T_\beta = 100$ samples, $\overline{T}_s = 30$ samples, $T_0 = 20$ samples, $s = 13$, $t = 9$, $r = 4$, and $\hat{q}_k^2 = 0$ ($k = 1, \ldots, 100$).

The value $\hat{q}_k^2$ ($k = 1, \ldots, 100$) is assumed to be zero since it is necessary to converge the estimated parameters first. Although the analysis results at the cross points of formants and antiformants are not accurate, formants and antiformants are accurately estimated except for these instants.

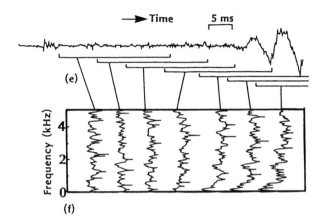

Figure 9-5(c) illustrates the MIS analysis for the synthetic waveform which is produced by the reference model represented in Fig. 9-5(a). The input is white Gaussian noise with zero mean and variance $\sigma_u^2 = 4.3$. The analysis conditions are the same as in the previous case. It is seen that the bandwidths of antiformants are not accurately estimated and that the estimation accuracy on the antiformants is less than that on the formants.

Next the analysis results for Japanese real speech are shown. The real speech was first filtered by a low-pass filter with 4.5 kHz cutoff frequency and sampled with 10 kHz by using a 12-bit A/D converter.

The MIS spectra for the Japanese real speech sound /tsu/ are shown in Fig. 9-6. The conditions used in Fig. 9-6(a), (b), and (c) are $T_s = 1$ (0.1 ms), $T_s = 30$ (3 ms), and $q^2 = 0$, respectively. The AR and MA orders of the estimation model and C_0 were 13, 6, and 0.01, respectively. The spectra estimated from k = 250 to 750 are shown. When $q^2 = 0$, we obtain spectra which are identical to spectra estimated by the modified MRAS [30].

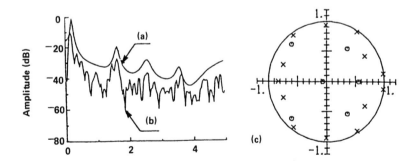

FIGURE 9-7 Analysis results: (a) MIS spectrum; (b) FFT spectrum; (c) poles and zeros of (a).

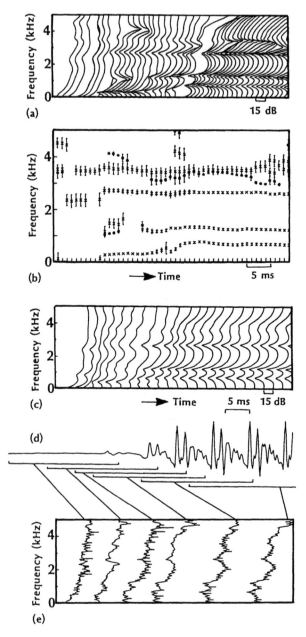

FIGURE 9-8 Analysis for real speech /ba/: (a) MIS spectra; (b) estimated formants and antiformants by the MIS method; (c) LPC spectra; (d) waveform of /ba/; (e) FFT spectra.

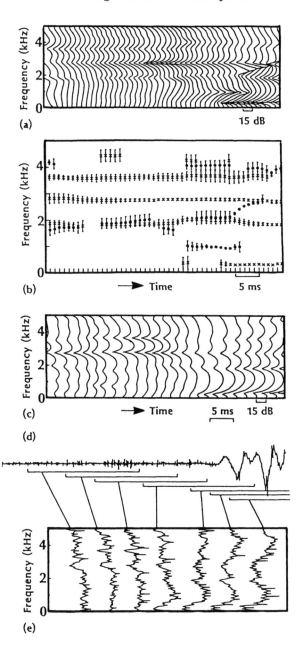

FIGURE 9-9 Analysis for real speech /shi/: (a) MIS spectra; (b) estimated formants and antiformants by the MIS method; (c) LPC spectra; (d) waveform of /shi/; (e) FFT spectra.

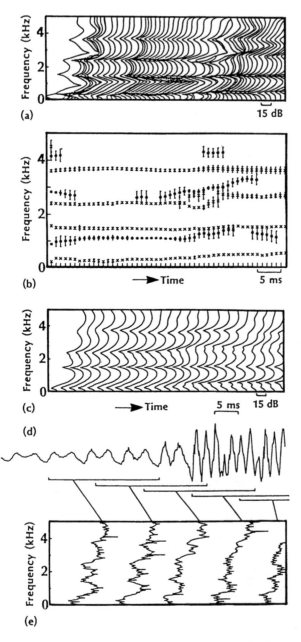

(a)

(b)

(c)

(d)

(e)

FIGURE 9-10 Analysis for real speech /re/: (a) MIS spectra; (b) estimated formants and antiformants by the MIS method; (c) LPC spectra; (d) waveform of /re/; (e) FFT spectra.

In addition, the LPC spectra and the FFT spectra are also shown. The LPC
analysis was applied with a 20-ms frame interval and the Kaiser window.
Each spectrum is represented at every 2 ms. Figure 9-6(e) represents the
frame intervals (256 samples) used in FFT. The Kaiser window was also
used to FFT. When $T_s = 1$, an accurate spectrum cannot be obtained because $q^2 \neq 0$. Thus the spectra obtained are over-charged. But when
$T_s = 30$, we can obtain the smoothed spectrum without this influence. It
turns out that the spectra in (b) represent the time-varying model more
precisely than do the spectra in (c). The LPC spectra provide the formants
with wide bandwidths since the LPC method smooths its parameters into
one frame.

Since the input influences are eliminated, the MIS method can estimate a reference model (i.e., a speech production model). Therefore, the
estimated ARMA model consists of a glottal model, a vocal tract model,
and a radiation model. However, it is very difficult to obtain three models
from the estimated ARMA model. For example, the estimated spectrum at
k = 600 in Fig. 9-6 is shown in Fig. 9-7(a). In Fig. 9-7(b), the FFT spectrum evaluated from the observed samples (k = 544 to 800) is also shown.
Figure 9-7(c) represents the poles and zeros of the MIS spectrum. Although
it can be considered that the poles which are close to a unit circle represent
the formants of the vocal tract model, it is not possible to identify which
poles and zeros represent the other models.

Figures 9-8, 9-9, and 9-10 illustrate the analysis results for Japanese
real speech sounds /ba/, /shi/, and /re/, respectively. The estimated
ARMA spectra and both estimated formants and antiformants are illustrated
from k = 100 in (a) and (b), respectively. The following conditions were
used in the MIS method: n + t = 13, m + t = 6, and $C_0 = 0.01$ for both /ba/
and /shi/, and n + t = 14, m + t = 8, and $C_0 = 0.01$ for /re/. The other
conditions were the same conditions as those used in the analysis above.
Since the ARMA orders were unknown in real speech, the orders of the
estimation model were empirically determined as high orders. Thus we
obtained formants and antiformants whose bandwidths were somewhat wide.
In view of this point, the formants and antiformants whose bandwidths are
greater than 500 Hz are not represented in (b). In addition, LPC spectra,
the observed real speech waveform, and FFT spectra are shown in (c), (d),
and (e), respectively. The AR order used in the LPC analysis is 15 for /ba/
and /shi/ and 16 for /re/. Note that the tracking ability of the MIS method
is superior to that of the conventional LPC analysis.

9-8 SUMMARY

In this chapter we have discussed several difficulties for the ARMA analysis
of speech. It has been shown especially that speech analysis needs to use
both first and second information. In addition, it has been mentioned that

the ARMA analysis becomes a nonlinear problem since the input signals of
the reference model cannot be observed. It can be solved by an adaptive
processing.

The MIS method proposed in this chapter consists of three algorithms:
the adaptive identification, the input estimation, and the model reduction.
The calculation cost of the adaptive identification is almost the same amount
as the Kalman filter. The input estimation has a very small amount of calculation cost since all values used in this algorithm are calculated in the adaptive identification. But the model reduction needs high calculation cost. If
the fast calculation algorithm will be used to the model reduction, it can be
reduced.

By comparing the MIS method with some conventional methods, we
have shown that the analysis algorithm must be changed by finding out
whether the input signal is the pulse series or white Gaussian process. In
addition, it has also been shown that the accurate parameters can be estimated by setting proper orders. As for the orders of the estimation model,
it is mentioned that the orders of the speech production model must be determined in order for the AR analysis and the ARMA analysis to get the
accurate estimates. In addition, from the analysis results for a synthesized
waveform and a real speech waveform, it has been shown that the antiformants should be estimated even when only formants are required.

REFERENCES

1. E. A. Robinson, "A historical perspective of spectrum estimation,"
   Proc. IEEE, vol. 70, no. 9, pp. 885-907 (Sept. 1982).

2. J. W. Cooley and J. W. Tukey, "An algorithm for the machine calculation of complex Fourier series," Math. Comput., vol. 19, pp. 297-
   301 (Apr. 1965).

3. P. Eykhoff, System Identification, Wiley-Interscience, Chichester,
   West Sussex, England, 1977.

4. P. Eykhoff, Trends and Progress in System Identification, Pergamon
   Press, Oxford, 1981.

5. E. A. Robinson, "Iterative least-squares procedure for ARMA spectral
   estimation," in Nonlinear Methods of Spectral Analysis, ed. S. Haykin,
   Springer-Verlag, Berlin, 1979.

6. J. Capon, "Maximum likelihood spectral estimation," in Nonlinear
   Methods of Spectral Analysis, ed. S. Haykin, Springer-Verlag, Berlin,
   1979.

7. R. N. McDonough, "Application of the maximum-likelihood method and
   the maximum entropy method to array processing," in Nonlinear Methods of Spectral Analysis, ed. S. Haykin, Springer-Verlag, Berlin, 1979.

8. P. R. Gutowski, E. A. Robinson, and S. Treitel, "Spectral estimation: fact or fiction," IEEE Trans. Geosci. Electron., vol. GE-16, pp. 80-84 (Apr. 1978).

9. J. P. Burg, "A new analysis technique for time series data," NATO Advanced Study Institute on Signal Processing, Aug. 12-23, 1968.

10. J. P. Burg, "Maximum entropy spectral analysis," 37th Meeting Society of Exploration Geophysicists, 1967.

11. T. J. Ulrych, "Maximum entropy spectral analysis and autoregressive decomposition," Rev. Geophys. Space Phys., vol. 13, pp. 183-200 (Feb. 1975).

12. P. C. Young, "An instrumental variable method for real-time identification of a noisy process," Automatica, vol. 6, pp. 271-287 (1970).

13. P. C. Young, "Comments on 'on-line identification of linear dynamic systems with applications to Kalman filter'," IEEE Trans. Autom. Control, vol. AC-17, pp. 269-270 (Apr. 1972).

14. I. D. Landau, "Unbiased recursive identification using model reference adaptive technique," IEEE Trans. Autom. Control, vol. AC-21, no. 1, pp. 194-202 (Apr. 1976).

15. I. D. Landau, "An addendum to 'unbiased recursive identification using model reference adaptive techniques'," IEEE Trans. Autom. Control, vol. AC-23, no. 1, pp. 97-99 (Feb. 1978).

16. P. Hagander and B. Wittenmark, "A self-tuning filter for fixed-lag smoothing," IEEE Trans. Inf. Theory, vol. IT-23, no. 3, pp. 377-384 (May 1977).

17. G. A. Dumont and P. R. Belanger, "Self-tuning control of a titanium dioxide kiln," IEEE Trans. Autom. Control, vol. AC-23, no. 4, pp. 532-538 (Aug. 1978).

18. B. Wittenmark, "A self-tuning predictor," IEEE Trans. Autom. Control, vol. AC-19, no. 6, pp. 848-851 (Dec. 1974).

19. J. Wieslander and B. Wittenmark, "An approach to adaptive control using real time identification," Automatica, vol. 7, pp. 211-217 (1971).

20. H. Matsuzawa, N. Ishii, A. Iwata, and N. Suzumura, "Adaptive identification by the time variable model of time series," Trans. IECE (Japan), vol. J61-A, No. 2, pp. 181-186 (Feb. 1978).

21. T. Bohlin, "Four cases of identification of changing systems," in System Identification: Advances and Case Studies, ed. R. K. Mehra and D. G. Lainiotis, Academic Press, New York, 1976.

22. H. Morikawa and H. Fujisaki, "A speech analysis-synthesis system

based on the SEARMA method," Monogr. Acoust. Soc. Jpn., vol. S79-47 (Nov. 1979).

23. H. Morikawa and H. Fukisaki, "A simplified method for simultaneous estimation of ARMA parameters and a method for adaptive analysis of speech," Monogr. Acoust. Soc. Jpn., vol. S79-05 (May 1979).

24. V. Strejc, "Least squares and regression methods," in Trends and Progress in System Identification, ed. P. Eykhoff, Pergamon Press, Oxford, 1981, pp. 103-144.

25. R. K. Mehra, "Choice of input signals," in Trends and Progress in System Identification, ed. P. Eykhoff, Pergamon Press, Oxford, 1981, pp. 308-366.

26. T. Tukabayashi and H. Suzuki, "Speech analysis by linear pole-zero model," Trans. IECE, vol. 58-A, pp. 270-277 (May 1975).

27. H. Morikawa and H. Fujisaki, "Speech analysis based on simultaneous estimation of autoregressive and moving-average parameters," Trans. IECE (Japan), vol. J61-A, no. 3, pp. 195-203 (Mar. 1978).

28. Y. Miyanaga, N. Miki, and N. Nagai, "An estimation method of auto-regressive and moving-average parameters of speech with pitch esti-mation," Trans. Inst. Elec. Commun. Eng. (Jpn.), vol. 63-A, no. 11, pp. 737-744 (Nov. 1980).

29. Y. Miyanaga, N. Miki, N. Nagai, and K. Hatori, "Adaptive simultane-ous estimation of time variable autoregressive and moving-average parameters," Trans. Inst. Elec. Commun. Eng. (Jpn.), vol. J64-D, no. 4, pp. 308-315 (Apr. 1981).

30. Y. Miyanaga, N. Miki, N. Nagai, and K. Hatori, "A speech analysis which eliminates the influence of pitch using the model reference adap-tive system," IEEE Trans. Acoust. Speech Signal Process., vol. ASSP-30, no. 1, pp. 88-96 (Feb. 1982).

31. Y. Miyanaga, N. Miki, and N. Nagai, "An optimum estimation of auto-regressive and moving-average parameters in a short time," Trans. Inst. Elec. Commun. Eng. (Jpn.), vol. J66-A, no. 1, pp. 1-8 (Jan. 1983).

10

ARMA Digital Lattice Filter Based on Linear Prediction Theory

YOSHIKAZU MIYANAGA

Hokkaido University, Sapporo, Japan

10-1 INTRODUCTION

In AR parameter estimation, the characteristics of the Levinson algorithm are well known in digital signal processing [1-3]. It is especially important that the design of an AR lattice filter be associated with the Levinson algorithm [3]. The AR lattice filter is applied to an equalizer and a filter design. For the application, the property of the lattice filter, that is, an orthogonal condition of each error produced at this filter element section, is employable. It has been shown in the application that the dynamic range of coefficients and the sensitivity are superior to those of other filters.

In ARMA model identification, several design methods for an ARMA lattice error filter have been presented [4-12]. In this chapter we introduce a new ARMA lattice filter. The lattice filter can be designed with an arbitrary AR order and an arbitrary MA order. The design method of the ARMA lattice filter is based on the four prediction errors yielded by a fast-calculation algorithm, also shown in this chapter. This design method is similar to the Levinson algorithm, by which the AR lattice error filter is built up.

First we define the four prediction errors, and then show their stochastic properties. From these errors, three recursions are derived. The ARMA lattice error filter is based on these recursions. The ARMA lattice filter is designed in the same way.

Next, as a special case, it will be assumed that the input signal of a filter is a white Gaussian process [12]. At that time the fast calculation algorithm for ARMA parameters is fairly simplified. This algorithm uses only three prediction errors. From these errors, a new ARMA lattice filter is derived.

10-2 FAST RECURSIVE ALGORITHM AND FOUR PREDICTION ERRORS

Let us show four prediction errors. The properties about each error are also discussed. Let the reference transfer functions be

$$H_a(z^{-1}) = Z[h_a(k)], \quad x(k) = con[h_a(k), u(k)]$$

$$H_b(z^{-1}) = Z[h_b(k)], \quad y(k) = con[h_b(k), u(k)]$$

$$(10\text{-}1)$$

where $con[\cdot]$ denotes convolution. The operator $Z[\ ]$ denotes the z-transfer. The variable $u(k)$ is a zero-mean white Gaussian process with variance σ_u^2. The stochastic variables $x(k)$ and $y(k)$ are an input signal and an output signal, respectively. The z-transfer functions $H_a(z^{-1})$ and $H_b(z^{-1})$ are rational transfer functions. It is shown in (10-1) that the input signal and the output signal can be represented as Gaussian process, that is, a canonical representation of innovations.

For the reference transfer functions in (10-1), the following estimation models are defined:

$$\hat{A}(z^{-1}) = 1 + \hat{a}_1 z^{-1} + \cdots + \hat{a}_s z^{-s}$$

$$\hat{B}(z^{-1}) = \hat{b}_0 + \hat{b}_1 z^{-1} + \cdots + \hat{b}_t z^{-t}$$

$$(10\text{-}2)$$

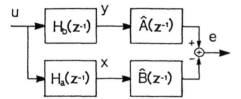

FIGURE 10-1 Block diagram of the least mean square estimation problem in (10-3).

In addition, let us define the following criterion:

$$V_{s,t} = \frac{\sigma_u^2}{2\pi j} \oint_{|z|=1} |H_b(z^{-1})\hat{A}(z^{-1}) - H_a(z^{-1})\hat{B}(z^{-1})|^2 \frac{dz}{z} \qquad (10\text{-}3)$$

To minimize the criterion in (10-3) is identical to the parameter estimation of $\hat{A}(z^{-1})$ and $\hat{B}(z^{-1})$, which minimizes the mean square of e(k) in Fig. 10-1. The parameters $\hat{a}_i$ and $\hat{b}_j$ ($i = 1, 2, \ldots, s$; $j = 0, 1, \ldots, t$) are obtained as follows:

$$[1 \ \hat{\underline{a}}_{s,t}^T \ \hat{b}_0 \ \hat{\underline{b}}_{s,t}^T] \underline{R}_{s+1, t+1} = [V_{s,t}^m \ 0 \ 0 \ \cdots \ 0] \qquad (10\text{-}4)$$

$$\min V_{s,t} = V_{s,t}^m \qquad (10\text{-}5)$$

where T denotes transpose. Each matrix and vector are

$$\hat{\underline{a}}_{s,t}^T = [\hat{a}_1 \ \cdots \ \hat{a}_s], \quad \hat{\underline{b}}_{s,t}^T = [\hat{b}_1 \ \cdots \ \hat{b}_t] \qquad (10\text{-}6)$$

$$\underline{R}_{s+1, t+1} = \begin{bmatrix} \underline{R}_{s+1}^y & -\underline{R}_{s+1, t+1}^{yx} \\ -(\underline{R}_{s+1, t+1}^{yx})^T & \underline{R}_{t+1}^x \end{bmatrix}$$

$$\underline{R}_{s+1}^y = \begin{bmatrix} r_0^y & r_1^y & \cdots & r_s^y \\ r_{-1}^y & r_0^y & \cdots & r_{s-1}^y \\ \vdots & \vdots & & \vdots \\ r_{-s}^y & r_{1-s}^y & \cdots & r_0^y \end{bmatrix}$$

$$
\underline{R}^x_{t+1} =
\begin{bmatrix}
r^x_0 & r^x_1 & \cdots & r^x_t \\
r^x_{-1} & r^x_0 & \cdots & r^x_{t-1} \\
\vdots & \vdots & & \vdots \\
r^x_{-t} & r^x_{1-t} & \cdots & r^x_0
\end{bmatrix}
\tag{10-7}
$$

$$
\underline{R}^{yx}_{s+1,\,t+1} =
\begin{bmatrix}
r^{yx}_0 & r^{yx}_{-1} & \cdots & r^{yx}_{-t} \\
r^{yx}_1 & r^{yx}_0 & \cdots & r^{yx}_{1-t} \\
\vdots & \vdots & & \vdots \\
r^{yx}_s & r^{yx}_{s-1} & \cdots & r^{yx}_{s-t}
\end{bmatrix}
$$

where $\underline{R}^y_{s+1}$, $\underline{R}^x_{t+1}$, and $\underline{R}^{yx}_{s+1,\,t+1}$ are Toeplitz matrices, and

$$
r^x_i = E[x(k)x(k-i) \mid g(k)]
$$

$$
r^y_i = E[y(k)y(k-i) \mid g(k)]
\tag{10-8}
$$

$$
r^{yx}_i = E[x(k)y(k-i) \mid g(k)]
$$

where $g(k)$ is a space spanned by $[y(i), x(i), i \le k]$. Equation (10-8) denotes conditional mean values on $g(k)$.

From (10-1), the reference transfer function is obtained as

$$
H(z^{-1}) = \frac{H_b(z^{-1})}{H_a(z^{-1})}
\tag{10-9}
$$

From (10-3), the following model is regarded as an estimation transfer function for (10-9):

$$
\hat{H}(z^{-1}) = \frac{\hat{B}(z^{-1})}{\hat{A}(z^{-1})}
\tag{10-10}
$$

A fast recursive algorithm estimates the ARMA parameters which approximate the reference transfer function [i.e., the ARMA parameters in (10-9)].

The four vector equations used in the fast recursive algorithm are defined as

$$\underline{R}_{s+1,t+1}{}^{\epsilon}_{s,t} = [-V^{\epsilon\nu}_{s,t}\ \underline{0} \mid V^{\epsilon}_{s,t}\ \underline{0}]^{T}$$

$$\epsilon^{T}_{s,t} = [0\ \hat{\underline{a}}^{\epsilon}_{s,t} \mid 1\ \hat{\underline{b}}^{\epsilon}_{s,t}]^{T} \tag{10-11}$$

$$\underline{R}_{s+1,t+1}{}^{\nu}_{s,t} = [V^{\nu}_{s,t}\ \underline{0} \mid -V^{\nu\epsilon}_{s,t}\ \underline{0}]^{T}$$

$$\nu^{T}_{s,t} = [1\ \hat{\underline{a}}^{\nu}_{s,t} \mid 0\ \hat{\underline{b}}^{\nu}_{s,t}]^{T} \tag{10-12}$$

$$\underline{R}_{s+1,t+1}{}^{\mathcal{H}}_{s,t} = [\underline{0}\ -V^{\gamma\xi}_{s,t} \mid \underline{0}\ V^{\gamma}_{s,t}]^{T}$$

$$\mathcal{H}_{s,t} = [\hat{\underline{a}}^{\gamma}_{s,t}\ 0 \mid \hat{\underline{b}}^{\gamma}_{s,t}\ 1]^{T} \tag{10-13}$$

$$\underline{R}_{s+1,t+1}{}^{\xi}_{s,t} = [\underline{0}\ V^{\xi}_{s,t} \mid \underline{0}\ -V^{\xi\gamma}_{s,t}]^{T}$$

$$\xi^{T}_{s,t} = [\hat{\underline{a}}^{\xi}_{s,t}\ 1 \mid \hat{\underline{b}}^{\xi}_{s,t}\ 0]^{T} \tag{10-14}$$

where $\underline{0}$ is a column zero vector, and

$$\epsilon^{T}_{s,t} = [0\ \hat{a}^{\epsilon}_{1}\ \cdots\ \hat{a}^{\epsilon}_{s} \mid 1\ \hat{b}^{\epsilon}_{1}\ \cdots\ \hat{b}^{\epsilon}_{t}]$$

$$\nu^{T}_{s,t} = [1\ \hat{a}^{\nu}_{1}\ \cdots\ \hat{a}^{\nu}_{s} \mid 1\ \hat{b}^{\nu}_{1}\ \cdots\ \hat{b}^{\nu}_{t}]$$

$$\gamma^{T}_{s,t} = [\hat{a}^{\gamma}_{0}\ \cdots\ \hat{a}^{\gamma}_{s-1}\ 0 \mid \hat{b}^{\gamma}_{0}\ \cdots\ \hat{b}^{\gamma}_{t-1}\ 1]$$

$$\xi^{T}_{s,t} = [\hat{a}^{\xi}_{0}\ \cdots\ \hat{a}^{\xi}_{s-1}\ 1 \mid \hat{b}^{\xi}_{0}\ \cdots\ \hat{b}^{\xi}_{t-1}\ 0]$$

From (10-11) to (10-14), the values in (10-4) and (10-5) are

$$\hat{\underline{a}}_{s,t} = \hat{\underline{a}}^{\nu}_{s,t} + \frac{\hat{\underline{a}}^{\epsilon}_{s,t} V^{\nu\epsilon}_{s,t}}{V^{\epsilon}_{s,t}}$$

$$\hat{b}_0 = \frac{V_{s,t}^{\epsilon\nu}}{V_{s,t}^{\nu}} \tag{10-15}$$

$$\hat{b}_{s,t} = b_{s,t}^{\nu} + \frac{\hat{b}_{s,t}^{\epsilon} V_{s,t}^{\nu\epsilon}}{V_{s,t}^{\epsilon}}$$

$$\min V_{s,t} = V_{s,t}^{\nu} - \frac{V_{s,t}^{\epsilon\nu} V_{s,t}^{\nu\epsilon}}{V_{s,t}^{\epsilon}} \tag{10-16}$$

Let us show the relation between the estimation models and the estimation errors.

Definition 10-1.

Forward estimation model for $x(k)$:

$$\hat{x}^{\epsilon}(k) = -\sum_{i=1}^{t} \hat{b}_i^{\epsilon} x(k-i) + \sum_{j=1}^{s} \hat{a}_j^{\epsilon} y(k-j) \tag{10-17}$$

Forward estimation model for $y(k)$:

$$\hat{y}^{\nu}(k) = -\sum_{j=1}^{s} \hat{a}_j^{\nu} y(k-j) + \sum_{i=1}^{t} \hat{b}_i^{\nu} x(k-i) \tag{10-18}$$

Backward estimation model for $x(k-t)$:

$$\hat{x}^{\gamma}(k-t) = -\sum_{i=0}^{t-1} \hat{b}_i^{\gamma} x(k-i) + \sum_{j=0}^{s-1} \hat{a}_j^{\gamma} y(k-j) \tag{10-19}$$

Backward estimation model for $y(k-s)$:

$$\hat{y}^{\xi}(k-s) = -\sum_{j=0}^{s-1} \hat{a}_j^{\xi} y(k-j) + \sum_{i=0}^{t-1} \hat{b}_i^{\xi} x(k-i) \tag{10-20}$$

From (10-17) to (10-20), the four estimation errors at time k are defined as

$$\epsilon_{s,t}(k) = x(k) - \hat{x}^{\epsilon}(k)$$

$$\nu_{s,t}(k) = y(k) - \hat{y}^{\nu}(k)$$

$$\gamma_{s,t}(k) = x(k-t) - \hat{x}^{\gamma}(k-t) \tag{10-21}$$

$$\xi_{s,t}(k) = y(k-s) - \hat{y}^{\xi}(k-s)$$

$\blacksquare$

Each estimation error is calculated on $g(k)$. Let us show that the parameters of these estimation models are obtained by the equations (10-11) to (10-14). From (10-11), $V_{s,t}^{\epsilon}$ is

$$
\begin{aligned}
V_{s,t}^{\epsilon} &= -\sum_{j=1}^{s} \hat{a}_j^{\epsilon} r_j^{yx} + r_0^x + \sum_{i=1}^{t} \hat{b}_i^{\epsilon} r_i^x \\
&= E[x(k)x(k) \mid g(k)] - E\left[x(k)\left(-\sum_{i=1}^{t} \hat{b}_i^{\epsilon} x(k-i) + \sum_{j=1}^{s} \hat{a}_j^{\epsilon} y(k-j) \right) \,\middle|\, g(k) \right] \\
&= E[x(k)(x(k) - \hat{x}^{\epsilon}(k)) \mid g(k)]
\end{aligned}
\tag{10-22}
$$

The ARMA parameters $\hat{a}_j^{\epsilon}$ and $\hat{b}_i^{\epsilon}$ ($j = 1, 2, \ldots, s$; $i = 0, 1, \ldots, t$) of the forward prediction model in (10-17) minimize

$$E[(x(k) - \hat{x}^{\epsilon}(k))^2 \mid g(k)] = E[(\epsilon_{s,t}(k))^2 \mid g(k)] \tag{10-23}$$

Differentiating (10-23) with respect to $\hat{a}_j^{\epsilon}$ and $\hat{b}_i^{\epsilon}$ ($j = 1, 2, \ldots, s$; $i = 0, 1, \ldots, t$), it turns out that these parameters satisfy (10-11). In addition, the minimum value is obtained in (10-22). The other equations can also be obtained by the similar way. Thus we get the following theorems.

Theorem 10-1. Every parameter in (10-17) to (10-20) satisfies the equations (10-11) to (10-14), respectively. In addition, the least mean square errors are obtained as

$$
\begin{aligned}
V_{s,t}^{\epsilon} &= E[x(k)\epsilon_{s,t}(k) \mid g(k)] \\[6pt]
V_{s,t}^{\nu} &= E[y(k)\nu_{s,t}(k) \mid g(k)] \\[6pt]
V_{s,t}^{\gamma} &= E[x(k-t)\gamma_{s,t}(k) \mid g(k)] \\[6pt]
V_{s,t}^{\xi} &= E[y(k)-s)\xi_{s,t}(k) \mid g(k)]
\end{aligned}
\tag{10-24}
$$

$\blacksquare$

Several orthogonal conditions for the estimation errors in (10-21) are derived.

<u>Theorem 10-2</u>.

Orthogonal relations on $\epsilon_{s,t}(k)$:

$$E[\epsilon_{s,t}(k)y(k-j) \mid g(k)] = 0, \qquad j = 1,2,\ldots,s$$

$$E[\epsilon_{s,t}(k)x(k-i) \mid g(k)] = 0, \qquad i = 1,2,\ldots,t \tag{10-25}$$

Orthogonal relations on $\nu_{s,t}(k)$:

$$E[\nu_{s,t}(k)y(k-j) \mid g(k)] = 0, \qquad j = 1,2,\ldots,s$$

$$E[\nu_{s,t}(k)x(k-i) \mid g(k)] = 0, \qquad i = 1,2,\ldots,t \tag{10-26}$$

Orthogonal relations on $\gamma_{s,t}(k)$:

$$E[\gamma_{s,t}(k)y(k-j) \mid g(k)] = 0, \qquad j = 0,1,\ldots,s-1$$

$$E[\gamma_{s,t}(k)x(k-i) \mid g(k)] = 0, \qquad i = 0,1,\ldots,t-1 \tag{10-27}$$

Orthogonal relations of $\xi_{s,t}(k)$:

$$E[\xi_{s,t}(k)y(k-j) \mid g(k)] = 0, \qquad j = 0,1,\ldots,s-1$$

$$E[\xi_{s,t}(k)x(k-i) \mid g(k)] = 0, \qquad i = 0,1,\ldots,t-1 \tag{10-28}$$

∎

<u>Proof</u> Let us show (10-25). From (10-21) we get

$$E[\epsilon_{s,t}(k)y(k-j) \mid g(k)] = r_j^{yx} + \sum_{i=1}^{t} \hat{b}_i^\epsilon r_{j-i}^{yx} - \sum_{k=1}^{s} \hat{a}_k^\epsilon r_{j-k}^{y} \tag{10-29}$$

From (10-11) we obtain

$$[\underline{R}_{s+1}^{y} \quad -\underline{R}_{s+1,t+1}^{yx}]\epsilon_{s,t} = [-V_{s,t}^{\epsilon\nu} \quad \underline{0}]^T$$

Thus (10-29) is zero at $j = 1, 2, \ldots, s$. In addition, from (10-21) we get

$$E[\epsilon_{s,t}(k)x(k-i) \mid g(k)] = r_i^{x} + \sum_{k=1}^{t} \hat{b}_k^\epsilon r_{i-k}^{x} - \sum_{j=1}^{s} \hat{a}_j^\epsilon r_{j-i}^{yx} \tag{10-30}$$

On the other hand, from (10-11) we obtain

$$[-(R^{yx}_{s+1,\,t+1})^T \; R^x_{t+1}]\epsilon_{s,\,t} = [V^\epsilon_{s,\,t} \; \underline{0}]^T$$

Thus (10-30) is zero at $i = 1, 2, \ldots, t$. The other orthogonal conditions can be proved in a similar way.

From Theorem 10-2, $V^{\epsilon\nu}_{s,\,t}$ in (10-11) is

$$V^{\epsilon\nu}_{s,\,t} = -\sum_{i=1}^{s} \hat{a}^\epsilon_i r^y_i + r^{yx}_0 + \sum_{j=1}^{t} \hat{b}^\epsilon_j r^{yx}_{-j}$$

$$= E[y(k)\epsilon_{s,\,t}(k) \mid g(k)] = E[\nu_{s,\,t}(k)\epsilon_{s,\,t}(k) \mid g(k)] \tag{10-31}$$

The value $V^{\nu\epsilon}_{s,\,t} = E[\nu_{s,\,t}(k)\epsilon_{s,\,t}(k) \mid g(k)]$ is also obtained. In addition, $V^{\gamma\xi}_{s,\,t}$ and $V^{\xi\gamma}_{s,\,t}$ are calculated as follows:

$$V^{\gamma\xi}_{s,\,t} = E[y(k-s)\gamma_{s,\,t}(k) \mid g(k)] = E[\xi_{s,\,t}(k)\gamma_{s,\,t}(k) \mid g(k)]$$

$$V^{\xi\gamma}_{s,\,t} = E[x(k-t)\xi_{s,\,t}(k) \mid g(k)] = E[\gamma_{s,\,t}(k)\xi_{s,\,t}(k) \mid g(k)] \tag{10-32}$$

Hence the following theorem is obtained:

<u>Theorem 10-3.</u> $V^{\epsilon\nu}_{s,\,t}$, $V^{\nu\epsilon}_{s,\,t}$, $V^{\gamma\xi}_{s,\,t}$, and $V^{\xi\gamma}_{s,\,t}$ are calculated as

$$V^{\epsilon\nu}_{s,\,t} = V^{\nu\epsilon}_{s,\,t} = E[\nu_{s,\,t}(k)\epsilon_{s,\,t}(k) \mid g(k)]$$

$$V^{\gamma\xi}_{s,\,t} = V^{\xi\gamma}_{s,\,t} = E[\gamma_{s,\,t}(k)\xi_{s,\,t}(k) \mid g(k)] \tag{10-33}$$

∎

If the following vector is defined:

$$h^T_{s+1,\,t+1}(k) = [y(k) \;\; y(k-1) \;\; \cdots \;\; y(k-s) \;\; -x(k) \;\; -x(k-1) \;\; \cdots \;\; -x(k-t)] \tag{10-34}$$

then from (10-17) to (10-21) the four prediction errors can be described as

$$\epsilon_{s,\,t}(k) = -h^T_{s+1,\,t+1}(k)\,\epsilon_{s,\,t}$$

$$\nu_{s,\,t}(k) = h^T_{s+1,\,t+1}(k)\,\nu_{s,\,t}$$

$$\gamma_{s,t}(k) = -\underline{h}_{s+1,t+1}(k)^T \underline{\gamma}_{s,t} \tag{10-35}$$

$$\xi_{s,t}(k) = \underline{h}_{s+1,t+1}(k)^T \underline{\xi}_{s,t}$$

10-3 ELEMENTARY SECTIONS OF ARMA DIGITAL LATTICE ERROR FILTER

Based on the fast recursive algorithm mentioned above, an ARMA lattice error filter is designed. The term "error" is used because of the following reason. When the design of an AR lattice filter is required, we construct the filter whose inputs are observed signals and whose outputs are estimation errors. Generally speaking, the observed signals are regarded as the output of a reference model. The first designed filter yields the estimation errors as its outputs where its inputs are the outputs of the reference model as shown in Fig. 10-2(a). Thus this filter is called the AR lattice

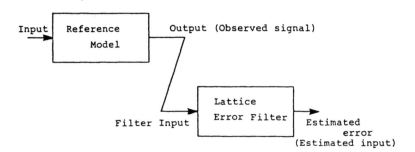

(a)

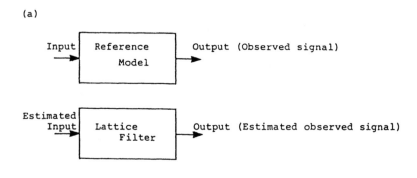

(b)

FIGURE 10-2 (a) AR lattice error filter; (b) AR lattice filter.

error filter. In addition, this filter is also called a whitening lattice filter.
On the other hand, by the same design method, we can design an AR lattice
filter whose inputs are the estimation errors [i.e., the zero-mean white
Gaussian processes, whose outputs are the observed signals shown in
Fig. 10-2(b)].

In this section we develop two elementary sections used in the ARMA
lattice error filter. One of the two elementary sections designs the filter
as the AR order of this filter increases by one. The other section designs
the filter as the MA order of this filter increases by one. Each section
satisfies the orthogonal conditions in Theorem 10-2. Since the input signal
y and the output signal x have some kind of correlation with each other, the
parts of the AR and MA models cannot be represented independently. We
first propose the recursive formulas, which increase AR order and MA
order independently. Then from the AR recursive formula, the AR elementary section is designed. In addition, the MA elementary section is
designed from the MA recursive formula.

10-3-1 AR Lattice Error Filter

Let us derive the AR recursive formula, which estimates the ARMA parameters with $s + 1$ AR order and t MA order under the condition that the values
in (10-11) to (10-14) with s AR order and t MA order have already been
given. First define two matrices $\underline{I}_1$ and $\underline{I}_2$ with $s + t + 3$ rows and $s + t + 2$
columns as

$$\underline{I}_1[w(1) \quad \cdots \quad w(s + 1) \mid w(s + 2) \quad \cdots \quad w(s + t + 2)]^T$$

$$= [w(1) \quad \cdots \quad w(s + 1) \mid w(s + 2) \quad \cdots \quad w(s + t + 2)]^T$$

$$\underline{I}_2[w(1) \quad \cdots \quad w(s + 1) \mid w(s + 2) \quad \cdots \quad W(s + t + 2)]^T \tag{10-36}$$

$$= [0 \quad w(1) \quad \cdots \quad w(s + 1) \mid 0 \quad w(s + 2) \quad \cdots \quad w(s + t + 1)]^T$$

Since $\underline{R}_{s+2}^y$, $\underline{R}_{t+1}^x$, and $\underline{R}_{s+2,t+1}^{yx}$ are Toeplitz matrices, (10-11) to (10-14)
are rewritten as

$$\underline{R}_{s+2,t+1}\underline{I}_1\underline{\epsilon}_{s,t} = [-V_{s,t}^{\epsilon\nu} \ \underline{0} \ -V_{s,t}^{\xi\epsilon} \mid V_{s,t}^{\epsilon} \ \underline{0}]^T$$

$$\underline{R}_{s+2,t+1}\underline{I}_1\underline{\nu}_{s,t} = [V_{s,t}^{\nu} \ \underline{0} \ V_{s,t}^{\xi\nu} \mid -V_{s,t}^{\epsilon\nu} \ \underline{0}]^T$$

$$\underline{R}_{s+2,t+1}\underline{I}_1\underline{\gamma}_{s,t} = [\underline{0} \ -V_{s,t}^{\gamma\xi} \ -V_{s,t}^{\gamma\gamma} \mid \underline{0} \ V_{s,t}^{\gamma}]^T \tag{10-37}$$

$$\underline{R}_{s+2,t+1}\underline{I}_1\underline{\xi}_{s,t} = [\underline{0} \ \ v^{\xi}_{s,t} \ \ v^{y\xi}_{s,t} \ \vdots \ \underline{0} \ \ -v^{\gamma\xi}_{s,t}]^T$$

$$\underline{R}_{s+2,t+1}\underline{I}_2\underline{\xi}_{s,t} = [v^{\xi\epsilon}_{s,t} \ \ \underline{0} \ \ v^{\xi}_{s,t} \ \vdots \ v^{\xi\nu}_{s,t} \ \ \underline{0}]^T$$

where

$$v^{\xi\epsilon}_{s,t} = -[r^y_{-s} \ r^y_{1-s} \ \cdots \ r^y_{-1}]\hat{\underline{a}}^{\epsilon}_{s,t} + r^{yx}_{-1-s} + [r^{yx}_{-s} \ r^{yx}_{1-s} \ \cdots \ r^{yx}_{t-1-s}]\hat{\underline{b}}^{\epsilon}_{s,t}$$

$$= E[\xi_{s,t}(k-1)\epsilon_{s,t}(k) \ | \ g(k)]$$

$$v^{\xi\nu}_{s,t} = r^y_{-1-s} + [r^y_{-s} \ r^y_{1-s} \ \cdots \ r^y_{-1}]\hat{\underline{a}}^{\nu}_{s,t} - [r^{yx}_{-s} \ r^{yx}_{1-s} \ \cdots \ r^{yx}_{t-s-1}]\hat{\underline{b}}^{\nu}_{s,t}$$

$$= E[\xi_{s,t}(k-1)\nu_{s,t}(k) \ | \ g(k)]$$

$$\hspace{10cm} (10\text{-}38)$$

$$v^{y\gamma}_{s,t} = -[r^y_{-s-1} \ r^y_{-s} \ \cdots \ r^y_{-2}]\hat{\underline{a}}^{\gamma}_{s,t} + [r^{yx}_{-1-s} \ r^{yx}_{-s} \ \cdots \ r^{yx}_{t-s-2}]\hat{\underline{b}}^{\gamma}_{s,t} + r^{yx}_{t-s-1}$$

$$= E[y(k-s-1)\gamma_{s,t}(k) \ | \ g(k)]$$

$$v^{y\xi}_{s,t} = [r^y_{-1-s} \ r^y_{-s} \ \cdots \ r^y_{-2}]\hat{\underline{a}}^{\xi}_{s,t} + r^y_{-1} - [r^{yx}_{-s-1} \ r^{yx}_{-s} \ \cdots \ r^{yx}_{t-s-2}]\hat{\underline{b}}^{\xi}_{s,t}$$

$$= E[y(k-s-1)\xi_{s,t}(k) \ | \ g(k)]$$

Thus from (10-36) and (10-37), an AR recursive formula is derived as

$$\underline{\epsilon}_{s+1,t} = \underline{I}_1\underline{\epsilon}_{s,t} + \mu_1\underline{I} \ \underline{\xi}_{s,t} \hspace{6cm} (10\text{-}39)$$

$$\underline{\nu}_{s+1,t} = \underline{I}_1\underline{\nu}_{s,t} - \mu_2\underline{I} \ \underline{\xi}_{s,t} \hspace{6cm} (10\text{-}40)$$

$$\underline{\gamma}_{s+1,t} = \underline{I}_1[\underline{\gamma}_{s,t} + \mu_3\underline{\xi}_{s,t}] \hspace{6cm} (10\text{-}41)$$

$$\underline{\xi}_{s+1,t} = \underline{I}_2\underline{\xi}_{s,t} + \underline{I}_1[\mu_{4,1}\underline{\epsilon}_{s,t} - \mu_{4,2}\underline{\nu}_{s,t} + \mu_{4,3}\underline{\gamma}_{s,t} + \mu_{4,4}\underline{\xi}_{s,t}] \hspace{1.5cm} (10\text{-}42)$$

$$\mu_1 = \frac{v^{\xi\epsilon}_{s,t}}{v^{\xi}_{s,t}}, \quad \mu_2 = \frac{v^{\xi\nu}_{s,t}}{v^{\xi}_{s,t}}, \quad \mu_3 = \frac{v^{\gamma\xi}_{s,t}}{v^{\xi}_{s,t}}$$

$$\begin{bmatrix} \mu_{4,1} \\ \mu_{4,2} \end{bmatrix} = \frac{-1}{v^{\epsilon\nu}_{s,t}v^{\epsilon\nu}_{s,t} - v^{\nu}_{s,t}v^{\epsilon}_{s,t}} \begin{bmatrix} v^{\nu}_{s,t} & -v^{\nu\epsilon}_{s,t} \\ -v^{\nu\epsilon}_{s,t} & v^{\epsilon}_{s,t} \end{bmatrix} \begin{bmatrix} v^{\xi\epsilon}_{s,t} \\ v^{\xi\nu}_{s,t} \end{bmatrix}$$

$$\mu_{4,3} = \frac{V_{s,t}^{\gamma\xi} V_{s,t}^{y\xi} + V_{s,t}^{\xi} V_{s,t}^{yy}}{V_{s,t}^{\xi} V_{s,t}^{\gamma} - (V_{s,t}^{\gamma\xi})^2}$$

$$\mu_{4,4} = \mu_{4,3}\mu_3 \tag{10-43}$$

$$V_{s+1,t}^{\nu} = V_{s,t}^{\nu} + V_{s,t}^{\xi\nu}\mu_2$$

$$V_{s+1,t}^{\epsilon\nu} = V_{s,t}^{\epsilon\nu} - V_{s,t}^{\xi\nu}\mu_1$$

$$V_{s+1,t}^{\epsilon} = V_{s,t}^{\epsilon} - V_{s,t}^{\xi\epsilon}\mu_3$$

$$V_{s+1,t}^{\xi} = V_{s,t}^{\xi} - \mu_{4,1}V_{s,t}^{\xi\epsilon} + \mu_{4,2}V_{s,t}^{\xi\nu} - \mu_{4,3}V_{s,t}^{yy} + \mu_{4,4}V_{s,t}^{y\xi}$$

$$V_{s+1,t}^{\gamma\xi} = V_{s,t}^{yy} - \mu_3 V_{s,t}^{y\xi}$$

According to (10-39) to (10-43), we can obtain $\epsilon_{s+1,t}$, $\nu_{s+1,t}$, $\gamma_{s+1,t}$, and $\xi_{s+1,t}$ from $\epsilon_{s,t}$, $\nu_{s,t}$, $\gamma_{s,t}$, and $\xi_{s,t}$. In addition, the residual power values with $s + 1$ AR order and t MA order are also calculated from the residual power values with s AR order and t MA order. The coefficient $\mu_{4,3}$ in (10-43) and the equations in (10-43) are obtained from Theorems 10-2 and 10-3.

From (10-43) the AR lattice error filter is built up. In other words, from (10-34) and (10-36) we get

$$\underline{h}_{s+2,t+1}(k)^T \underline{I}_1 = [y(k)\ y(k-1)\ \cdots\ y(k-s)\ \vdots\ -x(k)\ -x(k-1)\ \cdots\ -x(k-t)]$$

$$\underline{h}_{s+2,t+1}(k)^T \underline{I}_2 \tag{10-44}$$

$$= [y(k-1)\ y(k-2)\ \cdots\ y(k-s-1)\ \vdots\ -x(k-1)\ -x(k-2)\ \cdots\ -x(k-t)\ 0]$$

Thus by using (10-35) and (10-39), $\epsilon_{s+1,t}(k)$ is rewritten as

$$\epsilon_{s+1,t}(k) = -\underline{h}_{s+1,t}(k)^T \underline{\epsilon}_{s+1,t}$$

$$= -\underline{h}_{s+1,t}(k)^T \underline{I}_1\underline{\epsilon}_{s,t} - \mu_1\underline{h}_{s+1,t}(k)^T \underline{I}_2\underline{\xi}_{s,t} \tag{10-45}$$

$$= \epsilon_{s,t}(k) - \mu_1\xi_{s,t}(k-1)$$

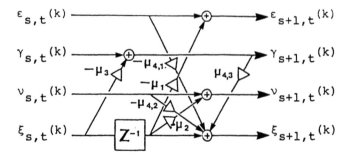

FIGURE 10-3 Lattice error filter of AR type.

In addition, we obtain

$$\nu_{s+1,t}^{(k)} = \nu_{s,t}^{(k)} - \mu_2 \xi_{s,t}^{(k-1)}$$

$$\gamma_{s,t}^{(k)} = \gamma_{s,t}^{(k)} - \mu_3 \xi_{s,t}^{(k)} \qquad (10\text{-}46)$$

$$\xi_{s+1,t}^{(k)} = \xi_{s,t}^{(k-1)} - \mu_{4,1} \epsilon_{s,t}^{(k)} - \mu_{4,2} \nu_{s,t}^{(k)} + \mu_{4,3} \gamma_{s,t}^{(k)} - \mu_{4,4} \xi_{s,t}^{(k)}$$

By using $\mu_{4,4} = \mu_{4,3} \mu_3$ in (10-43), $\xi_{s+1,t}$ is rewritten as

$$\xi_{s+1,t}^{(k)} = \xi_{s,t}^{(k-1)} - \mu_{4,1} \epsilon_{s,t}^{(k)} - \mu_{4,2} \nu_{s,t}^{(k)} + \mu_{4,3} \gamma_{s+1,t}^{(k)} \qquad (10\text{-}47)$$

From (10-45) to (10-47), the AR lattice error filter is designed in Fig. 10-3.

10-3-2 MA Lattice Error Filter

Let us derive an MA recursive formula for the ARMA parameters whose AR and MA orders increase from s and t to s and t + 1, respectively. Since its derivation is similar to the derivation in Sec. 10-3-1, let us show only final results.

The following matrices $\underline{I}_3$ and $\underline{I}_4$ are defined:

$$\underline{I}_3 [w(1) \; \cdots \; w(s+1) \; | \; w(s+2) \; \cdots \; w(s+t+2)]^T$$

$$= [w(1) \; \cdots \; w(s+1) \; | \; w(s+2) \; \cdots \; w(s+t+2) \; 0]^T$$

$$\underline{I}_4 [w(1) \; \cdots \; w(s+1) \; | \; w(s+2) \; \cdots \; w(s+t+2)]^T$$

$$= [0 \; w(1) \; \cdots \; w(s) \; | \; 0 \; w(s+2) \; \cdots \; w(s+t+2)]^T$$

$$(10\text{-}48)$$

By using the two matrices above, the MA recursive formula is derived as follows:

$$\epsilon_{s,t+1} = \underline{I}_3 \epsilon_{s,t} - \eta_1 \underline{I}_4 \gamma_{s,t} \tag{10-49}$$

$$\nu_{s,t+1} = \underline{I}_3 \nu_{s,t} + \eta_2 \underline{I}_4 \gamma_{s,t} \tag{10-50}$$

$$\gamma_{s,t+1} = \underline{I}_4 \gamma_{s,t} - \underline{I}_3 [\eta_{3,1} \epsilon_{s,t} - \eta_{3,2} \nu_{s,t} + \eta_{3,3} \gamma_{s,t} - \eta_{3,4} \xi_{s,t}] \tag{10-51}$$

$$\xi_{s,t+1} = \underline{I}_3 [\xi_{s,t} + \eta_4 \gamma_{s,t}] \tag{10-52}$$

$$\eta_1 = \frac{V_{s,t}^{\gamma\epsilon}}{V_{s,t}^{\gamma}}, \quad \eta_2 = \frac{V_{s,t}^{\gamma\nu}}{V_{s,t}^{\gamma}}$$

$$\begin{bmatrix} \eta_{3,1} \\ \eta_{3,2} \end{bmatrix} = \frac{-1}{\left(V_{s,t}^{\epsilon\nu}\right)^2 - V_{s,t}^{\nu} V_{s,t}^{\epsilon}} \begin{bmatrix} V_{s,t}^{\nu} & -V_{s,t}^{\epsilon\nu} \\ -V_{s,t}^{\epsilon\nu} & V_{s,t}^{\epsilon} \end{bmatrix} \begin{bmatrix} V_{s,t}^{\gamma\epsilon} \\ V_{s,t}^{\gamma\nu} \end{bmatrix}$$

$$\eta_{3,3} = -\eta_{3,4}\eta_4$$

$$\eta_{3,4} = \frac{V_{s,t}^{\gamma} V_{s,t}^{x\xi} - V_{s,t}^{\gamma\xi} V_{s,t}^{x\gamma}}{(V_{s,t}^{\epsilon\nu})^2 - V_{s,t}^{\nu} V_{s,t}^{\epsilon}}$$

$$\eta_4 = \frac{V_{s,t}^{\gamma\xi}}{V_{s,t}^{\gamma}}$$

$$V_{s,t+1}^{\nu} = V_{s,t}^{\nu} - V_{s,t}^{\gamma\nu}\eta_2$$

$$V_{s,t+1}^{\epsilon} = V_{s,t}^{\epsilon} + V_{s,t}^{\gamma\epsilon}\eta_1$$

$$V_{s,t+1}^{\epsilon\nu} = V_{s,t}^{\epsilon\nu} - V_{s,t}^{\gamma\nu}\eta_1 \tag{10-53}$$

$$V_{s,t+1}^{\gamma\xi} = \eta_{3,3} V_{s,t}^{\gamma\xi} - \eta_{3,4} V_{s,t}^{\xi}$$

$$V_{s,t+1}^{\gamma} = V_{s,t}^{\gamma} + \eta_{3,1} V_{s,t}^{\gamma\epsilon} - \eta_{3,2} V_{s,t}^{\gamma\nu} + \eta_{3,3} V_{s,t}^{x\gamma} - \eta_{3,4} V_{s,t}^{x\xi}$$

where

$$V^{\gamma\epsilon}_{s,t} = -[r^{yx}_t \, r^{yx}_{t-1} \, \cdots \, r^{yx}_{t-s+1}] \hat{\underline{a}}^{\epsilon}_{s,t} + r^x_{-t-1} + [r^x_{-t} \, r^x_{1-t} \, \cdots \, r^x_{-1}] \hat{\underline{b}}^{\epsilon}_{s,t}$$

$$V^{\gamma\nu}_{s,t} = r^{yx}_{t+1} + [r^{yx}_t \, r^{yx}_{t-1} \, \cdots \, r^{yx}_{t-s+1}] \hat{\underline{a}}^{\nu}_{s,t} + [r^x_{-t} \, r^x_{1-t} \, \cdots \, r^x_{-1}] \hat{\underline{b}}^{\nu}_{s,t}$$

$$V^{x\gamma}_{s,t} = -[r^{yx}_{t+1} \, r^{yx}_t \, \cdots \, r^{yx}_{t-s+2}] \hat{\underline{a}}^{\gamma}_{s,t} + [r^x_{-t-1} \, r^x_{-t} \, \cdots \, r^x_{-2}] \hat{\underline{b}}^{\gamma}_{s,t} + r^x_{-1}$$

$$V^{x\xi}_{s,t} = [r^{yx}_{t+1} \, r^{yx}_t \, \cdots \, r^{yx}_{t-s+2}] \hat{\underline{a}}^{\xi}_{s,t} - r^{yx}_{t-s+1} + [r^x_{1-t} \, r^x_{-t} \, \cdots \, r^x_{-2}] \hat{\underline{b}}^{\xi}_{s,t}$$

$$(10\text{-}54)$$

According to (10-49) to (10-53), we can obtain $\epsilon_{s,t+1}$, $\nu_{s,t+1}$, $\gamma_{s,t+1}$, and $\xi_{s,t+1}$ from $\epsilon_{s,t}$, $\nu_{s,t}$, $\gamma_{s,t}$, and $\xi_{s,t}$. In addition, the residual power values with s AR order and t + 1 MA order are also calculated from the residual power values with s AR order and t MA order.

By using (10-39) to (10-42) and (10-49) to (10-52), ARMA parameters with arbitrary ARMA orders can recursively be estimated from the given correlation data.

From (10-49) to (10-52), the MA elementary section of an ARMA lattice error filter is designed. From (10-34) and (10-48), we get

$$\underline{h}_{s,t+1}(k)^T \underline{I}_3 = [y(k) \; y(k-1) \; \cdots \; y(k-s) \; \vdots \; -x(k) \; -x(k-1) \; \cdots \; -x(k-t)]$$

$$(10\text{-}55)$$

$$\underline{h}_{s,t+1}(k)^T \underline{I}_4 = [y(k-1) \; y(k-2) \; \cdots \; y(k-s) \; 0 \; \vdots \; -x(k-1) \; -x(k-2) \; \cdots \; -x(k-t-1)]$$

Thus from (10-49) and (10-52), we obtain

$$\epsilon_{s,t+1}(k) = \epsilon_{s,t}(k) - \eta_1 \gamma_{s,t}(k-1)$$

$$\nu_{s,t+1}(k) = \nu_{s,t}(k) - \eta_2 \gamma_{s,t}(k-1)$$

$$(10\text{-}56)$$

$$\gamma_{s,t+1}(k) = \gamma_{s,t}(k-1) - \eta_{3,1}\epsilon_{s,t}(k) - \eta_{3,2}\nu_{s,t}(k) - \eta_{3,3}\gamma_{s,t}(k) + \eta_{3,4}\xi_{s,t}(k)$$

$$\xi_{s,t+1}(k) = \xi_{s,t}(k) - \eta_4 \gamma_{s,t}(k)$$

By using $\eta_{3,3} = \eta_{3,4}\eta_4$, $\gamma_{s,t+1}(k)$ is rewritten as

$$\gamma_{s,t+1}(k) = \gamma_{s,t}(k-1) - \eta_{3,1}\epsilon_{s,t}(k) - \eta_{3,2}\nu_{s,t}(k) + \eta_{3,4}\xi_{s,t+1}(k) \quad (10\text{-}57)$$

From (10-56) and (10-57) the MA lattice error filter is designed in Fig. 10-4.

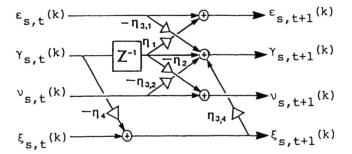

FIGURE 10-4 Lattice error filter of MA type.

10-3-3 ARMA Lattice Inverse Filter

Let us look at the ARMA recursive formula, which calculates the ARMA parameters with an AR order s + 1 and MA order t + 1 from the ARMA parameters with AR order s and MA order t. This formula is derived by using the AR recursive formula and the MA recursive formula at the same time. This formula is given as follows:

$$\underline{\epsilon}_{s+1,t+1} = \underline{I}_5 \underline{\epsilon}_{s,t} + \underline{I}_6 (\delta_{1,1} \underline{\gamma}_{s,t} - \delta_{1,2} \underline{\xi}_{s,t})$$

$$\begin{bmatrix} v^{\gamma}_{s,t} & v^{\gamma\xi}_{s,t} \\ v^{\gamma\xi}_{s,t} & v^{\xi}_{s,t} \end{bmatrix} \begin{bmatrix} \delta_{1,1} \\ \delta_{1,2} \end{bmatrix} = \begin{bmatrix} v^{\gamma\epsilon}_{s,t} \\ v^{\xi\epsilon}_{s,t} \end{bmatrix} \qquad (10\text{-}58)$$

$$\underline{\nu}_{s+1,t+1} = \underline{I}_5 \underline{\nu}_{s,t} + \underline{I}_6 (-\delta_{2,1} \underline{\gamma}_{s,t} + \delta_{2,2} \underline{\xi}_{s,t})$$

$$\begin{bmatrix} v^{\gamma}_{s,t} & v^{\gamma\xi}_{s,t} \\ v^{\gamma\xi}_{s,t} & v^{\xi}_{s,t} \end{bmatrix} \begin{bmatrix} \delta_{2,1} \\ \delta_{2,2} \end{bmatrix} = \begin{bmatrix} v^{\gamma\nu}_{s,t} \\ v^{\xi\nu}_{s,t} \end{bmatrix} \qquad (10\text{-}59)$$

$$\underline{\gamma}_{s+1,t+1} = \underline{I}_5 \underline{\gamma}_{s,t} + \underline{I}_6 (\delta_{3,1} \underline{\epsilon}_{s,t} - \delta_{3,2} \underline{\nu}_{s,t})$$

$$\begin{bmatrix} v^{\epsilon}_{s,t} & v^{\epsilon\nu}_{s,t} \\ v^{\epsilon\nu}_{s,t} & v^{\nu}_{s,t} \end{bmatrix} \begin{bmatrix} \delta_{3,1} \\ \delta_{3,2} \end{bmatrix} = \begin{bmatrix} v^{\gamma\epsilon}_{s,t} \\ v^{\gamma\nu}_{s,t} \end{bmatrix} \qquad (10\text{-}60)$$

$$\xi_{s+1,t+1} = \underline{I}_5 \xi_{s,t} + \underline{I}_6 (-\delta_{4,1} \epsilon_{s,t} + \delta_{4,2} \nu_{s,t})$$

$$\begin{bmatrix} v^{\epsilon}_{s,t} & v^{\epsilon\nu}_{s,t} \\ v^{\epsilon\nu}_{s,t} & v^{\nu}_{s,t} \end{bmatrix} \begin{bmatrix} \delta_{4,1} \\ \delta_{4,2} \end{bmatrix} = \begin{bmatrix} v^{\xi\epsilon}_{s,t} \\ v^{\xi\nu}_{s,t} \end{bmatrix} \qquad (10\text{-}61)$$

$$\begin{bmatrix} v^{\epsilon}_{s+1,t+1} & v^{\epsilon\nu}_{s+1,t+1} \\ v^{\epsilon\nu}_{s+1,t+1} & v^{\nu}_{s+1,t+1} \end{bmatrix} = \begin{bmatrix} v^{\epsilon}_{s,t} & v^{\epsilon\nu} \\ v^{\epsilon\nu}_{s,t} & v^{\nu}_{s,t} \end{bmatrix} - \begin{bmatrix} v^{\gamma\epsilon}_{s,t} & v^{\xi\epsilon}_{s,t} \\ v^{\gamma\nu}_{s,t} & v^{\xi\nu}_{s,t} \end{bmatrix} \begin{bmatrix} \delta_{1,1} & \delta_{2,1} \\ \delta_{1,2} & \delta_{2,2} \end{bmatrix}$$

$$\qquad\qquad\qquad\qquad\qquad\qquad\qquad\qquad\qquad\qquad (10\text{-}62)$$

$$\begin{bmatrix} v^{\gamma}_{s+1,t+1} & v^{\gamma\xi}_{s+1,t+1} \\ v^{\gamma\xi}_{s+1,t+1} & v^{\xi}_{s+1,t+1} \end{bmatrix} \begin{bmatrix} v^{\gamma}_{s,t} & v^{\gamma\xi}_{s,t} \\ v^{\gamma\xi}_{s,t} & v^{\xi}_{s,t} \end{bmatrix} \begin{bmatrix} v^{\gamma\epsilon}_{s,t} & v^{\gamma\nu}_{s,t} \\ v^{\xi\epsilon}_{s,t} & v^{\xi\nu}_{s,t} \end{bmatrix} \begin{bmatrix} \delta_{3,1} & \delta_{4,1} \\ \delta_{3,2} & \delta_{4,2} \end{bmatrix}$$

where $\underline{I}_5$ and $\underline{I}_6$ are defined as

$$\underline{I}_5 [w(1) \;\cdots\; w(s+1) \mid w(s+2) \;\cdots\; w(s+t+2)]^T$$

$$= [w(1) \;\cdots\; w(s+1) \; 0 \mid w(s+2) \;\cdots\; w(s+t+2) \; 0]^T$$

$$\qquad\qquad\qquad\qquad\qquad\qquad\qquad\qquad (10\text{-}63)$$

$$\underline{I}_6 [w(1) \;\cdots\; w(s+1) \mid w(s+2) \;\cdots\; w(s+t+2)]^T$$

$$= [0 \; w(1) \;\cdots\; w(s+1) \mid 0 \; w(s+2) \;\cdots\; w(s+t+2)]^T$$

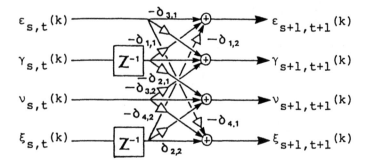

FIGURE 10-5 Lattice error filter of ARMA type.

The ARMA section of an ARMA lattice error filter is constructed by using (10-58) to (10-62). The relations among these prediction errors are

$$\epsilon_{s+1,t+1}^{(k)} = \epsilon_{s,t}^{(k)} - \delta_{1,1}\gamma_{s,t}^{(k-1)} - \delta_{1,2}\xi_{s,t}^{(k-1)}$$

$$\nu_{s+1,t+1}^{(k)} = \nu_{s,t}^{(k)} - \delta_{2,1}\gamma_{s,t}^{(k-1)} - \delta_{2,2}\xi_{s,t}^{(k-1)}$$

$$\gamma_{s+1,t+1}^{(k)} = \gamma_{s,t}^{(k-1)} - \delta_{3,1}\epsilon_{s,t}^{(k)} - \delta_{3,2}\nu_{s,t}^{(k)} \qquad (10\text{-}64)$$

$$\xi_{s+1,t+1}^{(k)} = \xi_{s,t}^{(k-1)} - \delta_{4,1}\epsilon_{s,t}^{(k)} - \delta_{4,2}\nu_{s,t}^{(k)}$$

Equation (10-64) is the recursive formula, which increases both the AR and MA orders by one simultaneously. The ARMA lattice error filter is shown in Fig. 10-5.

10-4 ELEMENTARY SECTIONS OF ARMA DIGITAL LATTICE FILTER

An ARMA lattice filter is derived from the ARMA lattice error filter. The lattice error filter calculates the forward prediction errors and the backward prediction errors for the observed input signal $x(k)$ and the observed output signal $y(k)$. In addition, it minimizes these error powers. The signals produced at each section are estimation errors. The orthogonal conditions in Theorem 10-2 are satisfied. The ARMA lattice filter [i.e., a synthesis filter whose input signals are $x(k)$'s and output signals are $y(k)$'s] is constructed by the same design process as that of the ARMA lattice error filter. Thus it can be designed by using the AR and MA elementary sections. The signals yielded at each section satisfy the orthogonal conditions.

The elementary units of the ARMA lattice filter are depicted in Figs. 10-6 to 10-8. The initial conditions at the first and last sections are

$$\nu_{0,0}^{(k)} = \xi_{0,0}^{(k)} = y(k)$$

$$\epsilon_{0,0}^{(k)} = \gamma_{0,0}^{(k)} = x(k) \qquad (10\text{-}65)$$

$$\nu_{n,m}^{(k)} - \frac{\epsilon_{n,m}^{(k)} V_{s,t}^{\nu\epsilon}}{V_{n,m}^{\epsilon}} = 0 \qquad (10\text{-}66)$$

if $\min V_{n,m} = 0$ at $s = n$ and $t = m$. Equation (10-66) implies that $\hat{H}(z^{-1})$ is exactly identical to $H(z^{-1})$. If (10-66) is used when $\min V_{s,t} \neq 0$, we get

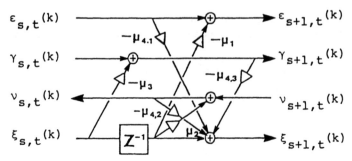

FIGURE 10-6 Lattice filter of AR type.

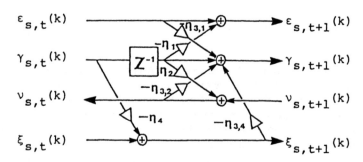

FIGURE 10-7 Lattice filter of MA type.

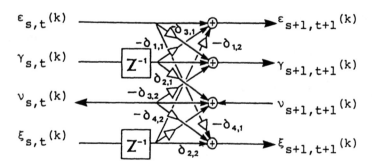

FIGURE 10-8 Lattice filter of ARMA type.

$$\nu_{s,t}(k) - \frac{\epsilon_{s,t}(k) V_{s,t}^{\nu\epsilon}}{V_{s,t}^{\epsilon}}$$

$$= y'(k) + \sum_{j=1}^{s} \left(\hat{a}_j^{\nu} + \frac{\hat{a}_j^{\epsilon} V_{s,t}^{\nu\epsilon}}{V_{s,t}^{\epsilon}} \right) y'(k-j) - \frac{x(k) V_{s,t}^{\nu\epsilon}}{V_{s,t}^{\epsilon}} - \sum_{i=1}^{t} \left(\hat{b}_i^{\nu} + \frac{\hat{b}_i^{\epsilon} V_{s,t}^{\nu\epsilon}}{V_{s,t}^{\epsilon}} \right) x(k-i)$$

$$= 0 \tag{10-67}$$

where $y'(k-i)$ $(i > 0)$ is synthesized on the condition of (10-66). By using (10-15) we obtain

$$y'(k) + \sum_{j=1}^{s} \hat{a}_j y'(k-j) - \hat{b}_0 x(k) - \sum_{i=1}^{t} \hat{b}_i x(k-i) = 0 \tag{10-68}$$

Since the output $y'(k)$ is synthesized by the ARMA parameters which are estimated by (10-4), it gives the optimal ARMA model when the AR and MA orders are s and t, respectively. Therefore, the ARMA lattice filter obtained by using (10-65) and (10-66) is a best approximation ARMA filter based on the least mean square criterion $V_{s,t}$ in (10-3). This filter is portrayed in Fig. 10-9.

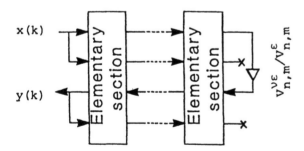

FIGURE 10-9 ARMA digital lattice filter.

10-5 ARMA DIGITAL LATTICE FILTER
WITH WHITE GAUSSIAN INPUT

If we assume an input sequence $x(k)$ as a white Gaussian process, a designed ARMA lattice filter is fairly simplified. Under the assumption, the vectors $\hat{\underline{a}}_{s,t}^{\epsilon}$ and $\hat{\underline{b}}_{s,t}^{\epsilon}$ become zero. Thus $\epsilon_{s,t}(k) = x(k)$. Since it is not necessary to calculate $\epsilon_{s,t}(k)$, we define the following new estimation error instead of $\epsilon_{s,t}(k)$ and $\nu_{s,t}(k)$:

$$\psi_{s,t}(k) = \nu_{s,t}(k) - \hat{b}_0 \epsilon_{s,t}(k) \tag{10-69}$$

where $\hat{b}_0$ is calculated by (10-15). The vector associated with $\psi_{s,t}(k)$ is also defined as

$$\psi_{s,t}(k) = \underline{h}_{s+1,t+1}^T(k)\underline{\psi}_{s,t} \tag{10-70}$$

where

$$\underline{\psi}_{s,t}^T = [1 \ \hat{a}_1^{\psi} \ \cdots \ \hat{a}_s^{\psi} \mid \hat{b}_0^{\psi} \ \cdots \ \hat{b}_t^{\psi}] \tag{10-71}$$

By using (10-69) three kinds of error recursions are given as:

AR elementary section:

$$\psi_{s+1,t}(k) = \psi_{s,t}(k) - \mu_2 \xi_{s,t}(k-1)$$

$$\gamma_{s+1,t}(k) = \gamma_{s,t}(k) - \mu_3 \xi_{s,t}(k) \tag{10-72}$$

$$\xi_{s+1,t}(k) = \xi_{s,t}(k-1) + \mu_4 \psi_{s,t}(k) - \mu_{4,3}\gamma_{s+1,t}(k)$$

where μ_4 is calculated as

$$\mu_4 = \frac{V_{s,t}^{\xi\psi}}{V_{s,t}^{\psi}}$$

$$V_{s,t}^{\xi\psi} = E[\xi_{s,t}(k)\psi_{s,t}(k) \mid g(k)]$$

$$V_{s,t}^{\psi} = E[\psi_{s,t}(k)\psi_{s,t}(k) \mid g(k)]$$

MA elementary section:

$$\psi_{s,t+1}{}^{(k)} = \psi_{s,t}{}^{(k)} - \eta_2 \gamma_{s,t}{}^{(k-1)}$$

$$\gamma_{s,t+1}{}^{(k)} = \gamma_{s,t}{}^{(k-1)} - \eta_3 \psi_{s,t}{}^{(k)} - \eta_{3,4} \xi_{s,t+1}{}^{(k)} \qquad (10\text{-}73)$$

$$\xi_{s,t+1}{}^{(k)} = \xi_{s,t}{}^{(k)} - \eta_4 \gamma_{s,t}{}^{(k)}$$

where η_3 is calculated as

$$\eta_3 = \frac{V_{s,t}^{\gamma\psi}}{V_{s,t}^{\psi}}$$

$$V_{s,t}^{\gamma\psi} = E[\gamma_{s,t}{}^{(k)} \psi_{s,t}{}^{(k)} \mid g(k)]$$

ARMA section:

$$\psi_{s+1,t+1}{}^{(k)} = \psi_{s,t}{}^{(k)} - \delta_{2,1} \gamma_{s,t}{}^{(k-1)} + \delta_{2,2} \xi_{s,t}{}^{(k-1)}$$

$$\gamma_{s+1,t+1}{}^{(k)} = \gamma_{s,t}{}^{(k-1)} - \delta_3 \psi_{s,t}{}^{(k)} \qquad (10\text{-}74)$$

$$\xi_{s+1,t+1}{}^{(k)} = \xi_{s,t}{}^{(k-1)} - \delta_4 \psi_{s,t}{}^{(k)}$$

where δ_3 and δ_4 are calculated as

$$\delta_3 = \frac{V_{s,t}^{\gamma\psi}}{V_{s,t}^{\psi}}$$

$$\delta_4 = \frac{V_{s,t}^{\xi\psi}}{V_{s,t}^{\psi}}$$

Each section is represented in Figs. 10-10 to 10-12. The ARMA digital lattice error filter can be constructed as the cascade of each section. The initial conditions are

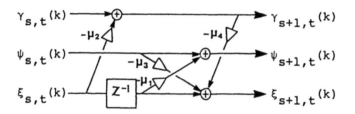

FIGURE 10-10 AR elementary section used in the lattice error filter.

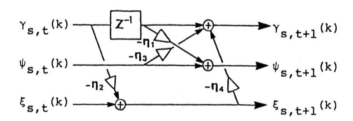

FIGURE 10-11 MA elementary section used in the lattice error filter.

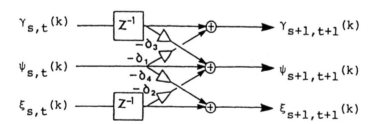

FIGURE 10-12 ARMA section used in the lattice error filter.

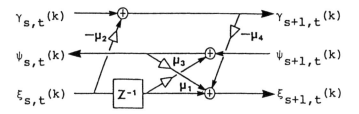

FIGURE 10-13 AR elementary section used in the lattice filter.

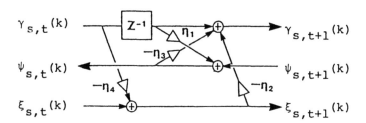

FIGURE 10-14 MA elementary section used in the lattice filter.

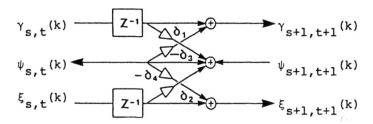

FIGURE 10-15 ARMA section used in the lattice filter.

$$\psi_{0,0}(k) = y(k) - \hat{b}_0 u(k), \quad \hat{b}_0 = \frac{r_c(0)}{\sigma_u^2}$$

$$\gamma_{0,0}(k) = u(k) \tag{10-75}$$

$$\xi_{0,0}(k) = y(k)$$

An ARMA lattice filter is designed from the ARMA lattice error filter. The input signal of the ARMA lattice filter is the observed input signal $u(k)$. The output signal is the estimated output signal $\hat{y}(k)$. The elementary sections are portrayed in Figs. 10-13 to 10-15. The conditions at the first section are the same as those in (10-75). The condition at the final section is

$$\psi_{s,t}(k) = 0 \tag{10-76}$$

where the power of an estimation error is assumed to be zero.

When the AR lattice filter is constructed by the Levinson algorithm, the output signal is synthesized from the filter excited by the estimation error at the final section as an input signal. In the ARMA lattice filter, (10-76) is used in the synthesizer. Thus the waveform $\hat{y}(k)$ is synthesized by the estimation model with AR and MA orders (s, t):

$$\hat{y}(k) = -\hat{a}_1 y(k-1) - \hat{a}_2 y(k-2) - \cdots - \hat{a}_{k-n} y(k-n)$$

$$+ \hat{b}_0 u(k) + \hat{b}_1 u(k-1) + \cdots + \hat{b}_m u(k-m) \tag{10-77}$$

Even if $v_{n,m} \neq 0$, (10-77) is obtained from (10-76). However, the synthesized signals in this case are not identical to the observed signals but to the optimum signals where the orders of ARMA estimation model are (n, m). Figure 10-16 shows the entire ARMA lattice filter.

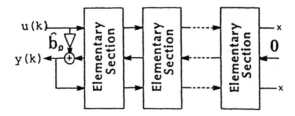

FIGURE 10-16   ARMA digital lattice filter.

10-6 EXPERIMENTAL RESULTS

By using the new fast recursive algorithm, we can design the ARMA lattice
filter from an input signal and an output signal. The observed waveform is
synthesized by the reference model whose resonance frequencies are 1.5
and 3 kHz, and whose antiresonance frequency is 2 kHz. Figure 10-17 shows
FFT spectrum of the observed output. The input of the reference model is a
white Gaussian process. With 800 samples of the input and the output data,
all correlation data are appropriately calculated. Every power value of
the estimation error with $s = 0, 1, \ldots, 8$ and $t = 0, 1, \ldots, 8$ is shown in
Fig. 10-18. Figure 10-18 represents the minimum value of $V_{s,t}$ at each
order. From (10-3) we get

$$V_{s,t} \geq 0 \qquad\qquad\qquad (10\text{-}78)$$

But since each correlation data point is calculated as

$$r_a(i) = \frac{1}{800} \sum_{k=1}^{800} y(k)y(k-i)$$

$$r_c(i) = \frac{1}{800} \sum_{k=1}^{800} y(k)u(k-i)$$

$$\sigma_u^2 = \frac{1}{800} \sum_{k=1}^{800} u(k)^2 \qquad\qquad (10\text{-}79)$$

$$y(j) = u(j) = 0 \quad (j \leq 0)$$

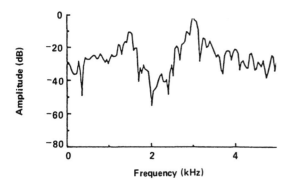

FIGURE 10-17 FFT spectrum of observed signals.

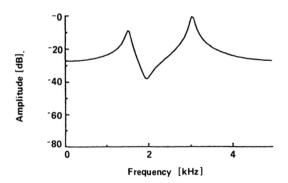

	0	1	2	3	4	5	6	7	8
0	147	146	142	133	132	107	102	89.9	84.7
1	140	140	136	129	129	102	99.8	84.5	82.0
2	52.9	51.8	48.7	45.8	43.1	33.8	33.8	33.3	33.0
3	38.5	37.9	36.6	36.5	32.1	27.0	26.6	25.6	24.5
4	20.4	14.7	0.276	0.009	-0.00	-0.44	-0.66	-1.00	-1.21
5	10.2	9.46	0.274	-20.1	-650	9050	8.46	3.73	2.43
6	9.97	8.05	0.070	-21.0	-6.74	-7.52	-7.76	-8.98	-13.4
7	7.29	5.85	-0.25	-60.3	-6.75	-5.17	-5.21	-5.21	-6.34
8	6.40	5.66	-0.26	21.0	-10.5	-6.13	0.657	-0.34	-0.36

order m (columns), order n (rows)

FIGURE 10-18 Least mean square estimation error power at each ARMA order.

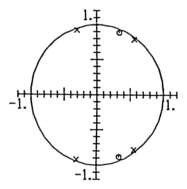

FIGURE 10-19 Spectrum of the estimation model.

FIGURE 10-20 Poles and zeros of the estimation model.

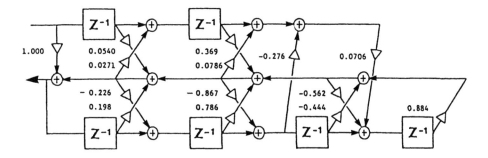

FIGURE 10-21   Estimated ARMA digital lattice filter.

we do not observe the exact correlation data. Thus $V_{s,t} \neq 0$ is obtained.
In Fig. 10-18, when the estimation model whose total number of parameters
$s + t + 1$ is minimum is selected under $V_{s,t} \cong 0$, we get $s = 4$ and $t = 2$.
Thus the estimation model with ARMA orders $(4, 2)$ is the minimal realiza-
tion model. Figure 10-19 shows the estimated spectrum with $s = 4$ and
$t = 2$. Figure 10-20 shows its poles and zeros in the Z-plane. From
Figs. 10-17 and 10-19, it is shown that a good approximation is obtained.
Figure 10-21 represents the estimated ARMA digital lattice filter with
$s = 4$ and $t = 2$.

10-7 SUMMARY

We have developed the ARMA lattice error filter based on a fast recursive
algorithm for ARMA parameter estimates. The lattice error filter consists
of two elementary sections: a section that increases the AR order by one
and another that increases the MA order by one. The signals yielded at each
elementary section have orthogonal conditions. The ARMA lattice filter is
also derived from the fast recursive algorithm. The lattice filter proposed
in this chapter is based on linear prediction theory, by which ARMA parameters can be estimated at a low cost; its properties have been explained.

We also discussed the lattice filter, where the input signal of the reference model is a white Gaussian process. A new fast recursive algorithm
is developed. In addition, the new ARMA lattice filter associated with this
algorithm has been represented. Its filter uses only three prediction errors.
Its calculation costs have been shown to be lower than those by previous
methods.

REFERENCES

1. N. Levinson, "A heuristic exposition of Wiener's mathematical theory of prediction and filtering," J. Math. Phy. (Cambridge, Mass.), vol. XXVI, no. 2, pp. 110-119 (July 1947).

2. H. Wakita, "Estimation of vocal tract shapes from acoustic data: its status of art," 9rth Meeting Acoustical Society of America, Miami Beach, Fla., Dec. 1978.

3. J. D. Markel and A. H. Gray, Jr., Linear Prediction of Speech, Springer-Verlag, Berlin, 1976.

4. M. D. Ortigueira and J. M. Tribolet, "On the double Levinson recursion formulation of ARMA spectral estimation," IEEE ICASSP Boston, pp. 1076-1079, 1983.

5. K. Ogino, "A fast algorithm for unmodified ARMA spectrum estimation," IEEE ICASSP, Boston, pp. 1106-1109, 1983.

6. D. T. L. Lee, M. Morf, and B. Friedlander, "Recursive least squares ladder estimation algorithms," IEEE Trans. Acoust. Speech Signal Process., vol. ASSP-29, no. 3, pp. 627-641 (June 1981).

7. D. T. Lee, B. Friedlander, and M. Morf, "Recursive ladder algorithms for ARMA lattice filters," IEEE Trans. Autom. Control, vol. AC-27, no. 4, pp. 753-764 (Aug. 1982).

8. B. Friedlander and S. Maitra, "Speech deconvolution by recursive ARMA lattice filters," IEEE ICASSP, pp. 343-346, 1981.

9. B. Friedlander, "System identification technique for adaptive signal processing," Circuits Syst. Signal Process, vol. 1, no. 1, pp. 3-41 (1982).

10. Y. Miyanaga, H. Watanabe, N. Miki, and N. Nagai, "A fast calculation algorithm for a parameter estimation of autoregressive and moving-average model," Trans. Inst. Elec. Commun. Eng. (Jpn.), vol. J66-A, no. 10, pp. 1000-1007 (Oct. 1983).

11. Y. Miyanaga, N. Miki, and N. Nagai, "ARMA digital lattice filter based on a linear prediction theory," Trans. Inst. Elec. Commun. Eng. (Jpn.), vol. J67-A, no. 5, pp. 487-494 (May 1984).

12. Y. Miyanaga, N. Nagai, and N. Miki, "ARMA digital lattice filter with white Gaussian input," Trans. IECE (Jpn.), vol. J67-A, no. 12, pp. 1270-1277 (Dec. 1985).

11

Wave Digital Filter Synthesized with the Darlington Procedure of Classical Circuit Theory

MASAKIYO SUZUKI and NOBUO NAGAI

Hokkaido University, Sapporo, Japan

11-1 INTRODUCTION

In this chapter we deal with wave digital realizations for scalar linear time-invariant digital filters with real coefficients. In the realization of digital filters, it is required that they have the properties of low sensitivity, suppressing zero-input parasitic oscillations and absence of delay-free loops. Since all the quantities in a digital filter are represented by finite word length when it is implemented by hardware or software, its characteristics are degraded. An important index for the degradation is the sensitivity to variations of the coefficients (parameters) contained in the digital filter. Therefore, the first property is required. The second property is required for internal stability, that is, for preventing the values of internal signals in the digital filter from getting unlimitedly big. The third property indicates that every feedback loop contained in the signal flow diagram describing the filter structure contains at least one delay element. This property guarantees that each response in the digital filter can be evaluated by a finite number of operations, that is, additions and multiplications.

Wave digital filters (WDFs), introduced by Fettweis [1] in 1971, are a class of digital filters made to imitate the behavior of conventional doubly terminated lossless filters. Wave digital ladder filters [3] are the translations from classical LC ladder filters. Wave digital Jauman [4] and lattice [5] filters are derived from symmetric LC filters. The WDFs above are realized as connections of several adapters, such as series, parallel, Jauman, and lattice (two-port) adapters, and delay elements. Each wave digital adapter represents wave flows in a portion of the electrical connections in circuits, such as series and parallel connection, and so on. The WDFs above are canonic in multipliers since the number of the multipliers is equal to the numbers of degrees of freedom in the transfer function to be realized. Hence they have corresponding analog filters even though they are realized by using finite-word-length multipliers. In other words, the variations of multiplier coefficients contained in the WDFs canonic in multipliers always correspond to the variations of the element values in analog filters. Consequently, WDFs canonic in multipliers inherit the property of low sensitivity of analog filters and are always stable internally due to the losslessness of analog filters [2]. Although connecting wave digital adapters yields delay-free loops, they can be avoided by using the well-known method introduced by Sedlmeyer and Fettweis [3], that is, choosing a suitable port resistance to make the port reflection-free. Furthermore, they have the canonic number of delays, that is, the number of the reactive elements. The typical WDFs above possess all the properties mentioned above. However, they can realize only restricted types of transfer functions because of the structural restrictions of the corresponding analog filters.

The Darlington synthesis method is a general method to realize an immittance function (positive-real function) as an input impedance of a reciprocal reactance two-port terminated in a single resistance. The reactance two-port to be synthesized is constructed as a cascade of certain sections, such as series, parallel, Brune, type C, and type D sections. These are reciprocal reactance two-ports. Youla [11] extended this method to the nonreciprocal case, introducing the type E section. In the nonreciprocal case, the Richards section, which was introduced by Richards in order to design distributed constant circuits, is also used. These sections are re-

ferred to as basic reactance sections. At the same time, the Darlington
realization for an impedance provides a transfer function of the two-ports.
This fact almost immediately suggests a way of designing a reactance two-
port for prescribed transfer characteristics. It is practically useful and
important.

To generalize the realizable transfer functions of WDFs, Nouta [6]
has shown that wave digital cascade synthesis is possible. It corresponds
to the Darlington-style synthesis method. For the cascade synthesis of
WDFs, some wave digital structures equivalent to the basic reactance sections have been introduced [6-10]. We call them basic wave digital filters
(basic WDFs). Nouta [6] has shown each basic WDF, which is the wave digital translation from a cascade of each basic reactance section and a unit
element. Fahmy [7] and Scanlan and Fagan [8] have introduced the use of
commensurate microwave circuits, that is, the Ikeno loop, the double-shunt
loop, and the equivalent of a type C section made with a two-wire line. The
Ideno loop is equivalent to a cascade of a Brune or type C section and a unit
element (UE). The double-shunt loop is equivalent to a cascade of a type D
section and a unit element. The basic WDFs [6-8] have more delays than
the canonic form does, except Fahmy's structure for a type C section.
Martens and Lê [9] have presented basic WDFs only for Brune and type C
sections without a unit element. Suzuki et al. [10] have derived basic WDFs
with a canonic number of delays for Brune, type C, type E, and type D sections. Since the basic WDFs shown in [6-10] are not canonic in multipliers
as shown in Table 11-1, passive realization for them is not always possible
if the multipliers are significantly quantized. It has been shown by computer
simulation, however, that in many cases they are still passive and have
good sensitivity properties.

In this chapter we show novel basic WDFs which are derived from the
distributed equivalents of the basic reactance sections. The basic WDFs
derived have fewer multipliers than the conventional basic WDFs (as shown
in Table 11-1) and are canonic in delays. Furthermore, they have low coefficient sensitivity properties.

This chapter is organized as follows. In Sec. 11-2 we show the scattering matrix of a distributed constant circuit which realizes a scalar transfer function of a digital filter. In Sec. 11-3 the Darlington synthesis method
is described briefly. The wave digital realization for series and parallel
sections is shown in Sec. 11-4. The derivation of a basic WDF for a Richards
section is described in Sec. 11-5. In Secs. 11-6 to 11-8, first we show the
distributed equivalent of a two-wire line, followed by the derivations of the
basic WDFs for the basic reactance sections other than the Richards section.
Finally, in Sec. 11-9, some results from time-domain simulations show
that WDFs realized with the proposed WDFs have good sensitivity properties.

TABLE 11-1 Numbers of Multipliers and Delays for Basic WDFs[a]

Basic WDF	Richards	Brune	Type C	Type E	Type D
Canonic number					
Multipliers	3 (2)	4 (3)	4 (3)	5 (4)	7 (6)
Delays	1	2	2	2	4
Nouta [6]	6 (5)	7 (6)	7 (6)	10 (9)	–
	2	3	3	3	–
Fahmy [7]	–	6 (5)	6 (5)	–	10 (9)
(Ikeno loop,	–	4	4	–	7
double shunt loop)					
Fahmy [7]	–	–	7 (6)	–	–
(two-wire line)	–	–	2	–	–
Scanlan and Fagan [8]	–	–	5 (4)	–	–
	–	–	3	–	–
Martens and Lê [9]	–	6 (5)	6 (5)	–	–
	–	2	2	–	–
Suzuki et al. [10]	–	6 (5)	6 (5)	10 (9)	12 (11)
	–	2	2	2	4
This chapter	6 (5)	5 (4)	5 (4)	9 (8)	9 (8)
	1	2	2	2	4

[a]Number in parentheses indicates the case in which a port in a basic WDF is made reflection-free.

11-2 COMMENSURATE DISTRIBUTED CONSTANT CIRCUIT WITH RATIONAL TRANSFER FUNCTIONS OF DIGITAL FILTERS

Let H(z) be a rational transfer function of z^{-1} with real coefficients, that is,

$$H(z) = \frac{B(z)}{A(z)} = \frac{b_0 + b_1 z^{-1} + \cdots + b_n z^{-n}}{1 + z_1 a^{-1} + \cdots + a_n z^{-n}} \qquad (11-1)$$

where $z \ (= e^{sT})$ is a complex variable used for the z-transform, s a usual complex frequency variable, and T the sampling period. A kind of digital systems is related to commensurate distributed constant circuits, which are composed of unit elements (UEs), ideal transformers, ideal transformer interconnections (ITIs), and ideal gyrators. The complex frequency variable used in the field of commensurate distributed constant circuits is given by

$$\lambda = \tanh\frac{sT}{2} = \frac{1 - z^{-1}}{1 + z^{-1}} \tag{11-2}$$

The transfer function $H(z)$ can correspond to one of several kinds of circuit functions in a suitably terminated reactance two-port, such as a voltage (or a current) transfer function, an open voltage (or a short current) transfer function, or a nondiagonal entry of a voltage (or a current) scattering matrix. If $|H(e^{j\theta})| \leq 1$ (a.e. θ) holds, then $H(z)$ can correspond to an entry of a scattering matrix or a diagonal entry of a voltage (or current) scattering matrix. In this section we discuss the case in which $H(z)$ corresponds to a $(2,1)$ entry of a scattering matrix.

The incident and reflected quantities at port i (i = 1, 2) of the circuit shown in Fig. 11-1 are defined as

$$\alpha_i = \frac{1}{2\sqrt{R_i}}(V_i + R_iI_i), \qquad \beta_i = \frac{1}{2\sqrt{R_i}}(V_i - R_iI_i) \qquad (i = 1, 2) \tag{11-3}$$

The relationships among incident and reflected quantities are described by a scattering matrix, that is,

$$\begin{bmatrix} \beta_1 \\ \beta_2 \end{bmatrix} = S(z)\begin{bmatrix} \alpha_1 \\ \alpha_2 \end{bmatrix} \tag{11-4}$$

Assuming that $|H(e^{j\theta})| \leq 1$ (a.e. θ), let $H(z)$ correspond to a $(2,1)$ entry of the scattering matrix of a doubly terminated reactance two-port. The scattering matrix is represented as follows:

$$S(z) = \frac{1}{A(z)}\begin{bmatrix} C(z) & \epsilon B_*(z) \\ B(z) & -\epsilon C_*(z) \end{bmatrix} \tag{11-5}$$

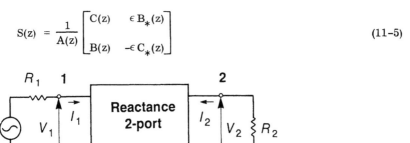

FIGURE 11-1 Doubly terminated reactance two-port.

where

$$A_*(z) = A\left(\frac{1}{z}\right)z^{-n} = a_n + a_{n-1}z^{-1} + \cdots + a_1 z^{-(n-1)} + z^{-n} \qquad (11\text{-}6)$$

$$\epsilon = \pm 1 \qquad (11\text{-}7)$$

and the polynomial $C(z)$ is determined by the spectral factorization of

$$C(z)C_*(z) = A(z)A_*(z) - B(z)B_*(z) \qquad (11\text{-}8)$$

Equation (11-8) represents the losslessness of the reactance two-port. Because of the condition $|H(e^{j\theta})| \leq 1$, it is guaranteed that the spectral factorization of (11-8) is accomplished. The input impedance at port 1 is given by

$$Z(z) = R_1 \frac{1 + S_{11}(z)}{1 - S_{11}(z)} = R_1 \frac{A(z) + C(z)}{A(z) - C(z)} \qquad (11\text{-}9)$$

where $S_{ij}(z)$, $(i, j = 1, 2)$ is an (i, j) entry of the scattering matrix.

11-3 DARLINGTON SYNTHESIS METHOD

According to the Darlington synthesis method, a reactance two-port terminated in a single resistance can be synthesized as a cascade of basic reactance sections as shown in Fig. 11-2. Each basic reactance section is extracted from the input impedance $Z(z)$. In this section the extraction procedures [11-13] are shown.

Theorem 11-1. A rational positive real function $W(\lambda)$, which has a positive real part for $\lambda > 0$, can be represented as

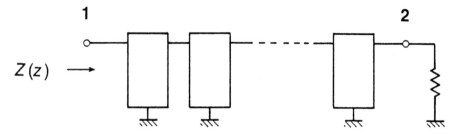

FIGURE 11-2 Circuit synthesized by the Darlington synthesis method.

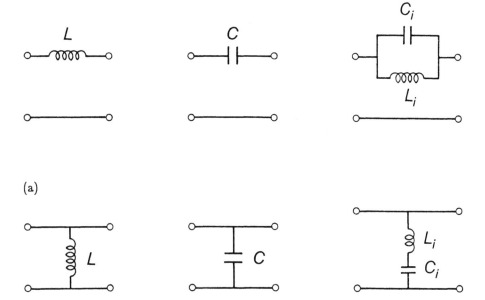

(a)

(b)

FIGURE 11-3 (a) Series sections; (b) parallel sections.

$$W(\lambda) = k_\infty \lambda + \frac{k_0}{\lambda} + \sum_{i=1}^{n} \frac{2k_i \lambda}{\lambda^2 + \omega_i^2} + W_1(\lambda) \qquad (11\text{-}10)$$

where

$$k_\infty = \lim_{\lambda \to \infty} \frac{W(\lambda)}{\lambda}, \quad k_0 = \lim_{\lambda \to 0} \lambda W(\lambda), \quad k_i = \lim_{\lambda \to j\omega_i} (\lambda - j\omega_i) W(\lambda) \qquad (11\text{-}11)$$

The function $W_1(\lambda)$ is also a rational positive real function not having poles at $\lambda = \infty$, $\lambda = 0$, or $\lambda = \pm j\omega_i$ $(i = 1, 2, \ldots, n)$. ∎

If $W(\lambda)$ is an impedance function, each term in (11-10) except $W_1(\lambda)$ represents one of the series sections shown in Fig. 11-3(a), where

$$L = k_\infty, \quad C = k_0, \quad L_i = \frac{2k_i}{\omega_i^2}, \quad C_i = \frac{1}{2k_i} \qquad (11\text{-}12)$$

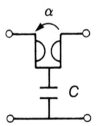

FIGURE 11-4 Richards section.

If $W(\lambda)$ is an admittance function, it represents one of the parallel sections shown in Fig. 11-3(b), where

$$C = k_\infty, \quad L = k_0, \quad C_i = \frac{2k_i}{\omega_i^2}, \quad L_i = \frac{1}{2k_i} \tag{11-13}$$

Extracting series and parallel sections by using (11-10) removes the poles and zeros on an imaginary axis in the complex λ-plane from the impedance function, respectively.

Theorem 11-2 (Richards). The following function $W_1(\lambda)$, obtained by transforming the positive real function $W(\lambda)$ with respect to a real point $\lambda = \sigma_0$ $(\sigma_0 > 0)$, is also a positive real function.

$$W_1(\lambda) = W_0 \frac{\sigma_0 W(\lambda) - \lambda W_0}{\sigma_0 W_0 - \lambda W(\lambda)} \tag{11-14}$$

where

$$W_0 = W(\sigma_0) \tag{11-15}$$

If $\lambda = \sigma_0$ is a zero of the para-even part of $W(\lambda)$, that is,

$$E(\lambda) = \frac{1}{2}\{W(\lambda) + W(-\lambda)\} \tag{11-16}$$

then the degree of $W_1(\lambda)$ is reduced to deg $W(\lambda) - 1$. Otherwise, $W_1(\lambda)$ has the same degree as $W(\lambda)$. ∎

The transformation (11-14) implies that the Richards section shown in Fig. 11-4 is extracted from the one-port having $W(\lambda)$ as its impedance. The element values in the Richards section is given by

$$\alpha = \pm W_0 , \quad C = \frac{1}{\sigma_0 \, W_0} \tag{11-17}$$

The Richards section realizes a single transmission zero at $\lambda = \sigma_0$ if $\alpha > 0$, or at $\lambda = -\sigma_0$ if $\alpha < 0$.

Theorem 11-3 (Darlington). The following function $W_2(\lambda)$ obtained by transforming the positive real function $W(\lambda)$ with respect to a real point $\lambda = \sigma_0$ ($\sigma_0 > 0$) is also a positive real function.

$$W_2(\lambda) = W_0 \, \frac{(\phi \lambda^2 + \sigma_0^2) \, W(\lambda) - \lambda(1 + \phi) \sigma_0 \, W_0}{(\phi^{-1} \lambda^2 + \sigma_0^2) \, W_0 - \lambda(1 + \phi^{-1}) \sigma_0 \, W(\lambda)} \tag{11-18}$$

where

$$\phi = \frac{W_0 - \sigma_0 \, W_0'}{W_0 + \sigma_0 \, W_0'} , \quad W_0 = W(\sigma_0), \quad W_0' = \left. \frac{dW(\lambda)}{d\lambda} \right|_{\lambda = \sigma_0} \tag{11-19}$$

If $\lambda = \sigma_0$ is the first-order real zero of the para-even part of $W(\lambda)$, the degree of $W_2(\lambda)$ is reduced to deg $W(\lambda)$ - 1. If $\lambda = \sigma_0$ is the second-order real zero of the para-even part of $W(\lambda)$, the degree of $W_2(\lambda)$ is reduced to det $W(\lambda)$ - 2. Otherwise, $W_2(\lambda)$ has the same degree as $W(\lambda)$. ∎

The transformation in (11-18) implies that the type C section (M < 0) shown in Fig. 11-5 is extracted from the one-port having $W(\lambda)$ as its impedance. The element values in the type C section are given by

$$L_1 = \frac{W_0 + \sigma_0 \, W_0'}{2\sigma_0} , \quad L_2 = \frac{(W_0 - \sigma_0 \, W_0')^2}{2\sigma_0 \, (W_0 + \sigma_0 \, W_0')}$$

$$M = -\sqrt{L_1 \, L_2} , \quad C = \frac{2}{W_0 - \sigma_0 \, W_0'} \tag{11-20}$$

The type C section realizes two real transmission zeros at $\lambda = \pm \sigma_0$.

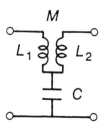

FIGURE 11-5 Reciprocal second-order section (type C section if M < 0, Brune section if M > 0).

Theorem 11-4 (Brune). If the positive real function $W(\lambda)$ satisfies

$$W(j\omega_0) + W(-j\omega_0) = 0 \tag{11-21}$$

the following function $W_2(\lambda)$ obtained by transforming $W(\lambda)$ with respect to the imaginary point $\lambda = j\omega_0$ is also a positive real function.

$$W_2(\lambda) = X_0 \frac{(\phi\lambda^2 + \omega_0^2)W(\lambda) - \lambda(1 - \phi)\omega_0 X_0}{(\phi^{-1}\lambda^2 + \omega_0^2)X_0 - \lambda(\phi^{-1} - 1)\omega_0 W(\lambda)} \tag{11-22}$$

where

$$\phi = \frac{X_0 - \sigma_0 X_0'}{X_0 + \sigma_0 X_0'}, \quad jX_0 = W(j\omega_0), \quad X_0' = \left.\frac{dW(\lambda)}{d\lambda}\right|_{\lambda=j\omega_0} \tag{11-23}$$

The degree of $W_2(\lambda)$ is reduced to deg $W(\lambda) - 2$.   ∎

The transformation (11-22) implies that the Brune section $(M > 0)$ shown in Fig. 11-5 is extracted from the one-port having $W(\lambda)$ as its impedance. The element values in the Brune section are given by

$$L_1 = \frac{\omega_0 X_0' + X_0}{2\omega_0}, \quad L_2 = \frac{(\omega_0 X_0' - X_0)^2}{2\omega_0(\omega_0 X_0' + X_0)}$$

$$M = \sqrt{L_1 L_2}, \quad C = \frac{2}{\omega_0(\omega_0 X_0' - X_0)} \tag{11-24}$$

The Brune section realizes two imaginary transmission zeros at $\lambda = \pm j\omega_0$.

Theorem 11-5 (Youla). The following function $W_2(\lambda)$ obtained by transforming the positive real function $W(\lambda)$ with respect to a complex point $\lambda = \lambda_0 = \sigma_0 + j\omega_0$ $(\sigma_0 > 0)$ is also a positive real function.

$$W_2(\lambda) = \frac{a(\lambda)W(\lambda) + b(\lambda)}{c(\lambda)W(\lambda) + d(\lambda)} \tag{11-25}$$

where

$$a(\lambda) = I_1\lambda^2 + |\lambda_0|^2, \quad b(\lambda) = -I_2\lambda,$$
$$c(\lambda) = -I_3\lambda, \quad d(\lambda) = I_4\lambda^2 + |\lambda_0|^2 \tag{11-26}$$

and

$$I_1 = \frac{R_0/\sigma_0 - X_0/\omega_0}{R_0/\sigma_0 + X_0/\omega_0}, \qquad I_2 = \frac{2|W_0|^2}{R_0/\sigma_0 + X_0/\omega_0}$$

$$I_3 = \frac{2}{R_0/\sigma_0 - X_0/\omega_0}, \qquad I_4 = I_1^{-1} \qquad \qquad (11\text{-}27)$$

$$W_0 = W(\lambda_0) = R_0 + jX_0$$

If $\lambda = \lambda_0$ is the complex zero of the para-even part of $W(\lambda)$, the degree of $W_2(\lambda)$ is reduced to deg $W(\lambda) - 2$. Otherwise, $W_2(\lambda)$ has the same degree as $W(\lambda)$.   ∎

The transformation in (11-25) implies that the type E section shown in Fig. 11-6 is extracted from the one-port having $W(\lambda)$ as its impedance. The element values in the type E section are given by

$$L_1 = \frac{1}{I_1 I_3}, \qquad L_2 = \frac{I_1}{I_3}, \qquad M = \frac{1}{I_3}$$

$$C = \frac{I_3}{|\lambda_0|^2}, \qquad \alpha = \pm \frac{2\sigma_0}{I_3} \qquad \qquad (11\text{-}28)$$

The type E section realizes two complex transmission zeros at $\lambda = \sigma_0 \pm j\omega_0$ if $\alpha > 0$, or at $\lambda = -\sigma_0 \pm j\omega_0$ if $\alpha < 0$.

A cascade of two type E sections extracted in succession from $W(\lambda)$ with respect to the same complex point $\lambda = \lambda_0 = \sigma_0 + j\omega_0$ can be transformed into a fourth-order section, called the type D section, which is shown in Fig. 11-7. One of the type E sections should realize the transmission zero at $\lambda = \sigma_0 \pm j\omega_0$, and the other at $\lambda = -\sigma_0 \pm j\omega_0$. The type D section to be extracted from $W(\lambda)$ can be identified as follows:

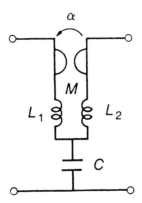

FIGURE 11-6 Type E section.

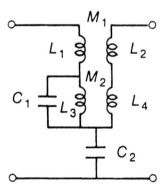

FIGURE 11-7 Type D section.

$$L_1 = \frac{I_4 J_4}{I_3 J_4 + I_1 J_3}, \qquad L_2 = \frac{I_1 J_1}{I_3 J_4 + I_1 J_3}, \qquad M_1 = \sqrt{L_1 L_2}$$

$$M_2 = -\frac{I_3^2 J_3^2 |\overline{W_2(\lambda_0)} + W(\lambda_0) I_1|^2}{|\lambda_0|^2 (I_3 J_4 + I_1 J_3)(I_3 + J_3)^2}, \qquad L_4 = -\frac{I_3 M_2}{J_3}, \qquad L_3 = \frac{M_2^2}{L_4}$$

$$C_1 = \frac{I_3 J_4 + I_1 J_3}{|\lambda_0|^2 (I_3 + J_3) L_3}, \qquad C_2 = \frac{I_3 + J_3}{|\lambda_0|^2} \tag{11-29}$$

where I_1, I_2, I_3, and I_4 are given by (11-27) with respect to $W(\lambda)$, and J_1, J_2, J_3, and J_4, which correspond to I_1, I_2, I_3, and I_4, respectively, are also given by (11-27) with respect to the positive real function $W_2(\lambda)$ transformed by (11-25).

Let $W_4(\lambda)$ denote the remaining impedance after extracting the type D section with respect to $\lambda = \lambda_0$. If $\lambda = \lambda_0$ is the first-order zero of the para-even part of $W(\lambda)$, the degree of $W_4(\lambda)$ is reduced to deg $W(\lambda) - 2$. If $\lambda = \lambda_0$ is the second-order zero of the para-even part of $W(\lambda)$, the degree of $W_4(\lambda)$ is reduced to deg $W(\lambda) - 4$. Otherwise, $W_4(\lambda)$ has the same degree as $W(\lambda)$. The type D section realizes four complex transmission zeros at $\lambda = \pm(\sigma_0 \pm j\omega_0)$.

11-4 WDFs FOR SERIES AND PARALLEL SECTIONS

In this section, wave digital realizations for the series and parallel sections are shown [1,3]. The inductance realizes the steady-state voltage-current relation, $V = \lambda L I$. The relationship between the incident and reflected voltage waves defined by

$$a = \frac{1}{2}(V + RI), \qquad b = \frac{1}{2}(V - RI), \qquad R = L \tag{11-30}$$

is given by

$$b = -z^{-1}a \qquad (11\text{-}31)$$

The capacitance realizes the steady-state voltage-current relation, $I = \lambda CV$. The relationship between the incident and reflected voltage waves defined by (11-30) with $R = 1/C$ instead of $R = L$ is given by

$$b = z^{-1}a \qquad (11\text{-}32)$$

Consider n ports, with ports i ($i = 1, 2, \ldots, n$). Let V_i and I_i represent the voltage and current at port i. The incident and reflected voltage waves a_i and b_i at port i are defined as

$$a_i = \frac{1}{2}(V_i + R_i I_i), \quad b_i = \frac{1}{2}(V_i - R_i I_i) \qquad (11\text{-}33)$$

where R_i is an arbitrary positive constant, called port resistance.
If these ports are interconnected in parallel, we have

$$V_1 = V_2 = \cdots = V_n$$
$$I_1 + I_2 + \cdots + I_n = 0 \qquad (11\text{-}34)$$

From (11-33) and (11-34), we obtain

$$b_i = a_0 - a_i, \quad a_0 = \sum_{i=1}^{n} \alpha_i a_i,$$

$$\alpha_i = \frac{2G_i}{G_1 + G_2 + \cdots + G_n}, \quad G_i = \frac{1}{R_i} \qquad (11\text{-}35)$$

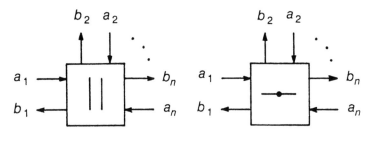

(a)   (b)

FIGURE 11-8 (a) Parallel adapter; (b) series adapter.

Equations (11–35) define a wave n–port called the parallel adapter.
    If these ports are interconnected in series, we have

$$I_1 = I_2 = \cdots = I_n$$
$$V_1 + V_2 + \cdots + V_n = 0 \qquad\qquad (11\text{-}36)$$

From (11–33) and (11–36) we obtain

$$b_i = a_i - \alpha_i a_0, \qquad a_0 = \sum_{i=1}^{n} a_i$$
$$\alpha_i = \frac{2R_i}{R_1 + R_2 + \cdots + R_n} \qquad\qquad (11\text{-}37)$$

Equations (11–37) define a wave n–port called the series adapter.
    In both cases, the α_i satisfies the relationships

$$\alpha_1 + \alpha_2 + \cdots + \alpha_n = 2 \qquad \alpha_i > 0 \qquad\qquad (11\text{-}38)$$

Due to the relationships in (11–38), the parallel and series adapters can be
realized by using n - 1 multipliers. These adapters are symbolized as shown
in Fig. 11–8.
    When two arbitrary adapters are interconnected directly, a closed
loop not containing any delay would normally appear. Then it would be impossible to find a sequence in which the values of signals can be computed.
Such delay-free loops, however, can be avoided by making a port in the
interconnection reflection-free. To make port n reflection-free, for the
parallel and series adapters, the port resistance at port n is chosen as
follows, respectively,

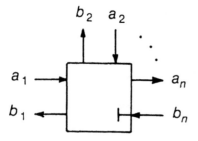

FIGURE 11-9 Adapter with a reflection-free port (port n).

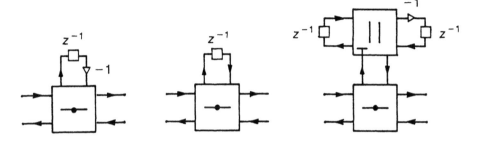

(a)

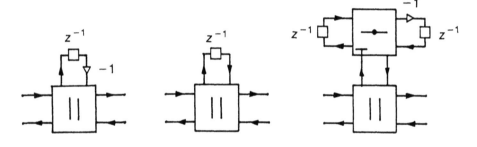

(b)

FIGURE 11-10 Wave digital realization for (a) series sections and (b) parallel sections.

$$G_n = G_1 + G_2 + \cdots + G_n \qquad\qquad (11\text{-}39)$$

$$R_n = R_1 + R_2 + \cdots + R_n \qquad\qquad (11\text{-}40)$$

Then one obtains

$$\alpha_n = 1 \qquad\qquad (11\text{-}41)$$

and a b_n for the parallel and series adapters, respectively, is given by

$$b_n = \alpha_1 a_1 + \alpha_2 a_2 + \cdots + \alpha_n a_n \qquad\qquad (11\text{-}42)$$

$$b_n = -a_1 - a_2 - \cdots - a_n \qquad\qquad (11\text{-}43)$$

Due to (11-41) the number of the independent multipliers in these adapters
is reduced to n - 2. An adapter having a reflection-free port (port n) is
shown in Fig. 11-9. WDFs for the series and parallel sections are realized
by using the parallel and series adapters and delays as shown in Fig. 11-10.

11-5 WDF FOR THE RICHARDS SECTION

The distributed equivalent of the Richards section is shown in Fig. 11-11(a),
where y_0 is the characteristic admittance of the transmission line and given
by $y_0 = C$. We first consider wave flows for the ideal gyrator in Fig. 11-11(a).
Let a_i and b_i ($i = 1, 2, 3$) indicate the incident and reflected voltage waves defined at port i with port resistance R_i ($R_3 = 1/y_0$). The voltage scattering
matrix among them is given by

$$
\begin{bmatrix} b_1 \\ b_2 \\ b_3 \end{bmatrix} = \begin{bmatrix} \mu_{11} & \mu_{12} & \mu_{13} \\ \mu_{21} & \mu_{22} & \mu_{23} \\ \mu_{31} & \mu_{32} & \mu_{33} \end{bmatrix} \begin{bmatrix} a_1 \\ a_2 \\ a_3 \end{bmatrix}
$$

$$\Delta = \alpha^2 + R_1 R_2 + R_2 R_3 + R_3 R_1$$

$$\mu_{11} = (\alpha^2 - R_1 R_2 + R_2 R_3 - R_3 R_1)/\Delta$$

$$\mu_{22} = (\alpha^2 - R_1 R_2 - R_2 R_3 + R_3 R_1)/\Delta$$

$$\mu_{33} = (\alpha^2 + R_1 R_2 - R_2 R_3 - R_3 R_1)/\Delta$$

$$\mu_{12} = 2R_1 (R_3 + \alpha)/\Delta, \quad \mu_{21} = 2R_2 (R_3 - \alpha)/\Delta$$

$$\mu_{13} = 2R_1 (R_2 - \alpha)/\Delta, \quad \mu_{31} = 2R_3 (R_2 - \alpha)/\Delta$$

$$\mu_{23} = 2R_2 (R_1 + \alpha)/\Delta, \quad \mu_{32} = 2R_3 (R_1 - \alpha)/\Delta$$

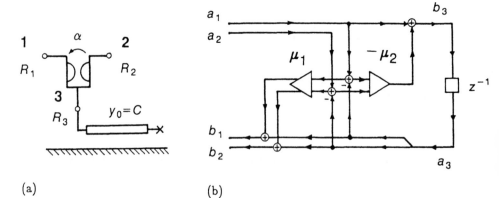

(a)   (b)

FIGURE 11-11 (a) Distributed equivalent of Richards section; (b) wave
digital realization for Richards section.

μ_{ij} (i, j = 1, 2, 3) satisfy the relationships

$$\mu_{i1} + \mu_{i2} + \mu_{i3} = 1 \tag{11-46}$$

Using the relationships in (11-46), the wave flow diagram for the ideal
gyrator can be realized as shown in Fig. 11-11(b) excluding the delay, where

$$\underline{\mu}_1 = \begin{bmatrix} \mu_{11} & \mu_{21} \\ \mu_{12} & \mu_{22} \end{bmatrix}, \quad \underline{\mu}_2 = [1 - \mu_{31} \quad -\mu_{32}] \tag{11-47}$$

Since the relationship between the incident and reflected waves in the
transmission line terminated in the open circuit shown in Fig. 11-11(a) is
represented as $b_3 = a_3 z^{-1}$, the basic WDF for the Richards section can be
depicted as shown in Fig. 11-11(b). It has six multipliers. This is the same
number as that of Nouta's structure [6] as shown in Table 11-1. The basic
WDF derived in this section, however, is canonic in delay.

When this basic WDF is interconnected with another wave digital
adapter or basic WDF, a delay-free loop appears in the interconnection
portion. However, it can be avoided by making port 1 or 2 reflection-free.
To make port 1 reflection-free, the port resistance R_1 is given as

$$R_1 = \frac{\alpha^2 + R_2 R_3}{R + R_3} \tag{11-48}$$

Then the multiplier μ_{11} vanishes, and the number of the multipliers is reduced to five. To make port 2 reflection-free, the port resistance R_1 is
given as

$$R_2 = \frac{\alpha^2 + R_1 R_3}{R_1 + R_3} \tag{11-49}$$

In this case, the multiplier μ_{22} vanishes.

11-6 EQUIVALENT CIRCUIT FOR TWO-WIRE LINE

In this section we show an equivalent circuit for a two-wire line, which is
used in succeeding sections. A two-wire line is shown in Fig. 11-12(a). Its
cascade matrix is given by

$$\begin{bmatrix} \underline{V}_0 \\ \underline{I}_0 \end{bmatrix} = \frac{1}{\sqrt{1 - \lambda^2}} \begin{bmatrix} \underline{1}_2 & \lambda\underline{\xi} \\ \lambda\underline{\eta} & \underline{1}_2 \end{bmatrix} \begin{bmatrix} \underline{V}_1 \\ \underline{I}_1 \end{bmatrix} \tag{11-50}$$

where $\underline{V}_i$ and $\underline{I}_i$ (i = 0, 1) are 2 × 1 vectors of voltage and current, respectively. η is a characteristic admittance matrix which is symmetric and sat-

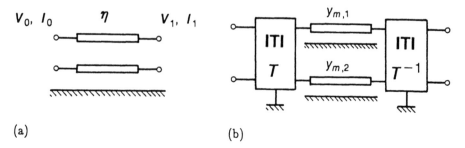

(a)   (b)

FIGURE 11-12 (a) Two-wire line; (b) equivalent circuit of (a).

isfies $\underline{\zeta} = \eta^{-1}$. $\underline{1}_2$ is a 2×2 identity matrix. Applying the method of modal decomposition [14] to the two-wire line, the cascade matrix can be represented as

$$
\begin{bmatrix} \underline{V}_0 \\ \underline{I}_0 \end{bmatrix} = \begin{bmatrix} \underline{T} & \underline{0} \\ \underline{0} & \underline{T}^{-1,T} \end{bmatrix} \frac{1}{\sqrt{1 - \lambda^2}} \begin{bmatrix} \underline{1}_2 & \lambda \underline{Y}_m^{-1} \\ \lambda \underline{Y}_m & \underline{1}_2 \end{bmatrix} \begin{bmatrix} \underline{T}^{-1} & \underline{0} \\ \underline{0} & \underline{T}^T \end{bmatrix} \begin{bmatrix} \underline{V}_1 \\ \underline{I}_1 \end{bmatrix}
$$

$$(11\text{-}51)$$

where the matrix $\underline{T}$ is a so-called voltage decomposition matrix, which diagonalizes the matrix η, that is,

$$
\underline{T}^T \underline{\eta} \, \underline{T} = \underline{Y}_m = \text{diag}(y_{m1}, y_{m2})
$$

$$(11\text{-}52)$$

Each of the first and third matrix factors in (11-51) indicates an ideal transformer interconnection (ITI) and the second factor represents uncoupled two transmission lines having y_{m1} and y_{m2} as their characteristic admittances. Hence the two-wire line is equivalently represented with the circuit shown in Fig. 11-12(b).

To simplify the succeeding calculation, we choose the voltage decomposition matrix $\underline{T}$ so that $\underline{Y}_m$ should be an identity matrix, that is,

$$
\underline{T} = \underline{\eta}^{-T/2} = \underline{\zeta}^{1/2}
$$

$$(11\text{-}53)$$

where $\underline{A}^{1/2}$ is a square root of the nonnegative definite matrix $\underline{A}$, $\underline{A}^{T/2} = (\underline{A}^{1/2})^T$, $\underline{A}^{-1/2} = (\underline{A}^{1/2})^{-1}$, and $\underline{A}^{-T/2} = (\underline{A}^{1/2})^{-1,T}$. $\underline{A}^{1/2}$ satisfies

$$
\underline{A}^{1/2} \underline{A}^{T/2} = \underline{A}
$$

$$(11\text{-}54)$$

11-7 WDFs FOR SECOND-ORDER SECTIONS

The distributed equivalents of the reciprocal and nonreciprocal second-
order sections are shown in Fig. 11-13(a) and (b), respectively, where the
characteristic impedance matrix of the two-wire line is given by

$$
\underline{\zeta} =
\begin{bmatrix}
L_1 + \dfrac{1}{C} & M + \dfrac{1}{C} \\[2ex]
M + \dfrac{1}{C} & L_2 + \dfrac{1}{C}
\end{bmatrix}
\tag{11-55}
$$

It can be considered that the reciprocal section shown in Fig. 11-13(a) is a
special case of the nonreciprocal section shown in Fig. 11-13(b), since the
former is obtained by giving $\alpha = 0$ to the latter. We consider the nonreciprocal section mainly.

Applying the method of the modal decomposition described in Sec. 11-6
to the two-wire line in Fig. 11-13(b), the equivalent of this circuit can be
obtained as shown in Fig. 11-14(a). The terminal conditions for $\underline{V}_L$ and $\underline{I}_L$
is given by

$$
\underline{V}_L = V_L[1, \ 1]^T, \quad \underline{I}_L = I_L[1, \ -1]^T
\tag{11-56}
$$

We choose the voltage decomposition matrix as

$$
\underline{T} = \frac{1}{\sqrt{\zeta_{11} + \zeta_{22} - 2\zeta_{12}}}
\begin{bmatrix}
\sqrt{\det \underline{\zeta}} & -(\zeta_{11} - \zeta_{12}) \\[2ex]
\sqrt{\det \underline{\zeta}} & \zeta_{22} - \zeta_{12}
\end{bmatrix}
\tag{11-57}
$$

The voltage vector $\underline{V}'_L$ and the current vector $\underline{I}'_L$ are represented as

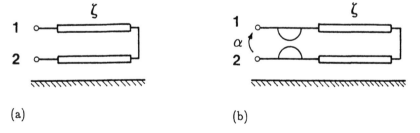

(a)   (b)

FIGURE 11-13 Distributed equivalents of second-order sections: (a) type C
and Brune section; (b) type E section.

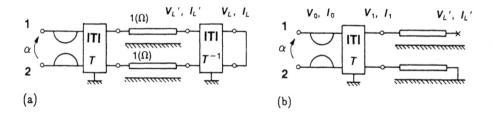

(a)   (b)

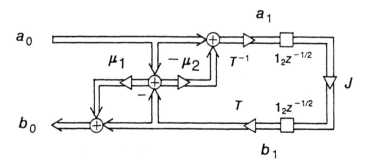

(c)

FIGURE 11-14 (a) Equivalent of Fig. 11-13(b); (b) final form of Fig. 11-13(b);
(c) wave flow diagram for (b).

$$\underline{V}'_L = \underline{T}^{-1}\underline{V}_L = [\underline{V}'_{L'}, \ 0]^T, \quad \underline{I}'_L = \underline{T}^T\underline{I}_L = [0, \ \underline{I}'_L]^T$$

$$\underline{V}'_L = \left(\frac{\zeta_{11} + \zeta_{22} - 2\zeta_{12}}{\det \zeta}\right)^{\frac{1}{2}}\underline{V}_L, \quad \underline{I}'_L = -(\zeta_{11} + \zeta_{22} - 2\zeta_{12})^{\frac{1}{2}}\underline{I}_L$$

(11-58)

The two-port shown in Fig. 11-14(a), therefore, can be represented equivalently as shown in Fig. 11-14(b).

The cascade matrix between $\underline{V}_0$, $\underline{I}_0$ and $\underline{V}_1$, $\underline{I}_1$ in Fig. 11-14(b) is
represented as

$$\begin{bmatrix} \underline{V}_0 \\ \underline{I}_0 \end{bmatrix} = \begin{bmatrix} \underline{1}_2 & \alpha \\ 0 & \underline{1}_2 \end{bmatrix}\begin{bmatrix} \underline{T} & 0 \\ 0 & \underline{T}^{-1,T} \end{bmatrix}\begin{bmatrix} \underline{V}_1 \\ \underline{I}_1 \end{bmatrix}$$

(11-59)

where

$$\underline{\alpha} = \begin{bmatrix} 0 & \alpha \\ -\alpha & 0 \end{bmatrix}$$

(11-60)

The incident and reflected voltage wave vectors for $\underline{V}_0$, $\underline{I}_0$ and $\underline{V}_1$, $\underline{I}_1$ are defined by

$$\underline{a}_0 = \frac{1}{2}(\underline{V}_0 + \underline{RI}_0), \quad \underline{b}_0 = \frac{1}{2}(\underline{V}_0 - \underline{RI}_0)$$

$$\underline{a}_1 = \frac{1}{2}(\underline{V}_1 + \underline{I}_1), \quad \underline{b}_1 = \frac{1}{2}(\underline{V}_1 - \underline{I}_1)$$

(11-61)

where $\underline{R} = \text{diag}(R_1, R_2)$, and R_1 and R_2 are port resistances at port 1 and port 2, respectively.

The voltage chain scattering matrix corresponding to (11-59) is given by

$$\begin{bmatrix} \underline{a}_0 \\ \underline{b}_0 \end{bmatrix} = \frac{1}{2} \begin{bmatrix} \underline{1}_2 + \underline{\alpha}\zeta^{-1} + \underline{R}\zeta^{-1} & \underline{1}_2 - \underline{\alpha}\zeta^{-1} - \underline{R}\zeta^{-1} \\ \underline{1}_2 + \underline{\alpha}\zeta^{-1} - \underline{R}\zeta^{-1} & \underline{1}_2 - \underline{\alpha}\zeta^{-1} + \underline{R}\zeta^{-1} \end{bmatrix} \begin{bmatrix} T & 0 \\ 0 & T \end{bmatrix} \begin{bmatrix} \underline{a}_1 \\ \underline{b}_1 \end{bmatrix}$$

(11-62)

The voltage scattering matrix is given by

$$\begin{bmatrix} \underline{b}_0 \\ \underline{a}_1 \end{bmatrix} = \begin{bmatrix} \underline{1}_2 & 0 \\ 0 & T^{-1} \end{bmatrix} \begin{bmatrix} \mu_1 & \underline{1}_2 - \mu_1 \\ \underline{1}_2 - \mu_2 & \mu_2 \end{bmatrix} \begin{bmatrix} \underline{1}_2 & 0 \\ 0 & T \end{bmatrix} \begin{bmatrix} \underline{a}_0 \\ \underline{b}_1 \end{bmatrix}$$

(11-63)

where

$$\mu_1 = (\underline{\zeta} + \underline{\alpha} - \underline{R})(\underline{\zeta} + \underline{\alpha} + \underline{R})^{-1}$$

$$\mu_2 = (\underline{R} + \underline{\alpha} - \underline{\zeta})(\underline{\zeta} + \underline{\alpha} + \underline{R})^{-1}$$

(11-64)

From (11-63) the voltage wave flow diagram of the circuit in Fig. 11-14(b) can be depicted as shown in Fig. 11-14(c), including the wave flows for a set of two transmission lines, where $\underline{J} = \text{diag}(1, -1)$. Since the multiplier matrix and the delay on the one-way path are commutative, the multiplier matrices $\underline{T}^{-1}$, $\underline{J}$, and $\underline{T}$ in Fig. 11-14(c) can be combined as follows.

$$\underline{\mu}_L = \underline{T}\underline{J}\underline{T}^{-1} = \begin{bmatrix} \mu_\ell & 1 - \mu_\ell \\ 1 + \mu_\ell & -\mu_\ell \end{bmatrix}$$

(11-65)

$$\mu_\ell = \frac{\zeta_{22} - \zeta_{11}}{\zeta_{11} + \zeta_{22} - 2\zeta_{12}}$$

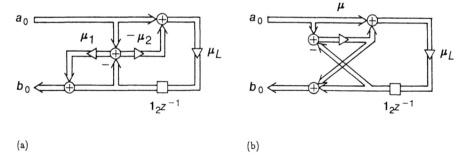

(a)   (b)

FIGURE 11-15 (a) Wave digital realization for type E section and (b) Brune and type C sections.

A pair of the delays of $T/2$ can be combined into one delay of T.

Accordingly, the structure of the basic WDF for the type E section can be obtained as shown in Fig. 11-15(a). It requires nine multipliers, since the wave flow diagram for μ_L can be realized by using one multiplier μ_ℓ. This number is less than that of any other basic WDF for the type E section shown in Table 11-1.

In order to make port 1 reflection-free, the port resistance R_1 is given by

$$R_1 = \zeta_{11} - \frac{\zeta_{12}^2 - \alpha^2}{\zeta_{22} + R_2} \tag{11-66}$$

In this case, the $(1,1)$ entry of $\underline{\mu}_1$ vanishes. The number of the multipliers, therefore, is reduced to eight. To make port 2 reflection-free, the port resistance R_2 is given by

$$R_2 = \zeta_{22} - \frac{\zeta_{12}^2 - \alpha^2}{\zeta_{11} + R_1} \tag{11-67}$$

In this case, the $(2,2)$ entry of $\underline{\mu}_1$ vanishes.

If α is zero, it holds that $\underline{\mu}_1 = -\underline{\mu}_2 \ (= \underline{\mu})$,

$$\underline{\mu} = (\underline{\zeta} - \underline{R})(\underline{\zeta} + \underline{R})^{-1} \tag{11-68}$$

Then we obtain the wave digital structure as shown in Fig. 11-15(b). It represents a wave flow diagram of the basic WDFs for reciprocal second-order sections, that is, Brune and type C sections. The basic WDF derived has only five multipliers, as does also the basic WDF proposed by Scanlan and Fagan [8]. This number is the least among those of the basic WDFs for the reciprocal second-order section shown in Table 11-1. The proposed basic

WDF is canonic in delays, whereas the Scanlan and Fagan structure is not. If either port 1 or port 2 is made reflection-free, since either the $(1, 1)$ or $(2, 2)$ entry of $\underline{\mu}$ vanishes, the number of the multipliers is reduced to four.

11-8 WDF FOR THE TYPE D SECTION

The distributed equivalent of a type D section is shown in Fig. 11-16(a). Let $\zeta_{ij}^{(k)}$ ($i, j, k = 1, 2$) represent the (i, j) entry of the characteristic impedance $\underline{\zeta}^{(k)}$ for the two-wire lines in Fig. 11-16(a), where $\zeta_{12}^{(k)} = \zeta_{21}^{(k)}$ holds. Each entry of $\underline{\zeta}^{(k)}$ is given by [10]

$$\zeta_{11}^{(1)} = L_1 + \frac{1}{C_2} + \frac{L_3}{1 + L_3 C_1}, \quad \zeta_{12}^{(1)} = M_1 + \frac{1}{C_2} + \frac{M_2}{1 + L_3 C_1},$$

$$\zeta_{22}^{(1)} = L_2 + \frac{1}{C_2} + \frac{L_4}{1 + L_3 C_1}$$

$$\zeta_{11}^{(2)} = (a_1 + L_3 a_2 - L_1 a_3)/b, \quad \zeta_{12}^{(2)} = (a_1 + M_2 a_2 - M_1 a_3)/b,$$

$$\zeta_{22}^{(2)} = (a_1 + L_4 a_2 - L_2 a_3)/b$$

$$(11\text{-}69)$$

where

$$a_1 = (1 + L_3 C_1)[(1 + L_3 C_1)\{L_a(1 + L_3 C_1) + L_{ab} C_2\} + L_b]$$

$$a_2 = L_3 C_1 C_2 \{L_a(1 + L_3 C_1) + L_{ab} C_2\}$$

$$a_3 = L_3 C_1 C_2 L_b (1 + L_3 C_1), \quad b = L_{ab} C_2^2 (1 + L_3 C_1)$$

$$(11\text{-}70)$$

$$L_a = L_1 + L_2 - 2M_1, \quad L_b = L_3 + L_4 - 2M_2, \quad L_{ab} = L_1 L_4 + L_2 L_3 - 2M_1 M_2$$

Let the voltage decomposition matrix $\underline{T}^{(k)}$ that diagonalizes the characteristic impedance $\underline{\zeta}^{(k)}$ be

$$\underline{T}^{(k)} = \frac{1}{\sqrt{\zeta_{11}^{(k)} + \zeta_{22}^{(k)} - 2\zeta_{12}^{(k)}}} \begin{bmatrix} \sqrt{\det \underline{\zeta}^{(k)}} & -(\zeta_{11}^{(k)} - \zeta_{12}^{(k)}) \\ \sqrt{\det \underline{\zeta}^{(k)}} & \zeta_{22}^{(k)} - \zeta_{12}^{(k)} \end{bmatrix} \quad (11\text{-}71)$$

Then the equivalent circuit of the type D section is obtained as shown in Fig. 11-16(b). The voltage wave vectors $\underline{a}_i$ and $\underline{b}_i$ ($i = 0, 1, 2, 3$) for $\underline{V}_i$ and $\underline{I}_i$ in Fig. 11-16(b) are defined as

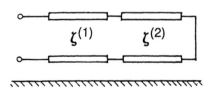

(a)

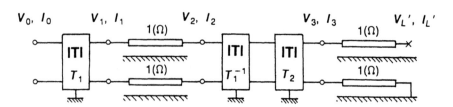

(b)

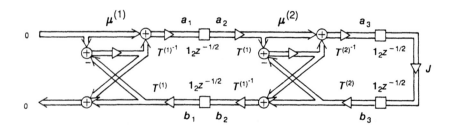

(c)

FIGURE 11-16 (a) Distributed equivalent of type D section; (b) equivalent circuit of (a); (c) wave flow diagram for (b).

$$\underline{a}_0 = \frac{1}{2}(\underline{V}_0 + \underline{RI}_0), \quad \underline{b}_0 = \frac{1}{2}(\underline{V}_0 - \underline{RI}_0),$$

$$\underline{a}_i = \frac{1}{2}(\underline{V}_i + \underline{I}_i), \quad \underline{b}_i = \frac{1}{2}(\underline{V}_i - \underline{I}_i) \quad (i = 1, 2, 3) \tag{11-72}$$

where $\underline{R} = \text{diag}(R_1, R_2)$, and R_1 and R_2 are port resistances at port 1 and port 2, respectively.

The voltage scattering matrix for $\underline{a}_0$, $\underline{a}_1$ and $\underline{b}_0$, $\underline{b}_1$ is given by

$$\begin{bmatrix} \underline{b}_0 \\ \underline{a}_1 \end{bmatrix} = \begin{bmatrix} 1_2 & 0 \\ 0 & \underline{T}^{(1)^{-1}} \end{bmatrix} \begin{bmatrix} \underline{\mu}^{(1)} & 1_2 - \underline{\mu}^{(1)} \\ 1_2 + \underline{\mu}^{(1)} & -\underline{\mu}^{(1)} \end{bmatrix} \begin{bmatrix} 1_2 & 0 \\ 0 & \underline{T}^{(1)} \end{bmatrix} \begin{bmatrix} \underline{a}_0 \\ \underline{b}_1 \end{bmatrix} \tag{11-73}$$

where

$$\underline{\mu}^{(1)} = (\underline{\zeta}^{(1)} - \underline{R})(\underline{\zeta}^{(1)} + \underline{R})^{-1} \tag{11-74}$$

The voltage scattering matrix for $\underline{a}_2$, $\underline{a}_3$ and $\underline{b}_2$, $\underline{b}_3$ is given by

$$\begin{bmatrix} \underline{b}_2 \\ \underline{a}_3 \end{bmatrix} = \begin{bmatrix} \underline{T}^{(1)^{-1}} & 0 \\ 0 & \underline{T}^{(2)^{-1}} \end{bmatrix} \begin{bmatrix} \underline{\mu}^{(2)} & 1_2 - \underline{\mu}^{(2)} \\ 1_2 + \underline{\mu}^{(2)} & -\underline{\mu}^{(2)} \end{bmatrix} \begin{bmatrix} \underline{T}^{(1)} & 0 \\ 0 & \underline{T}^{(2)} \end{bmatrix} \begin{bmatrix} \underline{a}_2 \\ \underline{b}_3 \end{bmatrix}$$

$$\tag{11-75}$$

where

$$\underline{\mu}^{(2)} = (\underline{\zeta}^{(2)} - \underline{\zeta}^{(1)})(\underline{\zeta}^{(2)} + \underline{\zeta}^{(1)})^{-1} \tag{11-76}$$

From (11-73) and (11-75), the wave flow diagram for the circuit in Fig. 11-16(b) is obtained as shown in Fig. 11-16(c), including the wave flow of the two sets of two transmission lines. Combining some of the multiplier matrices, we obtain the wave flow diagram shown in Fig. 11-17, where

$$\underline{\mu}_L = \underline{T}^{(2)} \underline{JT}^{(2)^{-1}} = \begin{bmatrix} \mu_\ell & 1 - \mu_\ell \\ 1 + \mu_\ell & -\mu_\ell \end{bmatrix}$$

$$\tag{11-77}$$

$$\mu_\ell = \frac{\zeta_{22}^{(2)} - \zeta_{11}^{(2)}}{\zeta_{11}^{(2)} + \zeta_{22}^{(2)} - 2\zeta_{12}^{(2)}}$$

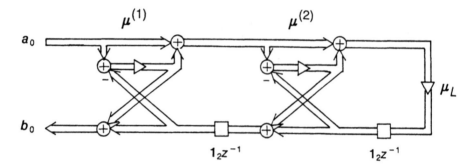

FIGURE 11-17 Wave digital realization for type D section.

and a pair of delays of T/2 are equivalently combined into a delay of T. This
represents the basic WDF for a type D section and requires nine multipliers.
This number is less than that of the other basic WDF in Table 11-1. The
basic WDF derived is also canonic in delays.

To make the port 1 reflection-free, R_1 is given by

$$R_1 = \zeta_{11}^{(1)} - \frac{\zeta_{12}^{(1)^2}}{\zeta_{22}^{(1)} + R_2} \tag{11-78}$$

In this case, since the $(1, 1)$ entry of $\underline{\mu}^{(1)}$ vanishes, the number of multipliers
is reduced to eight. To make the port 2 reflection-free, R_2 is given by

$$R_2 = \zeta_{22}^{(1)} - \frac{\zeta_{12}^{(1)^2}}{\zeta_{11}^{(1)} + R_1} \tag{11-79}$$

Then the $(2, 2)$ entry of $\underline{\mu}^{(1)}$ vanishes.

11-9 SOME EXAMPLES

In this section we show some results from time-domain simulations for WDFs
realized with the proposed basic WDFs. The transfer functions to be realized
are designed by using the method proposed by Fahmy and Rhodes [15, 16].
They are a class of even-degree transfer functions in the distributed domain
which exhibit flat amplitude and approximate a linear phase response over a
finite band. The general form of this transfer function is given by

$$S_{21}(\lambda) = \frac{A_n(\lambda) A_n(-\lambda) + \kappa^2 \lambda^2 A_{n-1}(\lambda) A_{n-1}(-\lambda)}{A_n^2(\lambda) + \kappa^2 \lambda^2 A_{n-1}^2(\lambda)} \qquad (11\text{-}80)$$

The $(1, 1)$ entry of the scattering matrix having $S_{21}(\lambda)$ as its $(2, 1)$ entry is given by

$$S_{11}(\lambda) = \frac{\kappa \lambda \{ A_n(\lambda) A_{n-1}(-\lambda) - A_n(-\lambda) A_{n-1}(\lambda) \}}{A_n^2(\lambda) + \kappa^2 \lambda^2 A_{n-1}^2(\lambda)} \qquad (11\text{-}81)$$

where κ is a constant and $A_n(\lambda)$ is a sequence of polynomials. They are defined by using three parameters, n, α, and θ_0, in [16].

We design three transfer functions by using the following sets of the parameters: (1) n = 2, α = 14, θ_0 = 0.06; (2) n = 2, α = 12, θ_0 = 0.06; and (3) n = 3, α = 8, θ_0 = 0.065. The WDFs that realize these transfer functions are synthesized by the usual cascade synthesis procedure described in Sec. 11-3, as shown in Fig. 11-18(a)-(e). We call them WDF(I), WDF(II), WDF(III), WDF(IV), and WDF(V), respectively. WDF(I) realizes the transfer function designed by using the parameters (1) and contains the basic WDF for a Brune section. WDF(II) realizes the transfer function designed by using the parameters (2) and contains the basic WDF for a type C section. WDF(III) realizes the same transfer function as WDF(II) and contains two basic WDFs for Richards sections instead of one basic WDF for a type C section. WDF(IV) realizes the transfer function designed by using the parameters (3) and contains the basic WDF for a type D section. WDF(V) realizes the same transfer function as WDF(IV) and contains two basic WDFs for type E sections instead of one basic WDF for a type D section.

In time-domain simulation, it is assumed that the word length of signals is long enough, and the multiplier coefficients are represented in a floating-point two's-complement whose mantissas are truncated with an appointed word length. The frequency responses are obtained from the fast Fourier transform of the unit impulse response.

Figure 11-19(a)-(e) show the attenuation characteristics in the passband of WDF(I) to WDF(V), respectively, when the multiplier coefficients are approximated in floating-point 6-bit and 8-bit form. The horizontal axis represents the frequency axis normalized by the sampling frequency, and attenuation at frequency zero is normalized to be zero. In Fig. 11-18, the order of the adapters and the location of the adapter not having a reflection-free port are adopted so that the sensitivities should be the lowest. From Fig. 11-19 it is found that WDFs realized with the proposed basic WDFs have very good sensitivities.

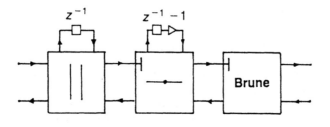

(a)

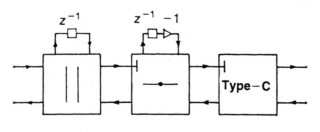

(b)

FIGURE 11-18 WDFs simulted as examples: (a) WDF(I); (b) WDF(II);
(c) WDF(III); (d) WDF(IV); (e) WDF(V).

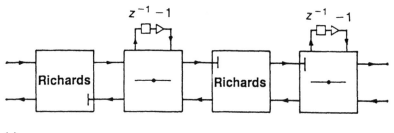

(c)

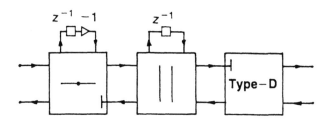

(d)

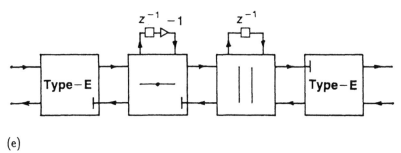

(e)

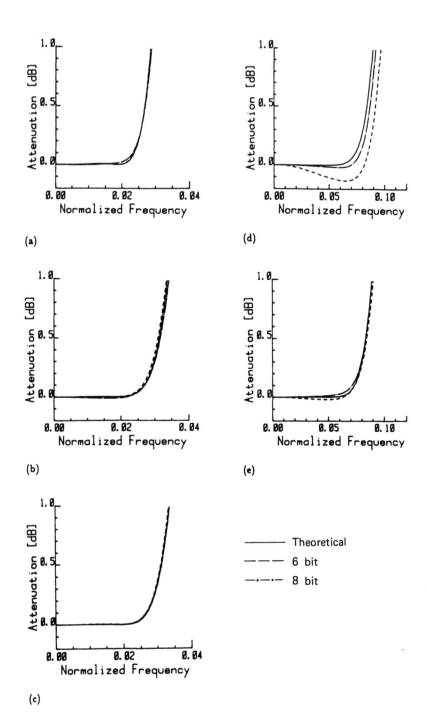

FIGURE 11-19 Attenuation characteristics of the WDFs with quantized multiplier coefficients: (a) WDF(I); (b) WDF(II); (c) WDF(III); (d) WDF(IV); (e) WDF(V).

11-10 CONCLUSIONS

In this chapter we have derived novel basic WDFs which are translated from the distributed equivalents of the basic reactance sections. The basic WDFs derived have fewer multipliers than the conventional basic WDFs and are canonic in delays. It has been shown by the time-domain simulation that WDFs realized with the proposed basic WDFs have low coefficient sensitivities.

REFERENCES

1. A. Fettweis, "Digital filter structures related to classical filter networks," AEÜ—Arch. Elektron. Übertragungstech., vol. 25, pp. 78-89 (Feb. 1971).

2. A. Fettweis, "Pseudo passivity, sensitivity, and stability of wave digital filters," IEEE Trans. Circuit Theory, vol. CT-19, pp. 668-678 (Nov. 1972).

3. A. Sedlmeyer and A. Fettweis, "Digital filter with true ladder configuration," Int. J. Circuit Theory Appl., vol. 1, pp. 5-10 (Mar. 1973).

4. R. Nouta, "Jauman structures in wave digital filters," Int. J. Circuit Theory Appl., vol. 2, pp. 163-174 (June 1974).

5. A. Fettweis, H. Levin, and A. Sedlmeyer, "Wave digital lattice filters," Int. J. Circuit Theory Appl., vol. 3 (Sept. 1975).

6. R. Nouta, "Wave digital cascade synthesis," Int. J. Circuit Theory Appl., vol. 3 (Sept. 1975).

7. M. F. Fahmy, "Digital realization of C- and D-sections," Int. J. Circuit Theory Appl., vol. 3, pp. 395-402 (Dec. 1975).

8. J. O. Scanlan and A. D. Fagan, "Wave digital equivalents of two-wire line and C-section," IEEE Trans. Circuits Syst., vol. CAS-24, pp. 422-428 (Aug. 1977).

9. G. O. Martens and H. H. Lê, "Wave digital adaptors for reciprocal second-order sections," IEEE Trans. Circuits Syst., vol. CAS-25, pp. 1077-1083 (Dec. 1978).

10. M. Suzuki, N. Miki, and N. Nagai, "New wave digital filters for basic reactance sections," IEEE Trans. Circuits Syst., vol. CAS-32, no. 4, pp. 337-348 (Apr. 1985).

11. D. C. Youla, "A new theory of cascade synthesis," IRE Trans. Circuit Theory, vol. CT-9, pp. 244-260 (Sept. 1961).

12. M. Saito, A Guide to Network Theory, University of Tokyo Press, Tokyo, 1967.

13. H. Watanabe, Theory and Design for Transmission Networks, Ohm
    Publishing, Tokyo, 1966.

14. N. Nagai, "Consideration of equivalent circuit representation and propa-
    gation modes for lossless multiwire lines," Trans. IECE (Japan),
    vol. J60-B, no. 5, pp. 305-312 (May 1977).

15. M. F. Fahmy and J. D. Rhodes, "Finite band approximation for distrib-
    uted and digital selective linear phase transfer functions," Int. J. Circuit
    Theory Appl., vol. 3, pp. 57-70 (Mar. 1975).

16. M. F. Fahmy and J. D. Rhodes, "The equidistant linear phase poly-
    nomial for distributed and digital networks," Int. J. Circuit Theory
    Appl., vol. 2, pp. 342-352 (Dec. 1974).

12

Complex Transmission-Line Circuit and Complex Wave Digital Filter

NOBUO NAGAI

Hokkaido University, Sapporo, Japan

12-1 INTRODUCTION

Dewilde et al. [1] have asserted that the Levinson realization algorithm for optimal linear predictors is a special case of Darlington network synthesis. They also suggested that ARMA process for a generalized form of optimal linear predictors can be synthesized by Darlington synthesis procedure.

343

Later, Dewilde et al. [2, 3] synthesized orthogonal digital filters (ODFs) with a Darlington synthesis procedure for ARMA process. Some of them can be regarded as complex digital filters.

To get a class of digital filters, Fettweis [4] has introduced wave digital filters (WDFs), which represent a class of digital filters that are closely related to classical lossless filters inserted between resistive terminations. Nagai et al. [6, 7] further described WDFs with canonical number of delays synthesized with the Darlington procedure. The WDFs are synthesized with the cascade connections of basic sections whose reference filters correspond to commensurate transmission-line (CTL) circuits [8, 9], which are a class of analog filters constituted with transmission lines and coupled transmission lines. Since analog filters are denoted with voltages and currents, the CTL circuit can be transformed into voltage WDF, current WDF, or power WDF.

Belevitch [10] shows the Darlington procedure for complex lumped-element networks. Complex networks may be free mathematical creations for analog circuits, but the theory is of practical interest concerning modulation problems and complex signal processing. The complex network is in our hands, when imaginary resistances are accepted as elements in a network. The imaginary resistance absorbs no real instantaneous power and is abstractly lossless.

Complex commensurate transmission-line (complex CTL) circuits can be obtained by admitting imaginary resistances as circuit elements to the usual CTL circuits. Since the imaginary resistance is an imaginary constant element, we cannot transform it alone to complex WDF. To get a complex WDF for a complex CTL network, we should create some elements with imaginary resistances.

In this chapter we attempt to transform complex ODFs to complex CTL circuits. For that purpose we give some new considerations about reflection coefficients for complex impedances in Sec. 12-2 to 12-6. In Sec. 12-7 WDFs are proved equivalent to CTL circuits. In Sec. 12-8 we describe passivity and losslessness for complex lumped, CTL, and digital circuits. From the passivity of complex circuits, the reflection coefficient for complex CTL circuits is determined to be the one defined for the conjugate matching condition. In Sec. 12-9 we define CUE (complex unit element), which is the most basic element for the complex CTL circuit. In Sec. 12-10 we extend the conventional gyrator to the one having complex linear relations, and in Sec. 12-11 we show that passive complex CTL circuits can be synthesized by using CUEs and complex gyrators and by utilizing the Darlington procedure.

12-2 ANALYSIS OF UNIFORM TRANSMISSION LINE

Now we consider a uniform transmission line described by the partial differential equations

$$-\frac{\partial v(x,t)}{\partial x} = Ri(x,t) + L\frac{\partial i(x,t)}{\partial t} \tag{12-1a}$$

$$-\frac{\partial i(x,t)}{\partial x} = Gv(x,t) + C\frac{\partial v(x,t)}{\partial t} \tag{12-1b}$$

where $v(x,t)$ and $i(x,t)$ are voltage and current, as a function of distance x along the line and time t. R, G, L, and C are per unit length series resistance, shunt conductance, series inductance, and shunt capacitance, respectively.

Let's use the Laplace transform with respect to time t in order to transform the partial differential equations into the ordinary differential equations and to obtain transient voltage and current responses. That is,

$$V(x,s) = \int_0^\infty v(x,t)e^{-st}\,dt \tag{12-2a}$$

$$I(x,s) = \int_0^\infty i(x,t)e^{-st}\,dt \tag{12-2b}$$

Then we arrive at the equations

$$-\frac{dV(x,s)}{dx} = Z(s)I(x,s) \tag{12-3a}$$

$$-\frac{dI(x,s)}{dx} = Y(s)V(x,s) \tag{12-3b}$$

where

$$Z(s) = (sL + R)$$

$$Y(s) = (sC + G)$$

The solutions for $V(x,s)$ and $I(x,s)$ are obtained as

$$V(x,s) = Ae^{-\gamma x} + Be^{\gamma x} \tag{12-4a}$$

$$I(x,s) = \frac{Ae^{-\gamma x} - Be^{\gamma x}}{Z_0} \tag{12-4b}$$

where γ is the propagation factor, which is given by

$$\gamma = \sqrt{Z(s)Y(s)} = \sqrt{(sL+R)(sC+G)} \tag{12-5a}$$

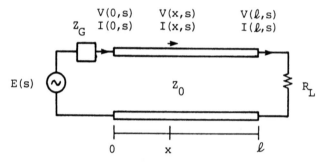

FIGURE 12-1 Transmission-line circuit represented by Laplace transform.

and Z_0 is the characteristic impedance, which is given by

$$Z_0 = \sqrt{\frac{Z(s)}{Y(s)}} = \sqrt{\frac{sL + R}{sC + G}} \qquad (12\text{-}5b)$$

We consider a circuit of the uniform transmission line shown in Fig. 12-1. The circuit is connected by the load resistance R_L at the end of the line at $x = \ell$, and by the voltage generator $E(s)$ and its internal impedance Z_G at the $x = 0$. The boundary conditions for the circuit shown in Fig. 12-1 are

$$V(0, s) + Z_G I(0, s) = E(s) \qquad (12\text{-}6a)$$

$$V(\ell, s) = R_L I(\ell, s) \qquad (12\text{-}6b)$$

Substituting the boundary condition (12-6) into Eq. (12-4), we obtain the following solutions for the voltage and current at x:

$$V(x, s) = \frac{E(s) Z_0 e^{-\gamma x}}{Z_G + Z_0} \frac{1 + m_L e^{-\gamma(\ell - x)}}{1 + m_L m_G e^{-2\gamma \ell}} \qquad (12\text{-}7a)$$

$$I(x, s) = \frac{E(s) e^{-\gamma x}}{Z_G + Z_0} \frac{1 - m_L e^{-\gamma(\ell - x)}}{1 + m_L m_G e^{-2\gamma \ell}} \qquad (12\text{-}7b)$$

where

$$m_G = \frac{Z_0 - Z_G}{Z_0 + Z_G} \tag{12-8a}$$

$$m_L = \frac{R_L - Z_0}{R_L + Z_0} \tag{12-8b}$$

$V(x, s)$ presented by Eq. (12-7a) may be written as

$$V(x, s) = \frac{E(s) Z_0}{Z_G + Z_0} \{ e^{-\gamma x} + m_L e^{-\gamma(2\ell-x)} - m_L m_G e^{-\gamma(2\ell+x)} + \cdots \} \tag{12-9}$$

The voltage $V(x, s)$ of Eq. (12-9) suggests that the voltage waves on the transmission line produce multiple reflections at $x = 0$ and $x = \ell$. That is, m_G and m_L show instantaneous voltage reflection coefficients for the transmission line at $x = 0$ and $x = \ell$, respectively.

If a voltage reflection coefficient is given by m_G of Eq. (12-8a), we show in Sec. 12-3 that the voltage transfer coefficient t_G is obtained as

$$t_G = \frac{2Z_0}{Z_G + Z_0} = 1 + m_G \tag{12-10}$$

As the result, Eq. (12-9) can be written

$$V(x, s) = \frac{E(s)}{2} t_G \{ e^{-\gamma x} + m_L e^{-\gamma(2\ell-x)} - m_L m_G e^{-\gamma(2\ell+x)} + \cdots \} \tag{12-11}$$

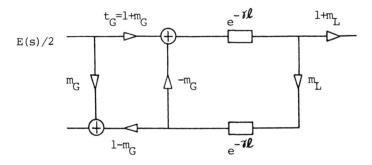

FIGURE 12-2 Voltage signal flow for the transmission-line circuit shown in Fig. 12-1.

Since multiple reflection and transfer waves appear only at x = 0 and
x = ℓ for the circuit shown in Fig. 12-1, the voltage signal flow for V(x, s)
is presented in Fig. 12-2.

As has been discussed above, a transmission-line circuit is analyzed
with reflection coefficients, which are related to impedance matching. So in
the next section we consider the problems of impedance matching and reflection coefficient defined between complex impedances.

12-3 DEFINITION OF VOLTAGE AND CURRENT WAVES

Since the reflected waves are produced at the connections of two transmission
lines or of a transmission line and a load, and so on, the signals travers-
ing the transmission line are formed by two waves: a forward wave and a
backward wave. These waves are represented by voltage and current. In
this section we discuss incident, reflected, and transfer waves, with each
wave described by voltage and current.

We consider a circuit of Fig. 12-3 to be loaded by a voltage source E_G
and a complex impedance

$$Z_G = R_G + jX_G \tag{12-12}$$

where $R_G > 0$. And the load is also assumed to be a complex impedance

$$Z_L = R_L + jX_L \tag{12-13}$$

where $R_L \geq 0$. We assume that the complex impedances Z_G and Z_L are
frequency-independent constants or are values at a single frequency.

To obtain the voltage and current reflection coefficients and voltage
and current transfer coefficients, each wave is defined as presented in
Fig. 12-3, where V_i and I_i are incident voltage and current waves, V_t and

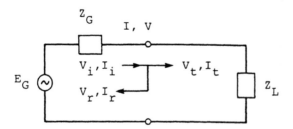

FIGURE 12-3 Representation of incident, reflected, and transfer voltage
and current waves.

I_t are transfer voltage and current waves, and V_r and I_r are reflected voltage and current waves.

We assume that each wave defined above satisfies the following definition [14].

Definition 12-1 (Definition for a Voltage Source).

$$V_t = V_i + V_r \tag{12-14a}$$

$$I_t = I_i - I_r \tag{12-14b}$$

$$\frac{V_t}{I_t} = Z_L \tag{12-14c}$$

$$\frac{V_r}{I_r} = Z_G \tag{12-14d}$$

$$E_G = (Z_G + Z_L)I_t \tag{12-14e}$$

∎

In the definition above we should notice that the impedance for the incident wave V_i/I_i is not involved.

At the arbitrary point on the uniform transmission line, no reflected wave is produced. This means that if the internal impedance Z_G and load impedance Z_L in the circuit shown in Fig. 12-3 are equal to each other, there is no reflected wave. We assume this fact to be a matching condition. That is,

Assumption 12-1 (Equal Matching Condition). (See Fig. 12-4.) If

$$Z_G = Z_L \tag{12-15a}$$

then

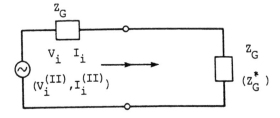

FIGURE 12-4 Equal impedance-matching circuit (conjugate impedance-matching circuit). Note: Marks in parentheses ($V_i^{(II)}$, $I_i^{(II)}$, and Z_G^*) apply to the "conjugate impedance-matching circuit."

$$V_r = I_r = 0 \qquad (12\text{-}15\text{b})$$

∎

Substituting Assumption 12-1 into Eq. (12-14) we get the following condition for the incident wave.

$$\frac{V_i}{I_i} = Z_G \qquad (12\text{-}16\text{a})$$

$$I_i = \frac{E_G}{2Z_G} \qquad (12\text{-}16\text{b})$$

Under Assumption 12-1, the voltage reflection coefficient (VRC) V_r/V_i and current reflection coefficient (CRC) I_r/I_i are obtained as follows:

$$\frac{V_r}{V_i} = \frac{I_r}{I_i} = \frac{Z_L - Z_G}{Z_L + Z_G} = m \qquad (12\text{-}17)$$

and the voltage transfer coefficient (VTC) V_t/V_i and current transfer coefficient (CTC) I_t/I_i are

$$\frac{V_t}{V_i} = \frac{2Z_L}{Z_L + Z_G} = 1 + m \qquad (12\text{-}18\text{a})$$

$$\frac{I_t}{I_i} = \frac{2Z_G}{Z_L + Z_G} = 1 - m \qquad (12\text{-}18\text{b})$$

By using Eqs. (12-14) to (12-18), we obtain the following relationship among each voltage and current:

$$V_r I_r + V_t I_t = V_i I_i \qquad (12\text{-}19)$$

When the voltage and current for a load impedance Z_L are given by V and I,

$$V = Z_L I \qquad (12\text{-}20\text{a})$$

the complex power W is defined as

$$W = VI^* \qquad (12\text{-}20\text{b})$$

where the superscript * denotes complex conjugate. In that case the average dissipated power of the load is represented by the real part of the complex power, or Re[VI*].

Let us consider that the wave propagates on a uniform transmission line whose characteristic impedance is given by Z_L. The voltage V and current I for the wave satisfy Eq. (12-20a). If the transmission line is connected to a load whose impedance is equal to the characteristic impedance Z_L, the wave propagating on the line matches the load, and the load dissipates the power given by Re[VI*]. Thus the wave can supply the average dissipated power Re[VI*] to the load. We can say that the active power of the wave is the real part of the complex power, or the average dissipated power.

$V_r I_r$, $V_t I_t$, and $V_i I_i$ represented in Eq. (12-19) do not denote the complex power for each wave defined in the foregoing. In other words, the quantity given by Eq. (12-19) is not related to the complex power or the average dissipated power for complex impedances.

Accordingly, if we use Eq. (12-17) as a reflection coefficient, it is notable that a reflection coefficient with an amplitude larger than unity has been obtained, although the two impedances correspond to a passive system [12]. Thus the coefficients given by Eqs. (12-17) and (12-18) are adopted to present voltage and current, but not to present the average dissipated power of transmission-line circuits.

It is well known that the maximum power transfer from a generator to a load is accomplished when the reactive components of Z_G and Z_L cancel and when the real components are equal. That is, the maximum power is supplied when a conjugate match of impedances is given. So let us assume that no reflection wave is produced when a conjugate match is achieved in the circuit shown in Fig. 12-3.

Assumption 12-2 (Conjugate Matching Condition). [See the ( )-presentation of Fig. 12-4.]

$$Z_L = Z_G^* \tag{12-21a}$$

then

$$V_r^{(II)} = I_r^{(II)} = 0 \tag{12-21b}$$

∎

Substituting Assumption 12-2 into Eq. (12-14), we obtain the following condition for the incident wave:

$$\frac{V_i^{(II)}}{I_i^{(II)}} = Z_G^* \tag{12-22a}$$

$$I_i^{(II)} = \frac{E_G}{2R_G} \qquad (12\text{-}22b)$$

Equation (12-22a) means that the impedance for the incident wave is given by the conjugate complex of the internal impedance of a generator, although such an impedance does not exist in the circuit shown in Fig. 12-3.

VRC and CRC, VTC and CTC, which are obtained in Eqs. (12-17) and (12-18) under Assumption 12-1, yield under Assumption 12-2:

$$\frac{V_r^{(II)}}{V_i^{(II)}} = \frac{Z_G}{Z_G^*} \frac{Z_L - Z_G^*}{Z_L + Z_G} \qquad (12\text{-}23a)$$

$$\frac{I_r^{(II)}}{I_i^{(II)}} = \frac{Z_L - Z_G^*}{Z_L + Z_G} \qquad (12\text{-}23b)$$

$$\frac{V_t^{(II)}}{V_i^{(II)}} = \frac{2R_G Z_L}{Z_G^*(Z_L + Z_G)} \qquad (12\text{-}23c)$$

$$\frac{I_t^{(II)}}{I_i^{(II)}} = \frac{2R_G}{Z_L + Z_G} \qquad (12\text{-}23d)$$

Since the incident voltage and current are given in Eq. (12-22), we can get the average dissipated power for the incident wave as follows:

$$\text{Re}(V_i^{(II)} I_i^{(II)*}) = \frac{|E_G|^2}{4R_G} \qquad (12\text{-}24)$$

This average dissipated power represents the maximum available power which is transferred from a generator to the load and which corresponds to conjugate matching.

By using Eqs. (12-22) to (12-24) we obtain the following relations for average dissipated power of waves.

$$\text{Re}(V_r^{(II)} I_r^{(II)*}) + \text{Re}(V_t^{(II)} I_t^{(II)*}) = \text{Re}(V_i^{(II)} I_i^{(II)*}) \qquad (12\text{-}25)$$

It should be noted that a reflection coefficient with an amplitude larger than unity is never obtained from the relation of Eq. (12-25).

12-4 COMPLEX VOLTAGE SCATTERING MATRIX

Fettweis has introduced wave digital filters (WDFs) [4, 5] as a class of digital filters to imitate the behavior of analog circuit. The analogy between a WDF and its reference analog filter is based on the wave quantity, and the description of a WDF is represented by a signal flow of voltage, current, or power waves. In this section we adopt mainly voltage wave quantities; that is, we define forward and backward voltage waves.

We can represent the behavior of a circuit with a scattering matrix, which is described by incident and reflected waves. These waves defined in a scattering matrix are in general normalized quantities and not pure voltages or currents [10]. However, we sometimes use voltage waves for scattering concepts [13, 14] and in that case we call it the voltage scattering matrix. WDFs described by voltage wave quantities can be represented by the voltage scattering matrix.

The forward and backward voltage waves α and β for WDF are defined by using a port voltage V, a port current I, and a port resistance R as follows:

$$\alpha = \frac{V + RI}{2} \tag{12-26a}$$

$$\beta = \frac{V - RI}{2} \tag{12-26b}$$

If a given port impedance is a complex number as is shown in Fig. 12-3 or Fig. 12-5, it is possible to define voltage waves in two different ways [13-15]. If we apply the definition described by Klein [15] to the circuit shown in Fig. 12-5, voltage waves are defined as follows:

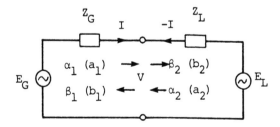

FIGURE 12-5 Definition of voltage waves (definition of normalized waves).

Definition 12-2a.   Incident voltage waves α_1 and α_2, and reflected voltage waves β_1 and β_2 for the circuit shown in Fig. 12-5 are defined as follows:

$$\alpha_1 = \frac{V + Z_G I}{2} \qquad (12\text{-}27a)$$

$$\beta_1 = \frac{V - Z_G I}{2} \qquad (12\text{-}27b)$$

$$\alpha_2 = \frac{V + Z_L(-I)}{2} \qquad (12\text{-}27c)$$

$$\beta_2 = \frac{V - Z_L(-I)}{2} \qquad (12\text{-}27d)$$

$\blacksquare$

If

$$E_L = 0 \qquad (12\text{-}28)$$

then we get

$$V = V_t \qquad (12\text{-}29a)$$

$$I = I_t \qquad (12\text{-}29b)$$

$$\alpha_2 = 0 \qquad (12\text{-}29c)$$

$\blacksquare$

In this case we can use the equations based on Assumption 12-1 [Eqs. (12-15) to (12-19)], and we obtain

$$\alpha_1 = Z_G I_i = V_i \qquad (12\text{-}30a)$$

$$\beta_1 = Z_G I_r = V_r \qquad (12\text{-}30b)$$

$$\beta_2 = Z_L I_t = V_t \qquad (12\text{-}30c)$$

That is, α_1, β_1, and β_2 are coincident with the incident, reflected, and transfer voltage waves, V_i, V_r, and V_t obtained under Assumption 12-1, respectively.

Next, if $E_L = 0$ and Eqs. (12-21) to (12-25) are substituted into Definition 12-2a, we get

$$\alpha_1 = Z_G I_i^{(II)} = V_i^{(II)} \tag{12-31a}$$

$$\beta_1 = \frac{(Z_G^* - Z_G) I_i^{(II)} + 2R_G I_r^{(II)}}{2} \tag{12-31b}$$

$$\alpha_2 = 0 \tag{12-31c}$$

$$\beta_2 = Z_L I_t^{(II)} = V_t^{(II)} \tag{12-31d}$$

In this case, α_1 and β_2 are represented by incident and transfer voltage waves, respectively, but β_1 involves the incident wave component. Therefore, the voltage waves given by Definition 12-2a conform only to the waves based on Assumption 12-1.

From Eqs. (12-27) to (12-30) and (12-15) to (12-18), we get a voltage scattering matrix as follows:

$$\begin{bmatrix} \beta_1 \\ \beta_2 \end{bmatrix} = \begin{bmatrix} m & 1 - m \\ 1 + m & -m \end{bmatrix} \begin{bmatrix} \alpha_1 \\ \alpha_2 \end{bmatrix} \tag{12-32}$$

where $m = (Z_L - Z_G)/(Z_L + Z_G)$, which is given by Eq. (12-17).

Since the electrical power for this signal flow can be represented by Eq. (12-19), which is not related to complex power, it is not so important to describe the signal flow with power. However, there is a description in Ref. 15 and we will refer to it later.

Definition 12-2b. Incident waves a_1 and a_2, and reflected waves b_1 and b_2 for the circuit shown in the ( )-presentation of Fig. 12-5 are defined as

$$a_1 = a_1^{(I)} = \frac{\alpha_1}{\sqrt{Z_G}} = \frac{V + Z_G I}{2\sqrt{Z_G}} \tag{12-33a}$$

$$b_1 = b_1^{(I)} = \frac{\beta_1}{\sqrt{Z_G}} = \frac{V - Z_G I}{2\sqrt{Z_G}} \tag{12-33b}$$

$$a_2 = a_2^{(I)} = \frac{\alpha_2}{\sqrt{Z_L}} = \frac{V + Z_L(-I)}{2\sqrt{Z_L}} \tag{12-33c}$$

$$b_2 = b_2^{(I)} = \frac{\beta_2}{\sqrt{Z_L}} = \frac{V - Z_L(-I)}{2\sqrt{Z_L}} \tag{12-33d}$$

∎

From this definition we get the following type of scattering matrix:

$$
\begin{bmatrix} b_1^{(I)} \\ b_2^{(I)} \end{bmatrix} = \begin{bmatrix} m & \sqrt{1-m^2} \\ \sqrt{1-m^2} & -m \end{bmatrix} \begin{bmatrix} a_1^{(I)} \\ a_2^{(I)} \end{bmatrix}
\tag{12-34}
$$

By using Eq. (12-19) we get the following power relation: If

$$
a_2^{(I)} = 0
\tag{12-35a}
$$

then

$$
b_1^{(I)2} + b_2^{(I)2} = a_1^{(I)2}
\tag{12-35b}
$$

We will use a chain scattering matrix [1-3] in Sec. 12-8 for synthesizing a passive complex circuit. The form of a chain scattering matrix is defined by making use of incident and reflected waves as

$$
\begin{bmatrix} b_2 \\ a_2 \end{bmatrix} = [\Theta] \begin{bmatrix} a_1 \\ b_1 \end{bmatrix}
\tag{12-36}
$$

Therefore, a scattering matrix given by Eq. (12-34) is transformed to

$$
\begin{bmatrix} b_2^{(I)} \\ a_2^{(I)} \end{bmatrix} = \frac{1}{\sqrt{1-m^2}} \begin{bmatrix} 1 & -m \\ -m & 1 \end{bmatrix} \begin{bmatrix} a_1^{(I)} \\ b_1^{(I)} \end{bmatrix}
\tag{12-37}
$$

12-5 COMPLEX SCATTERING MATRIX

Let's consider the other complex voltage wave flow. Following the definition described in Refs. 13 and 14, we can define voltage waves for the circuit shown in Fig. 12-5 as:

Definition 12-3a. Incident voltage waves α_1 and α_2, and reflected voltage waves β_1 and β_2 for the circuit shown in Fig. 12-5 are defined as

$$
\alpha_1 = \alpha_G = \frac{V + Z_G I}{2}
\tag{12-38a}
$$

$$\beta_1 = \beta_G = \frac{V - Z_G^* I}{2} \tag{12-38b}$$

$$\alpha_2 = \alpha_L = \frac{V + Z_L(-I)}{2} \tag{12-38c}$$

$$\beta_2 = \beta_L = \frac{V - Z_L^*(-I)}{2} \tag{12-38d}$$

∎

Assuming that Eq. (12-28) is satisfied, and substituting the equations based on Assumption 12-1 into Definition 12-3a, we get

$$\alpha_G = Z_G I_i = V_i \tag{12-39a}$$

$$\beta_G = \frac{(Z_G - Z_G^*)I_i + 2R_G I_r}{2} \tag{12-39b}$$

$$\beta_L = R_L I_t \tag{12-39c}$$

$$\alpha_L = 0 \tag{12-39d}$$

Therefore, the voltage waves given by Definition 12-3a does not suit the waves based on Assumption 12-1, because of Eq. (12-39b).

Next, substituting the equations based on Assumption 12-2 into Definition 12-3a, we obtain

$$\alpha_G = R_G I_i^{(II)} \tag{12-40a}$$

$$\beta_G = R_G I_r^{(II)} \tag{12-40b}$$

$$\alpha_L = 0 \tag{12-40c}$$

$$\beta_L = R_L I_t^{(II)} \tag{12-40d}$$

Although the incident and reflected waves obtained by the substitution are represented by voltages, they do not agree with the incident and reflected voltage waves $V_i^{(II)}$ and $V_r^{(II)}$ defined by Assumption 12-2, but each wave is presented by a current wave multiplied by the real part of impedance. This yields the fact that the waves defined by Eq. (12-38) are concerned with the average dissipated power in complex power. That is, we can obtain the following equations:

$$\frac{\alpha_G \alpha_G^*}{R_G} = \frac{|E_G|^2}{4R_G} = \text{Re}\,[V_i^{(\text{II})} I_i^{(\text{II})*}] \qquad\qquad (12\text{--}41\text{a})$$

$$\frac{\beta_G \beta_G^*}{R_G} = \left|\frac{Z_L - Z_G^*}{Z_L + Z_G}\right|^2 \frac{|E_G|^2}{4R_G} = \text{Re}\,[V_r^{(\text{II})} I_r^{(\text{II})*}] \qquad\qquad (12\text{--}41\text{b})$$

$$\frac{\beta_L \beta_L^*}{R_L} = \frac{R_L |E_G|^2}{|Z_L + Z_G|^2} = \text{Re}\,[V_t^{(\text{II})} I_t^{(\text{II})}] \qquad\qquad (12\text{--}41\text{c})$$

$$\alpha_L = 0 \qquad\qquad (12\text{--}41\text{d})$$

Now we define the following waves as

$$a_G = \frac{\alpha_G}{\sqrt{R_G}} = \sqrt{R_G}\, I_i^{(\text{II})} \qquad\qquad (12\text{--}42\text{a})$$

$$b_G = \frac{\beta_G}{\sqrt{R_G}} = \sqrt{R_G}\, I_r^{(\text{II})} \qquad\qquad (12\text{--}42\text{b})$$

$$b_L = \frac{\beta_L}{\sqrt{R_L}} = \sqrt{R_L}\, I_t^{(\text{II})} \qquad\qquad (12\text{--}42\text{c})$$

Then we get the following relations between the waves and the average dissipated power.

$$a_G a_G^* = \text{Re}\,[V_i^{(\text{II})} I_i^{(\text{II})*}] \qquad\qquad (12\text{--}43\text{a})$$

$$b_G b_G^* = \text{Re}\,[V_r^{(\text{II})} I_r^{(\text{II})*}] \qquad\qquad (12\text{--}43\text{b})$$

$$b_L b_L^* = \text{Re}\,[V_t^{(\text{II})} I_t^{(\text{II})*}] \qquad\qquad (12\text{--}43\text{c})$$

By using Eq. (12-25) we get the following relationship among the waves:

$$|b_L|^2 = |a_G|^2 - |b_G|^2 \qquad\qquad (12\text{--}44)$$

In this case, the waves are concerned with the average dissipated power, so we change Definition 12-3a to the following definition in order to fit a complex scattering matrix described in Refs. 13 and 14.

Definition 12-3b. Incident waves a_1 and a_2, and reflected waves b_1 and b_2 for the circuit shown in the ( )-presentation of Fig. 12-5 are defined as:

$$a_1 = \frac{V + Z_G I}{2\sqrt{R_G}} \qquad (12\text{-}45a)$$

$$b_1 = \frac{V - Z_G^* I}{2\sqrt{R_G}} \qquad (12\text{-}45b)$$

$$a_2 = \frac{V + Z_L(-I)}{2\sqrt{R_L}} \qquad (12\text{-}45c)$$

$$b_2 = \frac{V - Z_L^*(-I)}{2\sqrt{R_L}} \qquad (12\text{-}45d)$$

$\blacksquare$

From Definition 12-3b, we get the following complex scattering matrix:

$$\begin{bmatrix} b_1 \\ b_2 \end{bmatrix} = \begin{bmatrix} \dfrac{Z_L - Z_G^*}{Z_L + Z_G} & \dfrac{2\sqrt{R_G R_L}}{Z_L + Z_G} \\ \dfrac{2\sqrt{R_G R_L}}{Z_L + Z_G} & \dfrac{Z_G - Z_L^*}{Z_G + Z_L} \end{bmatrix} \begin{bmatrix} a_1 \\ a_2 \end{bmatrix} \qquad (12\text{-}46)$$

The $(1,1)$ element of this scattering matrix coincides with the CRC under assumption 12-2, which is given by Eq. (12-23b). Accordingly, we call the $(1,1)$ element or Eq. (12-23b) a reflection coefficient under Assumption 12-2 or we call it RC.

The complex scattering matrix above can be transformed to a chain scattering matrix defined in the form of Eq. (12-36) as

$$[\Theta] = \text{diag}\{d_1, d_2\} \frac{1}{\sqrt{1 - |\rho|^2}} \begin{bmatrix} 1 & \rho \\ \rho^* & 1 \end{bmatrix} \qquad (12\text{-}47)$$

where

$$\rho = \frac{Z_G - Z_L^*}{Z_G^* + Z_L^*}$$

$$d_1 = \frac{Z_G^* + Z_L^*}{|Z_G^* + Z_L^*|}$$

$$d_2 = \frac{Z_G + Z_L}{|Z_G + Z_L|}$$

12-6 RICCATI DIFFERENTIAL EQUATION AND THÉVENIN THEOREM

In the foregoing we have discussed that since the waves on a uniform transmission line with complex characteristic impedance satisfy Assumption 12-1, they are defined as Definition 12-2a or 12-2b (i.e., as a voltage scattering matrix). Since the waves represented by a voltage scattering matrix are not directly related to complex power, we cannot calculate the average dissipated power at a load by making use of the voltage scattering matrix derived from Definition 12-2b. In this section we consider an application of the complex scattering matrix in order to discuss the average dissipated power for transmission-line circuits.

We discussed in Sec. 12-2 that the voltage and current at an arbitrary point on the uniform transmission line shown in Fig. 12-1 are presented by Eq. (12-7). The equation shows that multiple reflections are produced, and the voltage on the line is figured out with a signal flow shown in Fig. 12-2. Since the voltage and current given by Eq. (12-7) indicate summations of multiple reflected waves, Eq. (12-7) shows the voltage and current at a steady state. We consider what circuit offers the voltage and current given by Eq. (12-7) first.

Since the voltage $V(x, s)$ and current $I(x, s)$ at an arbitrary point on a uniform transmission line are given by Eq. (12-4), the cascade matrix for the line of length ℓ is represented as

$$\begin{bmatrix} V(0, s) \\ I(0, s) \end{bmatrix} = \begin{bmatrix} \cosh \gamma \ell & Z_0 \sinh \gamma \ell \\ Z_0^{-1} \sinh \gamma \ell & \cosh \gamma \ell \end{bmatrix} \begin{bmatrix} V(\ell, s) \\ I(\ell, s) \end{bmatrix} \tag{12-48}$$

Therefore, we can obtain the impedance $Z_R(x)$ looking toward the load from the point x for the circuit shown in Fig. 12-1.

$$Z_R(x) = \frac{V(x, s)}{I(x, s)}$$

$$= \frac{R_L \cosh \gamma(\ell - x) + Z_0 \sinh \gamma(\ell - x)}{R_L Z_0^{-1} \sinh \gamma(\ell - x) + \cosh \gamma(\ell - x)} \tag{12-49}$$

Let us show that the impedance $Z_R(x)$ satisfies Riccati differential equation [16]. Equation (12-49) is rewritten as

$$V(x, s) = Z_R(x)I(x, s) \tag{12-50}$$

Differentiating with x, then,

$$\frac{dV(x, s)}{dx} = \frac{dZ_R(x)}{dx} I(x, s) + Z_R(x) \frac{dI(x, s)}{dx} \tag{12-51}$$

Substituting Eq. (12-3), then,

$$-(sL + R)I(x, s) = \frac{dZ_R(x)}{dx} I(x, s) - Z_R(x)(sC + G)V(x, s) \tag{12-52}$$

Dividing with $I(x, s)$, then,

$$\frac{dZ_R(x)}{dx} - (sC + G)Z_R^2(x) + (sL + R) = 0 \tag{12-53}$$

Thus $Z_R(x)$, which shows the impedance looking toward the load at a steady state, satisfies Riccati differential equation.

We can also obtain the Thévenin impedance $Z_S(x)$ looking toward the generator and the Thévenin equivalent generator $E_S(x)$ from the point x for the circuit shown in Fig. 12-1.

$$E_S(x) = \frac{E(s)Z_0}{Z_G \sinh \gamma x + Z_0 \cosh \gamma x} \tag{12-54a}$$

$$Z_S(x) = \frac{Z_G \cosh \gamma x + Z_0 \sinh \gamma x}{Z_G \sinh \gamma x + Z_0 \cosh \gamma x} \tag{12-54b}$$

Thus the equivalent circuit at the point x for the circuit shown in Fig. 12-1 is figured out in Fig. 12-6.

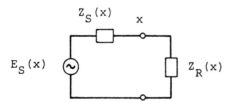

FIGURE 12-6 Equivalent-circuit representation at point x for the circuit
shown in Fig. 12-1.

The voltage and current at the port x for the circuit shown in **Fig. 12-6**
are given by $V(x, s)$ and $I(x, s)$ presented by Eq. (12-7). Thus at a steady
state we can use the Thévenin theorem for transmission-line circuits.

Now we consider power transfer for a generalized lossless two-port.
Figure 12-7(a) shows a lossless two-port inserted between resistive terminations. We assume that the cascade matrix of the lossless two-port is presented as

$$\begin{bmatrix} V_1 \\ I_1 \end{bmatrix} = \begin{bmatrix} A & B \\ C & D \end{bmatrix} \begin{bmatrix} V_2 \\ -I_2 \end{bmatrix} \qquad (12\text{-}55)$$

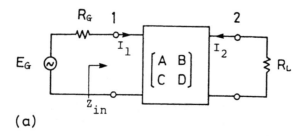

(a)

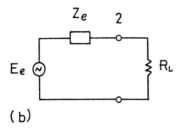

(b)

FIGURE 12-7 (a) A circuit with a lossless transmission two-port. (b) The
equivalent circuit at the port 2.

To make the two-port lossless, it is necessary and sufficient for the cascade matrix to satisfy the following condition [17].

Lossless Condition.

$$\begin{bmatrix} A^* & C^* \\ B^* & D^* \end{bmatrix} \begin{bmatrix} 0 & 1 \\ 1 & 0 \end{bmatrix} \begin{bmatrix} A & B \\ C & D \end{bmatrix} = \begin{bmatrix} 0 & 1 \\ 1 & 0 \end{bmatrix} \tag{12-56a}$$

Therefore, (12-56a) can be rewritten as

$$\begin{bmatrix} A & B \\ C & D \end{bmatrix}^{-1} = \begin{bmatrix} D^* & B^* \\ C^* & A^* \end{bmatrix} \tag{12-56b}$$

Thus the following condition for losslessness are obtained:

$$AD^* + BC^* = AD^* + B^*C = 1 \tag{12-56c}$$

$$Re[AB^*] = Re[AC^*] = Re[BD^*] = Re[CD^*] = 0 \tag{12-56d}$$

The reciprocity and the symmetry are presented below.

Reciprocity Condition. The reciprocity is represented by

$$AD - BC = 1 \tag{12-57}$$

Symmetry Condition. The symmetry is represented by

$$AD - BC = 1 \tag{12-58a}$$

and

$$A = D \tag{12-58b}$$

To obtain the current I_1 and voltage V_1 at port 1 for the circuit shown in Fig. 12-7(a), we should get the impedance Z_{in} looking toward the load from port 1.

$$Z_{in} = \frac{R_L A + B}{R_L C + D} \tag{12-59}$$

Next, to obtain the current I_2 and voltage V_2 at port 2 shown in Fig. 12-7(a), we use the Thévenin theorem. That is, the equivalent circuit at port 2 is shown in Fig. 12-7(b), and the voltage E_e and the internal impedance Z_e for the equivalent generator are

$$E_e = \frac{E_G}{A + R_G C} \tag{12-60a}$$

$$Z_e = \frac{B + R_G D}{A + R_G C} \tag{12-60b}$$

I_2 and V_2 are obtained by using Eq. (12-60) and R_L.

The maximum available power for the generator shown in Fig. 12-7(a) is $|E_G|^2/4R_G$, and the maximum available power for the equivalent generator shown in Fig. 12-7(b) is obtained from Eq. (12-60) as

$$\frac{|E_e|^2}{4\,\mathrm{Re}\,(Z_e)} = \frac{|E_G|^2}{4\,\mathrm{Re}\,(BA^* + R_G^2 DC^* + R_G(DA^* + BC^*))} \tag{12-61}$$

$$= \frac{|E_G|^2}{4R_G}$$

Thus the maximum available power for Thévenin equivalent generator is equal to the one for the original generator of Fig. 12-7(a), provided that the two-port is lossless.

Now we consider RC for the Thévenin equivalent circuit. RC at port 1 (we present with Γ_1) for the circuit shown in Fig. 12-7 is obtained by using R_G and Z_{in} as follows:

$$\Gamma_1 = \frac{Z_{in} - R_G}{Z_{in} + R_G} \tag{12-62a}$$

and RC at port 2 (Γ_2) is obtained by using Z_e and R_L as follows:

$$\Gamma_2 = \frac{R_L - Z_e^*}{R_L + Z_e} \tag{12-62b}$$

The absolute value of Γ_1 is

$$|\Gamma_1| = \frac{|R_L A + B - R_G R_L C - R_G D|}{\sqrt{R_L^2 AA^* + BB^* + R_L^2 R_G^2 CC^* + R_G^2 DD^* + 2R_G R_L}} \tag{12-63}$$

and we can prove

$$|\Gamma_2| = |\Gamma_1| \qquad\qquad\qquad (12\text{-}64)$$

Thus we have proved that the maximum available power for Thévenin equivalent generator is equal to the one for the given generator, and that the absolute value of RC for the Thévenin equivalent circuit is equal to the one at the
input port, provided that the two-port is lossless. If the circuit is at a
steady state after multiple reflections, these properties can be applied to
lossless transmission-line circuits, because a cascade matrix is a representation of the circuit at a steady state.

   We can also employ the Thévenin theorem to lossy two-port circuits.
For that case the absolute values of RC at port 1 and port 2 are not always
equal. Therefore, it is necessary that conjugate matching is usually achieved
at port 2, but not at port 1, in order that the load dissipates the maximum
available power it can dissipate. The conjugate matching at port 2 is achieved
when the load impedance is equal to conjugate complex of the internal impedance for the Thévenin equivalent generator.

12-7 COMMENSURATE TRANSMISSION-LINE CIRCUIT
AND WAVE DIGITAL FILTER

Among microwave circuits, there is a CTL circuit which is composed of
lengths of lossless transmission line that has commensurate one-way delay
and only propagates the transverse electromagnetic (TEM) mode. The CTL
section is lossless and uniform. That is, $R = G = 0$ in Eq. (12-1), and the
propagation constant and the characteristic impedance defined by Eq. (12-5)
are represented as follows:

$$\gamma = s\sqrt{LC} = \frac{s}{v_p} \qquad\qquad\qquad (12\text{-}65a)$$

where v_p denotes velocity of the propagation.

$$Z_0 = \sqrt{\frac{L}{C}} = R_0 \quad \text{(real)} \qquad\qquad\qquad (12\text{-}65b)$$

The complex variable λ for the CTL circuit is represented by

$$\lambda = \tanh \gamma \ell = \tanh \frac{s\ell}{v_p} = \tanh s\tau \qquad\qquad\qquad (12\text{-}66)$$

where τ is the one-way delay.

   The CTL section is called a unit element (UE). By using Eq. (12-48)
a cascade matrix of a UE can be written as

$$\frac{1}{\sqrt{1-\lambda^2}} \begin{bmatrix} 1 & \lambda R_0 \\ \dfrac{\lambda}{R_0} & 1 \end{bmatrix} \tag{12-67}$$

The input impedance of UE short- or open-circuited at the far end is

$$Z_{sh} = R_0 \lambda \tag{12-68a}$$

or

$$Z_{op} = \frac{R_0}{\lambda} \tag{12-68b}$$

This shows that Z_{sh} is the impedance of an inductance $L = R_0$ and Z_{op} is that of a capacitance $C = 1/R_0$.

Digital filters are characterized by the z-transform, and the complex variable is represented by a delay parameter z^{-1}. The frequency response is obtained by setting

$$z^{-1} = e^{-sT} \tag{12-69a}$$

or

$$z^{-1} = e^{-j\omega T} \tag{12-69b}$$

where T is the sampling period.
The bilinear transformation of z^{-1} is given by

$$\frac{1 - z^{-1}}{1 + z^{-1}} = \frac{1 - e^{-sT}}{1 + e^{-sT}} = \tanh \frac{sT}{2} \tag{12-70}$$

From Eqs. (12-66) and (12-69), if $T/2 = \tau$, then

$$\lambda = \frac{1 - z^{-1}}{1 + z^{-1}} \tag{12-71}$$

That is, some digital filters are equivalent to CTL circuits.
According to the Darlington classical theorem [18], any rational, positive-real function may be synthesized as the input impedance of a cascade of passive, lossless reciprocal two-ports terminated in a nonnegative resistor. The passive, lossless reciprocal two-ports are classified by zeros of the even part of the input impedance (i.e., transmission zero). That is, real-frequency, real-axis, and complex zeros are removed by Brune, type C, and type D sections, respectively. If ideal gyrators are permitted to be

used for the synthesis, the real-axis and complex zeros are removed by
Richards and type E sections, respectively. The element values of the various sections are obtained by Youla in close forms in terms of three or six
indexes [19].

The Darlington theory is applied to the CTL circuit, which is synthesized with UEs and n-wire lines without using ideal transformers and gyrators. That is, for the synthesis we can employ ideal transformers that the
equivalent circuit of an n-wire line contains, although the n-wire line is not
composed of ideal transformers.

In that case, the realizing condition for n-wire lines is given by the
hyperdominancy of the characteristic admittance matrix. Nevertheless, it
is shown that when a finite number of UEs are extracted, the remaining impedance can be realized with n-wire lines [9]. Thus a cascade synthesis
(Darlington synthesis) without using ideal transformers is possible due to
the extraction of UEs. But the resulting circuit may be far from canonic [9].

To get a class of digital filters, Fettweis [4] has introduced WDFs.
To every WDF corresponds a reference analog filter from which it is derived. Particularly, the WDFs described by Ref. 4 are digital filters constructed with series and parallel sections and with UEs, and the complex
variable for WDF is λ given by Eq. (12-71). Therefore, the WDF is considered to be synthesized with the CTL circuit theory.

The Darlington synthesis for CTL circuits described in the foregoing
is achieved without employing ideal transformers and gyrators, since the
synthesis is for microwave circuits and in microwave frequencies winding
transformers do not work well as ideal transformers. On the other hand, we
can freely use them in theoretical procedure, and the theoretical CTL circuit
is composed not only of UEs and n-wire lines, but also of ideal transformers
and gyrators.

As a result, we can synthesize basic sections of theoretical CTL circuits required for the Darlington method (e.g., Brune). And the basic sections for WDFs are obtained by converting the basic section mentioned above
[6,7]. In this case the transmission zeros on the λ-plane are mapped on the
z^{-1}-plane by the transformation of Eq. (12-71). For example, the real fre-
quencies on the λ-plane are mapped on the circle of $|z^{-1}| = 1$ on the z^{-1}-plane.

Dewilde et al. [1] described that the Levinson realization algorithm
for optimal linear predictors is a special case of the Darlington synthesis.
They also suggested that the ARMA process for a generalized form of optimal linear predictors can be synthesized by the Darlington procedure.
Later, Dewilde et al. [2,3] suggested that the synthesis be constituted with
ODFs. The theory borrows a number of classical network concepts, such
as scattering matrix, losslessness, and so on. It is easily shown that the
elementary sections of real ODFs are equivalent to the Richards, type E,
and Brune described in the Darlington synthesis. Therefore, it can be said
that the real ODFs are equivalent to WDFs and CTL circuits.

ODFs are basically represented by complex coefficients and can be regarded as complex digital circuits. The complex ODF can be applied to the interpolation known as the Nevanlinna-Pick problem, which has many applications, such as linear prediction, modeling filters, estimation problem, and so on.

Since the complex ODFs are a class of digital filters, it can be said that the Nevanlinna-Pick problem has been solved by the digital filter synthesis. If the Nevanlinna-Pick problem is solved by an analog circuit synthesis, it can be represented by circuit terminologies such as voltage, current, and impedance. In the following sections we consider a complex CTL circuit that belongs to a class of analog circuits.

12-8 PASSIVITY AND LOSSLESSNESS
FOR COMPLEX CIRCUIT

For a classical one-port circuit, passivity is represented by the positive-real function. On the other hand, the passivity for complex circuits is described by positive functions [10, 14]. The positive function for a lumped circuit is presented as follows:

Positive Condition. If a complex (linear, time-invariant) one-port circuit can be characterized by an impedance function $Z(s)$, a necessary condition for passivity is that $Z(s)$ be a positive function, that is,

$$\text{Re}\,[Z(s)] \geq 0 \tag{12-72a}$$

for almost all complex frequencies in the region

$$\text{Re}\,[s] \geq 0 \tag{12-72b}$$

The condition that impedance function $Z(\lambda)$ for the CTL circuit be a positive function can be also represented by

$$\text{Re}\,[Z(\lambda)] \geq 0 \quad \text{for} \quad \text{Re}\,[\lambda] \geq 0 \tag{12-73}$$

Thus, by using Eq. (12-71), the condition of a positive function for the digital filter is denoted by

$$\text{Re}\,[Z(z^{-1})] \geq 0 \quad \text{for} \quad |z^{-1}| \leq 1 \tag{12-74}$$

Belevitch [10] shows the Darlington procedure for the positive function of complex lumped circuits. Complex circuits may be a free mathematical creation for analog circuits, but the theory of complex circuits is of practical interest concerning modulation problems that adopt the Hilbert transform. The complex circuit is in our hands when imaginary resistances are accepted

as elements in a circuit. An imaginary resistance absorbs no real instantaneous power and is abstractly lossless.

If the complex variable s is replaced with λ, we obtain the circuit
function for CTL circuits. Moreover, we get complex CTL circuits by using
imaginary resistances as circuit elements to the usual CTL circuit.

Since an imaginary resistance is imaginary constant for all frequencies,
we cannot transform it alone to an element of the complex digital filter. We
should create some analog circuit elements with imaginary resistances to
consider a complex digital filter that can be converted from an analog circuit.

Dewilde et al. [1-3] proposed complex ODFs which are obtained by
applying the Darlington synthesis for the complex digital filter. The complex
ODF is described by the scattering matrix and chain scattering matrix. So
we briefly describe the passivity and losslessness for a complex scattering
matrix.

In Fig. 12-3, if the load impedance is given by an impedance function
Z, and the internal impedance is given by a resistor R, the voltage (and
current) reflection coefficient is given by

$$S = \frac{Z - R}{Z + R} \tag{12-75}$$

In this equation, if Z is a positive function, we call S a bounded function. Thus the bounded function for each complex variable is defined as
follows:

$$|S(s)| \leq 1 \quad \text{for } \text{Re}[s] \geq 0 \tag{12-76}$$

$$|S(\lambda)| \leq 1 \quad \text{for } \text{Re}[\lambda] \geq 0 \tag{12-77}$$

$$|S(z^{-1})| \leq 1 \quad \text{for } |z^{-1}| \leq 1 \tag{12-78}$$

To synthesize a bounded function by the Darlington method, we adopt
the lossless chain scattering matrix defined in Eq. (12-47). We describe
briefly the method based on Refs. 1 to 3.

Let

$$J = \begin{bmatrix} 1 & 0 \\ 0 & -1 \end{bmatrix} \tag{12-79}$$

Then Θ, as defined by Eq. (12-47), is J-lossless; that is, it is J-unitary on
$|z^{-1}| = 1$:

$$J = \Theta^*(e^{j\theta}) J \Theta(e^{j\theta}) = \Theta(e^{j\theta}) J \Theta^*(e^{j\theta}) \tag{12-80}$$

where Θ^* denotes the conjugate transpose of a matrix Θ, and J-contractive
in $|z^{-1}| \leq 1$; that is,

$$\Theta(z^{-1})J\Theta^*(z^{-1}) \leq J \qquad\qquad\qquad (12\text{-}81\text{a})$$

or equivalently,

$$\Theta^*(z^{-1})J\Theta(z^{-1}) \leq J \qquad\qquad\qquad (12\text{-}81\text{b})$$

where $A \geq B$ means that $A - B$ is nonnegative definite.

As described in Sec. 12-7, in employing the Darlington procedure, we synthesize the given circuit by extracting the elementary lossless sections, such as the Brune section. Thus the elementary lossless sections should satisfy the J-losslessness.

The elementary lossless sections are composed of constant sections and degree 1 sections. The general form of a constant J-lossless section is given by

$$\Theta = \text{diag}[d_1, d_2]H(\rho) \qquad\qquad\qquad (12\text{-}82\text{a})$$

and

$$H(\rho) = \frac{1}{(1 - |\rho|^2)^{\frac{1}{2}}}\begin{bmatrix} 1 & \rho \\ \rho^* & 1 \end{bmatrix} \qquad\qquad\qquad (12\text{-}82\text{b})$$

where

$$|d_1| = |d_2| = 1 \quad \text{and} \quad |\rho| < 1$$

The degree 1 J-lossless sections are related to a Blaschke factor, which is defined as

$$\xi_\omega(z^{-1}) = \frac{z^{-1} - \omega}{1 - \omega^* z^{-1}} \qquad\qquad\qquad (12\text{-}83)$$

where $|\omega| < 1$.

Degree 1 J-lossless sections have the following general form:

(i) When the pole of the section is in $|z^{-1}| > 1$, we use it as $1/\omega^*$, and Θ is given by

$$\Theta = H(\beta)\begin{bmatrix} \xi_\omega(z^{-1}) & 0 \\ 0 & 1 \end{bmatrix}H(\rho)\,\text{diag}[d_1, d_2] \qquad\qquad\qquad (12\text{-}84)$$

We call this the Schur I section.

(ii) When the pole is in $|z^{-1}| = 1$, which is denoted by α, then

$$\Theta = \left[1_2 - \frac{\sigma}{2} \frac{\alpha + z^{-1}}{\alpha - z^{-1}} JVV^* \right] H(\rho) \ \text{diag}\,[d_1, d_2] \qquad (12\text{-}85)$$

We call this the Brune section.

(iii) When the pole is in $|z^{-1}| < 1$, we use it as ω, and

$$\Theta = H(\beta) \begin{bmatrix} 1 & 0 \\ 0 & \zeta_\omega^{-1}(z^{-1}) \end{bmatrix} H(\rho) \ \text{diag}\,[d_1, d_2] \qquad (12\text{-}86)$$

We call this the Schur II section, where d_1, d_2, ρ, β, and σ are constants satisfying

$$|d_1| = |d_2| = 1, \quad |\rho| < 1, \quad |\beta| < 1, \quad \sigma > 0$$

and V is a constant vector satisfying

$$V = [V_1 \quad V_2]^T \quad \text{with} \quad |V_1| = |V_2| = 1$$

Now we describe briefly the degree reduction proved in Ref. 3. When a bounded function $S(z^{-1})$ is given, we can extract a Schur section; then the residual function $S_1(z^{-1})$ is represented by

$$S_1(z^{-1}) = \frac{S(z^{-1}) - S(\omega)}{1 - S^*(\omega) S(z^{-1})} \frac{1 - \omega^* z^{-1}}{z^{-1} - \omega} \qquad (12\text{-}87)$$

Since $S(z^{-1}) - S(\omega)$ and $z^{-1} - \omega$ are canceled, the degree of $S_1(z^{-1})$ is decreased by degree 1 if and only if

$$1 - S^*(\omega) S\left(\frac{1}{\omega^*}\right) = 0 \qquad (12\text{-}88)$$

Let us define

$$S_*(z^{-1}) = S^*(z^*) \qquad (12\text{-}89)$$

Then we can prove the reduction of the degree 1 in the following cases:

$$S_*(z^{-1}) \text{ has a pole at } \omega \text{ and } S(\omega) = 0 \qquad (12\text{-}90a)$$

$$1 - S_*(\omega) S(\omega) = 0 \qquad (12\text{-}90b)$$

A point where either of the equations above is satisfied is referred to as a point of local losslessness (PLL) of $S(z^{-1})$. The α that satisfies the following equation is a PLL,

$$1 - S^*(\alpha)\,S(\alpha) = 0 \tag{12-91}$$

12-9 COMPLEX UNIT ELEMENT

The most basic element for CTL circuits is UE, whose cascade matrix is given by Eq. (12-67). By using the transformation (12-71), it is represented as

$$\begin{bmatrix} V_1 \\ I_1 \end{bmatrix} = \frac{z^{\frac{1}{2}}}{2R_0} \begin{bmatrix} R_0(1 + z^{-1}) & R_0^2(1 - z^{-1}) \\ 1 - z^{-1} & R_0(1 + z^{-1}) \end{bmatrix} \begin{bmatrix} V_2 \\ -I_2 \end{bmatrix} \tag{12-92}$$

Referring to the properties described in Sec. 12-6, UE is a lossless, reciprocal, and symmetrical element.

To obtain the chain scattering matrix for UE, we define the following waves based on the circuit shown in Fig. 12-8:

$$a_1 = \frac{V_1 + R_1 I_1}{2\sqrt{R_1}} \tag{12-93a}$$

$$b_1 = \frac{V_1 + R_1 I_1}{2\sqrt{R_1}} \tag{12-93b}$$

$$a_2 = \frac{V_2 + R_2 I_2}{2\sqrt{R_2}} \tag{12-93c}$$

$$b_2 = \frac{V_2 - R_2 I_2}{2\sqrt{R_2}} \tag{12-93d}$$

Then we can obtain the following equation:

$$\begin{bmatrix} a_1 \\ b_1 \end{bmatrix} = H_a(m_1) \begin{bmatrix} z^{\frac{1}{2}} & 0 \\ 0 & z^{-\frac{1}{2}} \end{bmatrix} H_a(m_2) \begin{bmatrix} b_2 \\ a_2 \end{bmatrix} \tag{12-94a}$$

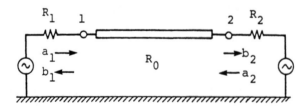

FIGURE 12-8 Normalized waves for a unit element.

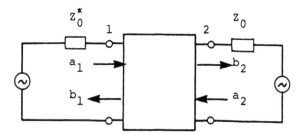

FIGURE 12-9 Circuit for obtaining a complex UE.

where

$$H_a(m_k) = \frac{1}{\sqrt{1 - m_k^2}} \begin{bmatrix} 1 & m_k \\ m_k & 1 \end{bmatrix} \qquad (12\text{-}94b)$$

$$m_1 = \frac{R_0 - R_1}{R_0 + R} \qquad (12\text{-}94c)$$

$$m_2 = \frac{R_2 - R_0}{R_2 + R_0} \qquad (12\text{-}94d)$$

Therefore, if $R_0 = R_1 = R_2$, then

$$\begin{bmatrix} a_1 \\ b_1 \end{bmatrix} = \begin{bmatrix} z^{\frac{1}{2}} & 0 \\ 0 & z^{-\frac{1}{2}} \end{bmatrix} \begin{bmatrix} b_2 \\ a_2 \end{bmatrix} \qquad (12\text{-}95)$$

Let us obtain a complex UE by using this chain scattering matrix. The general form of a constant J-lossless section is given by Eq. (12-82), and this equation is also given by Eq. (12-47), which is related to conjugate matching described in Assumption 12-2. Therefore, we take up a conjugate matching circuit shown in Fig. 12-9.

The waves represented in Eq. (12-95) are denoted in Fig. 12-9, and we define the following incident and reflected waves as

$$a_1 = \frac{V_1 + Z_0^* I_1}{2\sqrt{R_0}} \qquad (12\text{-}96a)$$

$$b_1 = \frac{V_1 - Z_0 I_1}{2\sqrt{R_0}} \qquad (12\text{-}96b)$$

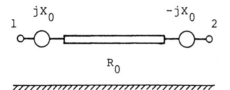

FIGURE 12-10 Circuit representation for the CTL circuit shown in
Fig. 12-9.

$$b_2 = \frac{V_2 - Z_0^* I_2}{2\sqrt{R_0}} \tag{12-96c}$$

$$a_2 = \frac{V_2 + Z_0 I_2}{2\sqrt{R_0}} \tag{12-96d}$$

where

$$\text{Re}[Z_0] = R_0$$

Substituting Eq. (12-96) into Eq. (12-95), we obtain the following cascade matrix:

$$\begin{bmatrix} V_1 \\ I_1 \end{bmatrix} = \frac{z^{\frac{1}{2}}}{2R_0} \begin{bmatrix} Z_0 + Z_0^* z^{-1} & Z_0 Z_0^* (1 - z^{-1}) \\ 1 - z^{-1} & Z_0^* + Z_0 z^{-1} \end{bmatrix} \begin{bmatrix} V_2 \\ -I_2 \end{bmatrix} \tag{12-97}$$

Belevitch [10] shows that a complex circuit can be obtained by adding imaginary resistances. The element given by Eq. (12-97) can also be represented by adding imaginary resistances to UE. That is, the element is shown in Fig. 12-10, where jX_0 and $-jX_0$ denote imaginary resistances, and

$$Z_0 = R_0 + jX_0 \tag{12-98a}$$

$$Z_0^* = R_0 - jX_0 \tag{12-98b}$$

A complex filter can be derived through transform on a usual real filter [20]. The complex filter is given by replacing z^{-1} with $z^{-1}e^{j\phi}$ in the complex variable of a real filter. This method is effective in getting a complex degree 1 section by transforming a degree 1 real section (e.g., the Richards section). Thus we consider an element whose cascade matrix is obtained by replacing z^{-1} with $z^{-1}e^{j\phi}$ in Eq. (12-97).

$$\frac{z^{1/2}e^{-j\phi/2}}{2R_0}\begin{bmatrix} Z_0 + Z_0^*e^{j\phi}z^{-1} & Z_0Z_0^*(1 - e^{j\phi}z^{-1}) \\ 1 - e^{j\phi}z^{-1} & Z_0^* + Z_0e^{j\phi}z^{-1} \end{bmatrix} \qquad (12\text{-}99)$$

This element can be obtained by the following way. Let us consider a uniform, lossless transmission line with which per-unit-length series imaginary resistance jr and parallel imaginary conductance jg are added to the usual series L and parallel C as shown in Fig. 12-11(a). In this case, $Z(s)$ and $Y(s)$ defined by Eq. (12-3) are given by

$$Z(s) = jr + sL \qquad (12\text{-}100a)$$

$$Y(s) = jg + sC \qquad (12\text{-}100b)$$

Since the characteristic impedance $\sqrt{Z(s)/Y(s)}$ for this line is generally a function of s, and changes with frequencies, we try to get a lossless, uniform line which satisfies the following condition:

$$\frac{L}{C} = \frac{r}{g} = R_0^2 \qquad (12\text{-}101)$$

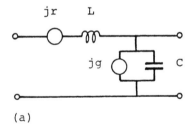

(a)

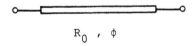

(b)

FIGURE 12-11 (a) Per-unit-length section, including imaginary resistances for a uniform, lossless line; (b) CUE with real characteristic impedance R_0.

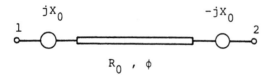

FIGURE 12-12   Representation of CUE (complex unit element).

Then the propagation constant γ is

$$\gamma = j\sqrt{rg} + s\sqrt{LC} \tag{12-102}$$

We define that the length ℓ for this element satisfies

$$e^{-(j\sqrt{rg}+s\sqrt{LC})\ell} = e^{-j\sqrt{rg}\,\ell}e^{-sT/2} = e^{j\phi/2}z^{-1/2} \tag{12-103}$$

Then we get the following cascade matrix with real characteristic impedance R_0.

$$\frac{z^{1/2}e^{-j\phi/2}}{2R_0}\begin{bmatrix} R_0(1 + e^{j\phi}z^{-1}) & R_0^2(1 - e^{j\phi}z^{-1}) \\ 1 - e^{j\phi}z^{-1} & R_0(1 + e^{j\phi}z^{-1}) \end{bmatrix} \tag{12-104}$$

This element is depicted in Fig. 12-11(b). Now we consider an element cascaded with a series imaginary resistance jX_0, the element shown in Fig. 12-11(b), and a series imaginary resistance $-jX_0$ as shown in Fig. 12-12. The cascade matrix for the element shown in Fig. 12-12 is represented by Eq. (12-99).

Since this element is one of expansions of the usual UE, we call it complex unit element (CUE). The CUE has the following properties:

    (i) It is lossless.
    (ii) It is reciprocal.
    (iii) It has transmission zeros at $z^{-1} = 0$ and $z^{-1} = \infty$.
    (iv) The impedance looking right from port 1 is Z_0 if we connect an impedance Z_0 at the load (port 2). This means that the impedance for a forward wave is given by Z_0. The impedance looking left from port 2 is Z_0^* if we connect an impedance Z_0^* at the input port (port 1). This means that the impedance for a backward wave is given by Z_0^*.

Thus the impedances of the forward and the backward waves for CUE are conjugate complex at each other. Therefore, we can use the conjugate matching assumed in Assumption 12-2 for CUE. On the other hand, the characteristic impedances of the forward and the backward waves for the usual transmission line are equal. Therefore, we cannot use the conjugate matching for the forward and the backward waves on the usual transmission line. But we can use it at a steady state as described in Sec. 12-6.

12-10 CONSTANT CHAIN SCATTERING MATRIX

Dewilde et al. [2, 3] synthesized digital filters with Darlington method, and named them ODFs. The objective of this chapter is to show that we can synthesize a complex CTL circuit, which is a kind of analog circuit, with the same method that is applied to ODFs. This means that the synthesized complex CTL circuit is sure to be converted to an ODF whose synthesis method is described in Ref. 3. Thus we omit to present the synthesis method of digital filters. In this section we consider some basic properties of constant scattering matrices necessary for synthesizing complex CTL circuits.

First we consider again the constant chain scattering matrix described by Definition 12-3b. That is, incident waves a_1 and a_2 and reflected waves b_1 and b_2 for the circuit shown in Fig. 12-5 are defined as

$$a_1 = \frac{V + Z_G I}{2\sqrt{R_G}} \tag{12-105a}$$

$$b_1 = \frac{V - Z_G^* I}{2\sqrt{R_G}} \tag{12-105b}$$

$$a_2 = \frac{V + Z_L(-I)}{2\sqrt{R_L}} \tag{12-105c}$$

$$b_2 = \frac{V - Z_L^*(-I)}{2\sqrt{R_L}} \tag{12-105d}$$

By using the equations above, we obtain a constant chain scattering matrix of the following form, which is an inverse matrix of the equation defined by Eq. (12-36):

$$\begin{bmatrix} a_1 \\ b_1 \end{bmatrix} = H(Z_G, Z_L) \begin{bmatrix} b_2 \\ a_2 \end{bmatrix} \tag{12-106a}$$

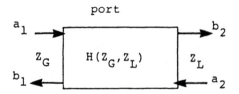

FIGURE 12-13 Representation of $H(Z_G, Z_L)$, which is a constant chain scattering matrix.

where

$$H(Z_G, Z_L) = \frac{1}{2\sqrt{R_G R_L}} \begin{bmatrix} Z_L + Z_G & Z_L^* - Z_G \\ Z_L - Z_G^* & Z_L^* + Z_G^* \end{bmatrix} \tag{12-106b}$$

Since the wave signal flow for the ()-presentation of Fig. 12-5 is given by Eq. (12-106), we figure out the wave signal flow as shown in Fig. 12-13.

$H(Z_G, Z_L)$ has the following properties:

$$H(Z_G, Z_G^*) = 1_2 \tag{12-107a}$$

where

$$1_2 = \begin{bmatrix} 1 & 0 \\ 0 & 1 \end{bmatrix} \tag{12-107b}$$

This means that incident waves are not reflected at conjugate matching, and this property is depicted in Fig. 12-14.

$$H^{-1}(Z_G, Z_L) = H(Z_L^*, Z_G^*) \tag{12-108}$$

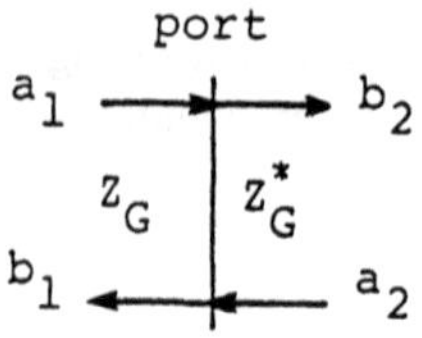

FIGURE 12-14 Representation for conjugate matching waves.

In other words, the inverse of this matrix is not the one in which the two impedances are replaced by each other. The matrix whose two impedances are interchanged is represented as

$$H(Z_L, Z_G) = J_2 H^{-1}(Z_G, Z_L) J_2 \tag{12-109a}$$

where

$$J_2 = \begin{bmatrix} 0 & 1 \\ 1 & 0 \end{bmatrix} \tag{12-109b}$$

and

$$J_2^{-1} = J_2 \tag{12-109c}$$

The following product is obtained:

$$H(Z_1, Z_2) H(Z_2^*, Z_3) = H(Z_1, Z_3) \tag{12-110}$$

Some constant scattering matrices are different from the one formed at a port interconnecting two impedances. The series and the parallel connections may be important among them. Since the series and parallel adapters for their connections can be discussed by the method described by Fettweis [4,5], we omit describing them here.

A complex ideal transformer described in Ref. 13 has a constant scattering matrix. The cascade matrix for a complex ideal transformer is defined as

$$\begin{bmatrix} V_1 \\ I_1 \end{bmatrix} = \begin{bmatrix} d^{-1} & 0 \\ 0 & d^* \end{bmatrix} \begin{bmatrix} V_2 \\ -I_2 \end{bmatrix} \tag{12-111}$$

A gyrator is a nonreciprocal two-port and has a constant scattering matrix. The linear relations between the terminal voltages and currents are obtained by the following cascade matrix [13]:

$$\begin{bmatrix} V_1 \\ I_1 \end{bmatrix} = \begin{bmatrix} 0 & R \\ R^{-1} & 0 \end{bmatrix} \begin{bmatrix} V_2 \\ -I_2 \end{bmatrix} \tag{12-112}$$

The number R is real and has a role somewhat analogous to the turns ratio of an ideal transformers.

In the Darlington synthesis, the usual gyrator is used in the Richards and type E sections [13]. The Richards section realizes a real transmission

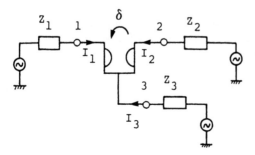

FIGURE 12-15 Three-port complex gyrator.

zero and degree 1 lossless basic section. In complex circuit synthesis, we
need degree 1 and lossless basic sections which realize complex transmission
zeros as expansions of the Richards section, which is described in Sec. 12-8.
The basic section is named the Schur section. To get a Schur section, it is
necessary to extend the usual gyrator to a gyrator with complex linear relations.

We consider a cascade connection of a complex ideal transformer and
a unit gyrator. The cascade matrix for the cascaded circuit is given by

$$\begin{bmatrix} d^{-1} & 0 \\ 0 & d^* \end{bmatrix} \begin{bmatrix} 0 & 1 \\ 1 & 0 \end{bmatrix} = \begin{bmatrix} 0 & d^{-1} \\ d^* & 0 \end{bmatrix} \tag{12-113a}$$

We represent the cascade matrix above as follows:

$$\begin{bmatrix} 0 & \delta \\ \delta^{-1*} & 0 \end{bmatrix} \tag{12-113b}$$

The complex gyrator is also a three-port, symbolized as shown in
Fig. 12-15. The linear relations among the port voltages and currents are

$$V_1 - V_3 = \delta(-I_2) \tag{12-114a}$$

$$\delta^* I_1 = V_2 - V_3 \tag{12-114b}$$

$$I_1 + I_2 + I_3 = 0 \tag{12-114c}$$

The incident waves a_k and the reflected waves b_k at each port are
defined as

$$a_k = \frac{V_k + Z_k I_k}{2\sqrt{R_k}} \tag{12-115a}$$

$$b_k = \frac{V_k - Z_k^* I_k}{2\sqrt{R_k}} \tag{12-115b}$$

$$Z_k = R_k + jX_k \quad (k = 1, 2, 3) \tag{12-115c}$$

Then we obtain the following scattering matrix for the complex gyrator:

$$\begin{bmatrix} b_1 \\ b_2 \\ b_3 \end{bmatrix} = \begin{bmatrix} S_{11} & S_{12} & S_{13} \\ S_{21} & S_{22} & S_{23} \\ S_{31} & S_{32} & S_{33} \end{bmatrix} \begin{bmatrix} a_1 \\ a_2 \\ a_3 \end{bmatrix} \tag{12-116}$$

where

$$\Delta = Z_1 Z_2 + Z_1 Z_3 + Z_2 Z_3 + Z_3(\delta - \delta^*) + \delta\delta^*$$
$$S_{11}\Delta = Z_2 Z_3 - Z_1^*(Z_2 + Z_3) + Z_3(\delta - \delta^*) + \delta\delta^*$$
$$S_{12}\Delta = 2\sqrt{R_1 R_2}\,(Z_3 - \delta)$$
$$S_{13}\Delta = 2\sqrt{R_1 R_3}\,(\delta + Z_2)$$
$$S_{21}\Delta = 2\sqrt{R_2 R_1}\,(Z_3 + \delta^*)$$
$$S_{22}\Delta = Z_1 Z_3 - Z_2^*(Z_1 + Z_3) + Z_3(\delta - \delta^*) + \delta\delta^*$$
$$S_{23}\Delta = 2\sqrt{R_2 R_3}\,(Z_1 - \delta^*)$$
$$S_{31}\Delta = 2\sqrt{R_3 R_1}\,(Z_2 - \delta^*)$$
$$S_{32}\Delta = 2\sqrt{R_3 R_2}\,(Z_1 + \delta)$$
$$S_{33}\Delta = Z_1 Z_2 - Z_3^*(Z_1 + Z_2) + Z_3^*(\delta^* - \delta) + \delta\delta^*$$

The scattering matrix above is simplified when the following conditions are satisfied:
(i) When

$$Z_1 = Z_2 = \delta^* \tag{12-117a}$$

then

$$\begin{bmatrix} 0 & S_{12} & S_{13} \\ 1 & 0 & 0 \\ 0 & S_{32} & S_{33} \end{bmatrix} \tag{12-117b}$$

where

$$S_{12} = \frac{Z_3 - \delta}{Z_3 + \delta*}$$

$$S_{13} = \frac{2\sqrt{R_1 R_3}}{Z_3 + \delta*}$$

$$S_{32} = \frac{2\sqrt{R_3 R_2}}{Z_3 + \delta*}$$

$$S_{33} = \frac{\delta* - Z_3^*}{Z_3 + \delta*}$$

The scattering matrix given by Eq. (12-117) can be divided into the following two equations:

$$b_2 = a_1 \qquad\qquad (12\text{-}118a)$$

$$\begin{bmatrix} b_1 \\ b_3 \end{bmatrix} = \begin{bmatrix} S_{12} & S_{13} \\ S_{32} & S_{33} \end{bmatrix} \begin{bmatrix} a_2 \\ a_3 \end{bmatrix} \qquad\qquad (12\text{-}118b)$$

Equation (12-118b) can be converted to the following constant chain scattering matrix represented by the form of Eq. (12-106):

$$\begin{bmatrix} a_2 \\ b_1 \end{bmatrix} = \frac{1}{2\sqrt{R_3 R_2}} \begin{bmatrix} Z_3 + \delta* & Z_3 - \delta* \\ Z_3 - \delta & Z_3 + \delta \end{bmatrix} \begin{bmatrix} b_3 \\ a_3 \end{bmatrix}$$

$$= H(\delta*, Z_3) \begin{bmatrix} b_3 \\ a_3 \end{bmatrix} \qquad\qquad (12\text{-}119)$$

Thus, the signal flow of these waves is indicated as shown in Fig. 12-16(a).

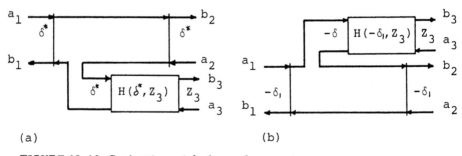

(a)   (b)

FIGURE 12-16 Conjugate matched complex gyrators.

(ii) When

$$Z_1 = Z_2 = -\delta \qquad (12\text{-}120\text{a})$$

then

$$\begin{bmatrix} 0 & 1 & 0 \\ S_{21} & 0 & S_{23} \\ S_{31} & 0 & S_{33} \end{bmatrix} \qquad (12\text{-}120\text{b})$$

where

$$S_{21} = \frac{Z_3 + \delta^*}{Z_3 - \delta}$$

$$S_{23} = \frac{2\sqrt{R_2 R_3}}{Z_3 - \delta}$$

$$S_{31} = \frac{2\sqrt{R_3 R_1}}{Z_3 - \delta}$$

$$S_{33} = \frac{\delta + Z_3^*}{\delta - Z_3}$$

The scattering matrix given by Eq. (12-120) can also be divided into

$$b_1 = a_2 \qquad (12\text{-}121\text{a})$$

$$\begin{bmatrix} b_2 \\ b_3 \end{bmatrix} = \begin{bmatrix} S_{21} & S_{23} \\ S_{31} & S_{33} \end{bmatrix} \begin{bmatrix} a_1 \\ a_3 \end{bmatrix} \qquad (12\text{-}121\text{b})$$

The equation above can be converted to

$$\begin{bmatrix} a_1 \\ b_2 \end{bmatrix} = \frac{1}{2\sqrt{R_3 R_1}} \begin{bmatrix} Z_3 + (-\delta) & Z_3^* - (-\delta) \\ Z_3 - (-\delta^*) & Z_3^* + (-\delta^*) \end{bmatrix} \begin{bmatrix} b_3 \\ a_3 \end{bmatrix} = H(-\delta, Z_3) \begin{bmatrix} b_3 \\ a_3 \end{bmatrix}$$

$$(12\text{-}122)$$

Thus the signal flow of these waves is indicated as shown in Fig. 12-16(b).

12-11 DEGREE 1 LOSSLESS SECTIONS

In this section we consider a synthesis method of complex CTL circuits for degree 1 J-lossless sections represented by Eqs. (12-84) to (12-86). The core is how to constitute a Blaschke factor given by Eq. (12-83) with complex CTL circuits. For realizing the factor, we consider the circuit shown in Fig. 12-17. Suppose that the cascade matrix for CUE shown in Fig. 12-17 is given by Eq. (12-104), and the boundary condition of far end for CUE is open-circuited, the impedance Z_3 connected to port 3 is represented by an impedance looking right from port 3.

$$Z_3 = \frac{Z_0 + Z_0^* e^{j\phi} z^{-1}}{1 - e^{j\phi} z^{-1}} \tag{12-123}$$

If $\mathrm{Re}[\delta] > 0$ for the gyrator shown in Fig. 12-17, we can use Eq. (12-117a) and S_{12} in Eq. (12-117) yields

$$S_{12} = \frac{e^{j\phi}(Z_0^* + \delta)}{Z_0 + \delta^*} \frac{z^{-1} - \omega}{1 - \omega^* z^{-1}} \tag{12-124a}$$

where

$$-\omega = \frac{e^{-j\phi}(Z_0 - \delta)}{Z_0^* + \delta} \tag{12-124b}$$

Thus if

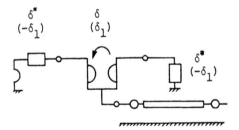

FIGURE 12-17   Realization of a Schur II section (Schur I section).

$$e^{j\phi} = \frac{Z_0 + \delta^*}{Z_0^* + \delta}$$ (12-125a)

$$Z_0 = \frac{\delta - \omega\delta^*}{1 + \omega}$$ (12-125b)

then S_{12} is represented by the following Blaschke factor:

$$S_{12} = \frac{z^{-1} - \omega}{1 - \omega^* z^{-1}}$$ (12-126)

If we present the relation above with a chain scattering matrix, it can be shown by Eq. (12-86). Thus the circuit shown in Fig. 12-17 is a Schur II for complex CTL circuits.

Next we consider a circuit shown in the ()-presentation of Fig. 12-17. If $\mathrm{Re}[\delta_1] < 0$ for the gyrator shown in Fig. 12-17, we can use Eq. (12-120a) and S_{21} in Eq. (12-120b) yields

$$S_{21} = \frac{Z_3 + \delta_1^*}{Z_3 - \delta_1} = \frac{e^{j\phi}(Z_0^* - \delta_1^*)}{Z_0 - \delta_1} \frac{z^{-1} - \omega}{1 - \omega^* z^{-1}}$$ (12-127a)

where

$$-\omega = \frac{Z_0 + \delta_1^*}{Z_0 - \delta_1}$$ (12-127b)

Thus if

$$e^{j\phi} = \frac{Z_0 - \delta_1}{Z_0^* - \delta_1^*}$$ (12-128a)

$$Z_0 = \frac{\omega\delta_1 - \delta_1^*}{1 + \omega}$$ (12-128b)

then S_{21} is

$$S_{21} = \frac{z^{-1} - \omega}{1 - \omega^* z^{-1}}$$ (12-129)

If we present the relation above with a chain scattering matrix, it can be

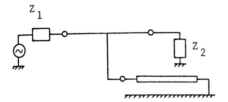

FIGURE 12-18 Brune section.

shown by Eq. (12-84). Thus the circuit shown in the ( )-presentation in
Fig. 12-17 is a Schur I for complex CTL circuits.

    Now we consider the circuit shown in Fig. 12-18 for a Brune section
of complex CTL circuits. Since the boundary condition of the far end for
CUE is short-circuited, the cascade matrix of the circuit shown in Fig. 12-18
is

$$\begin{bmatrix} 1 & 0 \\ \dfrac{Z_0 + Z_0^* e^{j\phi} z^{-1}}{Z_0 Z_0^* (1 - e^{j\phi} z^{-1})} & 1 \end{bmatrix} \tag{12-130}$$

In this case, Z_0 for CUE is real, that is,

$$Z_0 = R_0 = \frac{1}{G_0} \tag{12-131}$$

The chain scattering matrix Θ is represented by

$$\Theta = \frac{1}{2\sqrt{R_1 R_2}} \begin{bmatrix} Z_1^* + Z_2^* - Z_1^* Z_2^* B & Z_1 - Z_2^* - Z_1 Z_2^* B \\ Z_1^* - Z_2 + Z_1^* Z_2 B & Z_1 + Z_2 + Z_1 Z_2 B \end{bmatrix} \tag{12-132a}$$

where

$$B = \frac{G_0 (e^{j\phi} + z^{-1})}{e^{j\phi} - z^{-1}} \tag{12-132b}$$

The equation above can be rewritten as

$$\left[1_2 - \frac{G_0}{2} \frac{e^{-j\phi} + z^{-1}}{e^{-j\phi} - z^{-1}} \, JVV^* \right] H(\rho) \; \mathrm{diag}\,[d_1, d_2] \tag{12-133a}$$

where

$$\rho = \frac{Z_1 - Z_2^*}{Z_1 + Z_2} \qquad (12\text{-}133b)$$

$$d_1 = \frac{Z_1^* + Z_2^*}{|Z_1^* + Z_2^*|} \qquad (12\text{-}133c)$$

$$d_2 = \frac{Z_1 + Z_2}{|Z_1 + Z_2|} \qquad (12\text{-}133d)$$

$$J = \text{diag}[1, -1] \qquad (12\text{-}133e)$$

$$v^T = \left[\frac{Z_1}{\sqrt{R_1}} \quad \frac{Z_1^*}{\sqrt{R_1}} \right] \qquad (12\text{-}133f)$$

Thus the circuit shown in Fig. 12-18 represents Brune section given by Eq. (12-85).

Since the circuits shown in Figs. 12-17 and 12-18 are degree 1 J-lossless sections for complex CTL circuits, complex CTL circuits can clearly be synthesized with the Darlington procedure.

This book has no mention of the design methods described in the recently published Ref. 11. Although it presents some efficient design methods of multidimensional filters and doubly-complementary filters, they are not applicable to the Nevanlinna-Pick problem.

REFERENCES

1. P. Dewilde, A. C. Vieira, and T. Kailath, "On a generalized Szego-Levinson realization algorithm for optimal linear predictors based on a network synthesis approach," IEEE Trans. Circuits Syst., vol. CAS-25, no. 9, pp. 663-675 (Sept. 1978).

2. P. Dewilde and H. Dym, "Lossless inverse scattering digital filters, and estimation theory," IEEE Trans. Inf. Theory, vol. IT-30, no. 4, pp. 644-662 (July 1984).

3. E. Deprettere, P. Dewilde, and C. V. K. Prabhakara Rao, "Cascade orthogonal filter design," Delft University of Technology, 1984.

4. A. Fettweis, "Digital filter networks related to classical filter networks," AEÜ—Arch. Elektron. Übertragungstech., vol. 25, pp. 5-10 (1971).

5. A. Fettweis, "Wave digital filter: theory and practice," Proc. IEEE, vol. 74, no. 2, pp. 270-327 (Feb. 1986).

6. M. Suzuki, N. Miki, and N. Nagai, "New wave digital filters for basic reactance sections," IEEE Trans. Circuits Syst., vol. CAS-32, no. 4, pp. 337-348 (Apr. 1985).

7. N. Nagai, M. Suzuki, M. Suzuki, and N. Miki, "A synthesis on a generalized wave digital filter based on a distributed-constant network," Proc. Int. Symp. Circuits and Systems, Kyoto, pp. 519-522, 1985.

8. A. Matsumoto (ed.), Microwave Filters and Circuits, Academic Press, New York, 1970.

9. J. O. Scanlan, "Theory of microwave coupled-line networks," Proc. IEEE, vol. 68, no. 2, pp. 209-231 (Feb. 1980).

10. V. Belevitch, Classical Network Theory, Holden-Day, San Francisco, 1968.

11. "Special section on complex signal processing," IEEE Trans. Circuits Syst., vol. CAS-34, no. 4, pp. 337-399 (1987).

12. F. E. Gardiol, Lossy Transmission Lines, Artech House, Norwood, Mass., 1987.

13. H. J. Carlin and A. B. Giordano, Network Theory, Prentice-Hall, Englewood Cliffs, N.J., 1964.

14. E. S. Kuh and R. A. Rohrer, Theory of Linear Active Networks, Holden-Day, San Francisco, 1967.

15. W. Klein, Grundlagen der Theorie elektrischer Schaltungen, Akademie-Verlag, East Berlin, 1961.

16. W. T. Reid, Riccati Differential Equations, Academic Press, New York, 1972.

17. A. V. Efimov and V. P. Potapov, "J-expanding matrix functions and their role in the analytical theory of electrical circuits," Usp. Mat. Nauk, pp. 65-130 (1973).

18. S. Darlington, "Synthesis of reactance 4-poles," J. Math. Phys. (Cambridge, Mass.), vol. 18, pp. 257-353 (Sept. 1939).

19. D. C. Youla, "A new theory of cascade synthesis," IRE Trans. Circuit Theory, vol. CT-8, pp. 244-260 (Sept. 1961).

20. T. H. Crystal and L. Ehrman, "The design and applications of digital filters with complex coefficients," IEEE Trans. Audio Electroacoust., vol. AU-16, no. 3, pp. 315-320 (Sept. 1968).

13

Application of ARMA Digital Lattice Filter to Speech Analysis and Reconstruction

NOBUHIRO MIKI

Hokkaido University, Sapporo, Japan

13-1 SPEECH PRODUCTION MODEL

In human communication and person-machine communication, speech signal processing is required for transmission by data compression, and for recognition and synthesis. Recently, we have seen a great deal of digital signal processing (DSP) of speech as efficient digital codings, but we would inquire of the efficiency of the digital coding for speech waveforms. Since speech signal is a one-dimensional signal depending on time, it would seem that DSP theories and techniques could easily be applied for the speech signal processing: for example, speech synthesis from parameters or speech recognition with computers. In actuality, there are many difficulties because, although speech signals have unique properties, we have only a little knowledge about the speech production process. Thus it becomes very important to know what is essential for the processing of speech signals. For this reason a vocal tract model based on circuit theory is proposed, and this model and approach is applied to the speech synthesizer as a vocal tract simulator and to the estimation algorithm of vocal tract shapes.

13-2 VOCAL TRACT MODEL WITH DIGITAL LATTICE FILTER

The vocal tract is regarded as a connection of some different acoustic cylindrical pipes with different cross sections. It is known that the equivalent circuit for each section can be constructed by a distributed constant transmission line. In this transmission-line model it is necessary to account for the propagation loss caused by viscous friction, heat conduction of air, and th the wall vibration at the wall of the vocal tract. The equivalent-circuit model for a short length is shown as Fig. 13-1. In this model some of the equations related to physical constants are given as

$$L_i = \frac{\rho}{A_i} \tag{13-1a}$$

$$C_i = \frac{A_i}{\rho c^2} \tag{13-1b}$$

$$r_i(\omega) = \frac{S_i}{A_i^2} \sqrt{\frac{\omega\rho\mu}{2}} \tag{13-1c}$$

$$g_i(\omega) = \frac{0.4S_i}{\rho c^2} \sqrt{\frac{\omega\lambda}{2C_p\rho}} \tag{13-1d}$$

$$y_{w_i}(\omega) = \frac{1}{1400/S_i + 1.6j\omega/S_i} \tag{13-1e}$$

$$z_i(\omega) = r_i(\omega) + j\omega L_i \tag{13-1f}$$

$$y_i(\omega) = g_i(\omega) + y_{w_i}(\omega) + j\omega C_i \tag{13-1g}$$

where ρ is the air density, c the sound velocity, μ the viscosity coefficient of air, λ the coefficient of heat condition, C_p the specific heat of air at

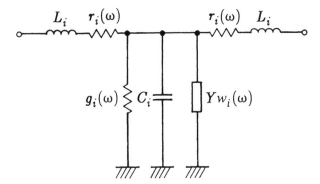

FIGURE 13-1 Equivalent circuit of vocal tract as an acoustic tube of very short length.

constant pressure, S_i the pipe circumference, and A_i the pipe area. In the equations above $r_i(\omega)$, $g_i(\omega)$, and y_{w_i} represent viscous loss, heat loss, and the acoustic admittance of the vocal tract wall, respectively, for unit length. If we denote

$$Z_i(\omega) = \left[\frac{z_i(\omega)}{y_i(\omega)} \right]^{1/2} \tag{13-2a}$$

and

$$\gamma_i^2(\omega) = z_i(\omega) y_i(\omega) \tag{13-2b}$$

the propagation equation in the frequency domain, in general, includes the propagation by wall vibration of the vocal tract, given by

$$-\frac{dV_i}{dx} = Z_i(\omega) U_i \tag{13-3a}$$

$$-\frac{dU_i}{dx} = Y_i(\omega) V_i \tag{13-3b}$$

where V_i is the sound pressure and U_i is the volume velocity.

The cascade matrix F in the ith section with length ℓ can be given from the equations above by

$$F_i = \begin{bmatrix} \cosh \gamma_i \ell & Z_i \sinh \gamma_i \ell \\ \dfrac{1}{Z_i} \sinh \gamma_i \ell & \cosh \gamma_i \ell \end{bmatrix} \qquad (13\text{-}4)$$

In this matrix the variable ω is omitted for simplicity.

We define traveling waves in the ith section as U_i^+ for the direction to the lips and U_i^- for the direction to the glottis.

$$U_i^+ = \frac{1}{2}\left(U_i + \frac{V_i}{Z_i}\right) \qquad (13\text{-}5a)$$

$$U_i^- = \frac{1}{2}\left(U_i - \frac{V_i}{Z_i}\right) \qquad (13\text{-}5b)$$

and

$$\begin{bmatrix} U_i^+ \\ U_i^- \end{bmatrix} = \begin{bmatrix} A_{w_i} & B_{w_i} \\ C_{w_i} & D_{w_i} \end{bmatrix} \begin{bmatrix} U_{i+1}^+ \\ U_{i+1}^- \end{bmatrix} \qquad (13\text{-}5c)$$

From this definition and the parameters of the cascade matrix for wave flow, we obtain

$$A_{w_i} = \frac{1}{2}[Z_i(A_i Z_{i+1} + B_i) + (C_i Z_{i+1} + D_i)] \qquad (13\text{-}6a)$$

$$B_{w_i} = \frac{1}{2}[Z_i(A_i Z_{i+1} - B_i) + (C_i Z_{i+1} - D_i)] \qquad (13\text{-}6b)$$

$$C_{w_i} = \frac{1}{2}[Z_i(A_i Z_{i+1} + B_i) + (C_i Z_{i+1} + D_i)] \qquad (13\text{-}6c)$$

$$D_{w_i} = \frac{1}{2}[Z_i(A_i Z_{i+1} - B_i) + (C_i Z_{i+1} - D_i)] \qquad (13\text{-}6d)$$

Using the relations above, the relation between traveling waves can be expressed as

$$\begin{bmatrix} U_i^+ \\ U_i^- \end{bmatrix} = \frac{e^{\gamma_i \ell}}{1 + \rho_i} \begin{bmatrix} 1 & -\rho_i \\ -\rho_i e^{-2\gamma_i \ell} & e^{-2\gamma_i \ell} \end{bmatrix} \begin{bmatrix} U_{i+1}^+ \\ U_{i+1}^- \end{bmatrix} \qquad (13\text{-}7)$$

where $\rho_i = (Z_i - Z_{i+1})/(Z_i + Z_{i+1})$.

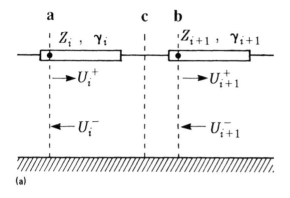

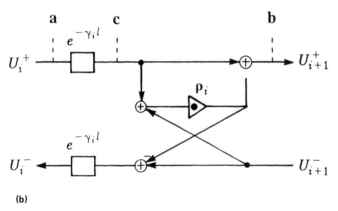

(b)

FIGURE 13-2 (a) Distributed line model of vocal tract; (b) equivalent digital lattice filter for the circuit (a).

In (13-6) the characteristic impedance Z_i and the propagation constant γ_i are complex variables depending on frequency and, consequently, the reflection coefficient ρ_i of the above equation is also a complex variable depending on frequency. It is seen, however, that (13-7) is the general form, which is the same as in a lossless case. From (13-7) and the above discussion, the equivalent expression with wave flow for the transmission line model of vocal tracts as in Fig. 13-2(a) is given as in Fig. 13-2(b). In this figure $e^{-\gamma_i \ell}$ and ρ_i can be realized as filters.

Consider the case where a sound source is located in the vocal tract (fricative, for example). The transmission line model then is of the form in Fig. 13-3(a). The volume velocity produced by the source V_s is given by $U_s = V_s/(Z_i + Z_{i+1})$. By the principle of superposition, the volume velocity wave at point b is given by adding (13-7) and U_s:

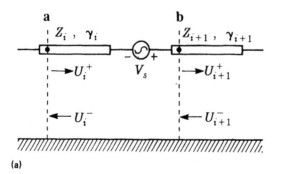

(a)

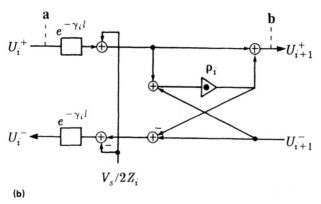

(b)

FIGURE 13-3 (a) Distributed line model with sound source in vocal tract; (b) equivalent digital lattice filter for the circuit (a).

$$U_i^+ e^{-\gamma_i \ell} + \frac{V_s}{Z_i + Z_{i+1}} - U_i^- e^{\gamma_i \ell} = U_{i+1}^+ - U_{i+1}^- - \frac{V_s}{Z_i + Z_{i+1}} \qquad (13\text{-}8a)$$

$$Z_i \left(U_i^+ e^{-\gamma_i \ell} + \frac{V_s}{Z_i + Z_{i+1}} + U_i^- e^{\gamma_i \ell} \right) = Z_{i+1} \left(U_{i+1}^+ - U_{i+1}^- - \frac{V_s}{Z_i + Z_{i+1}} \right)$$

$$(13\text{-}8b)$$

Rearranging the expression above, the following equation is obtained:

$$\begin{bmatrix} U_i^+ \\ U_i^- \end{bmatrix} = \frac{e^{\gamma_i \ell}}{1 + \rho_i} \begin{bmatrix} 1 & -\rho_i \\ -\rho_i e^{-2\gamma_i \ell} & e^{-2\gamma_i} \end{bmatrix} \begin{bmatrix} U_{i+1}^+ \\ U_{i+1}^- \end{bmatrix} - \frac{e^{\gamma_i \ell}}{2Z_i} \begin{bmatrix} 1 \\ e^{-2\gamma_i \ell} \end{bmatrix}^T V_s \quad (13\text{-}9)$$

From (13-8) and (13-9), the wave flow for the section with a sound source in the vocal tract is given as in Fig. 13-3(b). As shown in this figure, the filter structure in this case differs from that in Fig. 13-2(b) only in the presence of the sound source V_s. Consequently, in the synthesis of the consonant, where V_s is produced, the structure of the lattice need not be modified and the volume velocity wave $V_s/2Z_i$ should be added to the section (b) corresponding to narrowing of the vocal tract. As to the automatic generation of V_s for the narrow portion of the vocal tract, a circuit is proposed to generate the source from computation of the dc component in the vocal tract [5], which can be utilized in this case.

Nasal sounds are characterized by a branch in the middle of the vocal tract. Wave flow expression for a network with a branch is given by Fettweis [15], which is here applied to the circuit consisting of transmission lines with a branch. Define the pressure V_k and the volume velocity U_k as in Fig. 13-4(a) and assume that each transmission line is terminated by its characteristic impedance Z_i. From the continuity of current and pressure at junction b, the following relation is obtained:

$$U_i^+ e^{-\gamma_i \ell} - U_i^- e^{\gamma_i \ell} = (U_{i+1}^+ - U_{i+1}^-) + (U_{i'+1}^+ - U_{i'+1}^-) \qquad (13\text{-}10a)$$

$$(U_i^+ e^{-\gamma_i \ell} + U_i^- e^{\gamma_i \ell})Z_i = (U_{i+1}^+ + U_{i+1}^-)Z_{i+1}$$

$$= (U_{i'+1}^+ + U_{i'+1}^-)Z_{i+1} \qquad (13\text{-}10b)$$

Denoting the characteristic admittance of the transmission line in each section by G_i, the reflection coefficient is defined as follows:

$$\rho_i = \frac{G_{i+1} - G_{i'+1} - G_i}{G_{i+1} + G_{i'+1} + G_i} \qquad (13\text{-}11a)$$

$$\rho_{i'} = \frac{G_{i'+1} - G_{i+1} - G_i}{G_{i+1} - G_{i'+1} - G_i} \qquad (13\text{-}11b)$$

Using the definition above, we can modify (13-10) as follows:

$$\begin{bmatrix} U_i^+ \\ U_i^- \end{bmatrix} = \frac{e^{\gamma_i \ell}}{1+\rho_i} \begin{bmatrix} 1 & -\rho_i \\ -\rho_i e^{-2\gamma_i \ell} & e^{-2\gamma_i \ell} \end{bmatrix} \begin{bmatrix} U_{i+1}^+ \\ U_{i+1}^- \end{bmatrix} + e^{\gamma_i \ell} \begin{bmatrix} 0 & -1 \\ -e^{-2\gamma_i \ell} & 0 \end{bmatrix} \begin{bmatrix} U_{i'+1}^+ \\ U_{i'+1}^- \end{bmatrix}$$

$$\qquad (13\text{-}12a)$$

$$U_{i'+1}^+ = (1 + \rho_{i'+1})(e^{-\gamma_i \ell} U_i^+ + U_{i'+1}^-) + \rho_{i'+1} U_{i'+1}^- \qquad (13\text{-}12b)$$

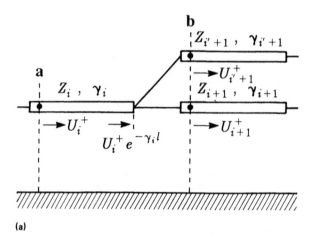

(a)

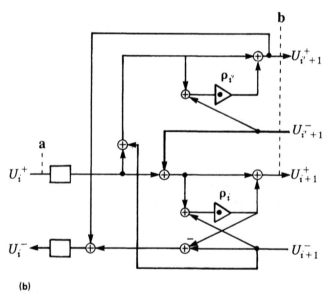

(b)

FIGURE 13-4 (a) Distributed line model with nasal branch in vocal tract;
(b) equivalent digital lattice filter for the circuit (a).

The equation above determines the wave matrix between points a and b
of Fig. 13-4(a). Determining the wave flow from the equation above, the
lattice filter of Fig. 13-4(b) is obtained. From this figure it is seen that the
wave flow wave flow in the case of branching is given by adding the reflection
of $\rho_{i'+1}$ to the lattice filter of the vowel-type section given in Fig. 13-2(b).

In the same manner as in the discussion above, we obtain a lattice filter in Fig. 13-5(b), which is the model of the branches for lateral consonants.

In the generation of consonants, in general, the vocal tract can be represented by a circuit with a branch in the vocal tract, as in the case of

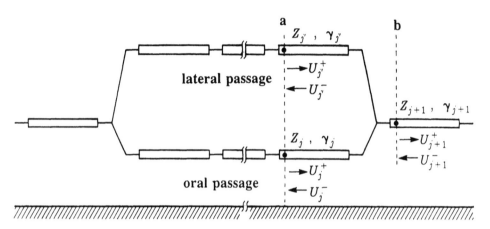

(a)

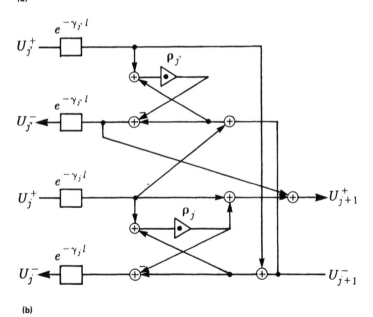

(b)

FIGURE 13-5 (a) Distributed line model with lateral passage in vocal tract; (b) equivalent digital lattice filter for the circuit (a).

nasals or lateral constants, and a circuit with a source in the vocal tract or at either end. By constructing lattice filters for those circuits, the general form of wave flow is determined as in Figs. 13-2(b) to 13-5(b) for the case with loss in the vocal tract. These structures are the same as in the lossless form, the only difference being that the coefficient ρ_i and the delay factor $e^{-\gamma_i \ell}$ depend on the frequency.

   Thus it becomes necessary to approximate these frequency-dependent elements by digital filters. A method has been proposed for approximation by lumped-constant circuits and transmission lines for a vocal tract system, including the loss due to wall vibration [1]. This is discussed in the next section.

13-3 LATTICE FORM FOR MIXED MODEL
AND SIMPLIFIED LATTICE

The losses for sound propagation are distributed within the vocal tract: serial resistance $r(\omega)$, parallel conductance $g(\omega)$, and wall admittance $y_w(\omega)$. It is known that $r(\omega)$ and $g(\omega)$ have an influence on the transfer function of vocal tracts, but it is less than that of $y_w(\omega)$. Thus the losses can be regarded as lumped elements since the influence of $y_w(\omega)$ is dominant in the low-frequency region. We propose a mixed model for unit sections of the vocal tract as illustrated in Fig. 13-6. In this model the loss factor of wave propagation is realized by the three lumped elements, and the propagation delay is realized by the lossless line. $R_i(\omega)$, $G_i(\omega)$, and $Yw_i(\omega)$ are regarded as a concentrated loss in the ith section.

   For the wave flow expression, we obtain

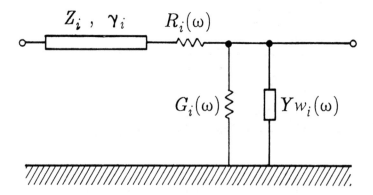

FIGURE 13-6 Mixed model of lossy vocal tract.

$$\begin{bmatrix} U_i^+ \\ U_i^- \end{bmatrix} = e^{sT} \begin{bmatrix} Aw_i & Bw_i \\ Cw_i e^{-2sT} & Dw_i e^{-2sT} \end{bmatrix} \begin{bmatrix} U_{i+1}^+ \\ U_{i+1}^- \end{bmatrix}$$ (13-13a)

$$Aw_i^{-1} = 2 \left\{ \frac{Z_{i+1} + Z_i + R_i(\omega)}{Z_i} + G_i(\omega)\left[\frac{R_i(\omega)Z_{i+1}}{Z_i} + Z_{i+1} \right] \right.$$

$$\left. + \left[\frac{R_i(\omega Z_{i+1}}{Z_i} + Z_{i+1} \right] Yw_i(\omega) \right\}^{-1}$$ (13-13b)

$$Bw_i = \frac{1}{2} \left\{ \frac{Z_{i+1} - Z_i - R_i(\omega)}{Z_i} + G_i(\omega)\left[\frac{R_i(\omega)Z_{i+1}}{Z_i} + Z_{i+1} \right] \right.$$

$$\left. + \left[\frac{R_i(\omega)Z_{i+1}}{Z_i} + Z_{i+1} \right] Yw_i(\omega) \right\}$$ (13-13c)

$$Cw_i = \frac{1}{2} \left\{ \frac{Z_{i+1} - Z_i + R_i(\omega)}{Z_i} + G_i(\omega)\left[\frac{R_i(\omega)Z_{i+1}}{Z_i} - Z_{i+1} \right] \right.$$

$$\left. + \left[\frac{R_i(\omega)Z_{i+1}}{Z_i} - Z_{i+1} \right] Yw_i(\omega) \right\}$$ (13-13d)

$$Dw_i = \frac{1}{2} \left\{ \frac{Z_{i+1} + Z_i - R_i(\omega)}{Z_i} + G_i(\omega)\left[\frac{R_i(\omega)Z_{i+1}}{Z_i} - Z_{i+1} \right] \right.$$

$$\left. + \left[\frac{R_i(\omega)Z_{i+1}}{Z_i} - Z_{i+1} \right] Yw_i(\omega) \right\}$$ (13-13e)

where Z_i is the characteristic impedance of the ith section's line. The section of the mixed model can be translated into an equivalent lattice filter form as in Fig. 13-7(a). The multiplier symbol $-\triangleright-$ represents a filter, and the square box represents a unit delay.

If the wall admittance is expressed with $Yw_i(\omega) = [r_i + j\omega\ell_i]^{-1}$ and the frequency variable of the lumped elements is adequately fixed for approximation, the filter can be realized with a first-order digital filter. Furthermore, this first-order digital filter can be approximated by a real constant multiplier, and we have a simplified lattice form as shown in Fig. 13-7(b). To investigate the accuracy of the mixed model, the maximum error of approximation is evaluated for x-ray data of vocal tract shape. The length of

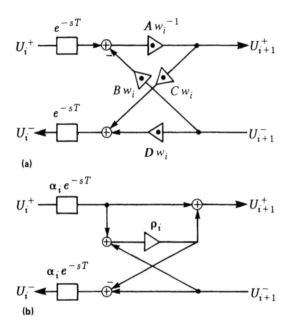

(a)

(b)

FIGURE 13-7 (a) Realization of the mixed model using four filters; (b) approximated digital lattice filter of filter (a).

one section of the vocal tract is set at 1.5 cm, and each section is approximated by a cascaded mixed model of several blocks. The approximation accuracy for vowels is sufficiently under a 1-dB error, even using only one block for one section.

In simplified approximation, we need to determine some parameter values. One is a fixed angular frequency ω_α for the resistance $R_i(\omega)$ and conductance $G_i(\omega)$. Another is a fixed angular frequency ω_β for admittance $Yw_i(\omega)$. These parameter values are determined so that the vocal tract transfer function is best approximated. Using these parameter values, we can approximate as

$$\rho_i = \text{sign}|\ \rho(\omega_\alpha, \omega_\beta)|\ \tag{13-14a}$$

$$\alpha_i = |\ \exp[-\gamma_i(\omega_\alpha, \omega_\beta)]|\ \tag{13-14b}$$

$$\alpha_i \exp(-sT) \approx \exp[-\gamma_i(\omega)\ell]\ \tag{13-14c}$$

where sign represents a plus sign or -1.

In the equations above ρ_i is a real constant, which replaces the complex coefficient $\rho(\omega)$; and the real constant α_i and the delay factor $\exp(-sT)$

are used as the attenuation and delay elements. Using (13-14) we can obtain the simplified lattice form as shown in Fig. 13-7(b). This simplified method is applicable to the other lattice model of constants.

13-4 REALIZATION OF BOUNDARY CONDITION
AT GLOTTIS AND LIPS

The two-mass model is well known as a model for self-oscillation of the vocal cord [8]. By this model the equivalent circuit for acoustic approximation can be expressed as a time-varying resistance and a time-varying inductance as shown in Fig. 13-8(a). In this figure the source Ps is the pressure at the lung, and the relation of the pressure and the volume velocity is obtained as

$$P_s + R_g(U_g^- - U_g^+) + \frac{d}{dt}[L_g(U_g^- - U_g^+)] = Z_1(U_g^+ + U_g^-) \qquad (13\text{-}15)$$

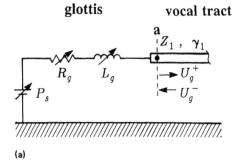

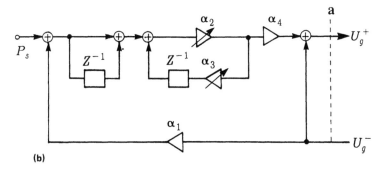

FIGURE 13-8 (a) Circuit model for the boundary at glottis; (b) digital filter realization of the circuit (a).

If the differential operator d/dt is approximated by the trapezoidal rule,

$$\frac{d}{dt} \approx \frac{2}{T}(1 - z^{-1})(1 + z^{-1})^{-1}$$

we obtain the following digital filter realization:

$$U_g = U_g^- + \alpha_4(\alpha_2^{-1} - z^{-1}\alpha_3)^{-1}(1 + z^{-1})(P_s + \alpha_1 U_g^-) \tag{13-16}$$

where

$$\alpha_1 = -2Z_1$$

$$\alpha_2 = \left(\frac{Z_1 + R_g}{L_g} + \frac{2}{T}\right)^{-1}$$

$$\alpha_3 = \frac{2}{T} - \frac{Z_1 + R_g}{L_g} \tag{13-17}$$

$$\alpha_4 = \frac{1}{L_g}$$

T is the sampled interval and z^{-1} is the time shift operator of T.

In the derivation above, noncommutative multiplication of z^{-1} and time-varying variables are used. The realization of (13-16) is shown in Fig. 13-8.

In most digital synthesizers, the driving point radiation impedance has been treated as zero or a real constant for simplicity. In vocal-tract modeling, the driving point radiation impedance must be treated as the termination of acoustic circuits. A radiation model with frequency dependency has also been developed recently [1-3]. Based on our experimental measurements of radiation characteristics [3], the dominant characteristics of amplitude and phas can be approximated as those of a spherical baffle model, and the difference component is compensated with an additional linear phase. Thus if the reflection coefficient at the lips is represented as

$$\mu_L(\omega) = \frac{z_0 - Z_L(\omega)}{z_0 + Z_L(\omega)} \tag{13-18}$$

where z_0 is the characteristic impedance of the uniform acoustic tube, $Z_L(\omega)$ the radiation impedance at the lips, and $\mu_L(\omega)$ is regarded as a complex reflection coefficient.

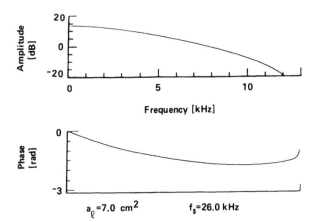

$a_\ell = 7.0$ cm^2 $f_s = 26.0$ kHz

FIGURE 13-9   Example of lip reflection coefficient model $\mu_\ell(z)$.

We can approximate Eq. (13-18) by a filter as follows:

$$\mu_L(\omega) \approx \mu_\ell(z) = \hat{\mu}_b(z) e^{-j\phi} \tag{13-19}$$

where

$$\hat{\mu}_b(z) = b \; \frac{\Pi^{nb}_{i=1} (1 - b_i z^{-1})}{\Pi^{na}_{i=1} (1 - a_i z^{-1})}$$

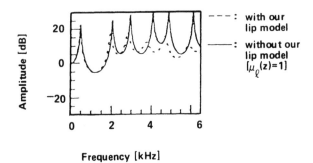

--- : with our
       lip model

——— : without our
       lip model
       $[\mu_\ell(z) = 1]$

FIGURE 13-10   Effects of lip model on transfer function of vocal tract.

and

$$e^{-j\phi} \approx \frac{\sum_{i=0}^{nc} c_{nc-1} z^{-i}}{\sum_{i=0}^{nc} c_i z^{-1}}$$

In the approximation above, we can get excellent approximation by setting the orders as na = 1, nb = 2, and nc = 2. For example, Fig. 13-9 shows a computation result of $|\mu_\ell(z)|$ from the spherical baffle model, and we see that the magnitude of reflection decreases along with frequency. The transfer functions of vocal tract for Japanese vowel /i/ are shown in Fig. 13-10, and we see that the function is greatly influenced with the lip impedance.

13-5 TIME-DOMAIN SIMULATION OF SPEECH PRODUCTION

Using the proposed vocal tract simulator with the digital lattice filter, speech waveforms can be obtained easily from the vocal tract shape, the vocal cord parameters, and the lung pressure P_s. In the synthesis of speech waveforms as a time-domain simulation of the equivalent circuit, the simulation is performed as a computation of sound propagation in the vocal tract.

When the proposed simulator is applied for the real speech synthesizer, it is important to reduce the information for vocal tract areas. The dynamical movement of vocal tract shape must be represented as a three-dimen-

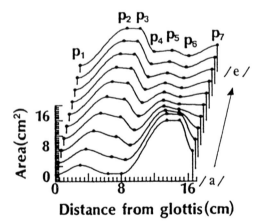

FIGURE 13-11 Reconstruction of dynamic vocal tract movement from vocal tract parameters represented by dots.

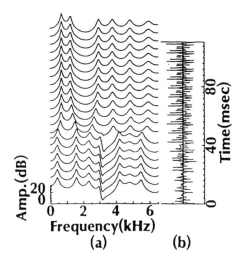

FIGURE 13-12 Computational results of dynamic vocal tract transfer function and synthesized speech /la/ from vocal tract parameters.

sional area function with a tract length axis, time axis, and area axis. As
shown in Fig. 13-11, by using the vocal tract parameters represented by
the dots, and interpolation for the space and time axes, the information reduction can be performed. In this example of the figure, the whole area

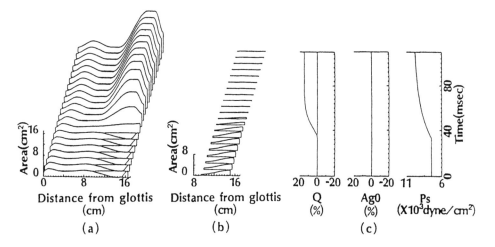

FIGURE 13-13 Computational results of reconstruction of dynamic vocal
tract shape and some control parameters.

function is computed from only one pair of parameters for /a/ and /e/, the pair of seven dots. In the space interpolation, Hermite interpolation is used, and in the time axis each parameter is interpolated by using a second-order filter. Control of the parameter of the second-order filter can be used for the variation of utterance speed in the synthesizer. The synthesized speech /la/, as an example, is shown in Figs. 13-12 and 13-13. Figure 13-13(a) represents the dynamic movement of the vocal tract area of the channel on the tongue, and (b) represents that of the lateral channel. The control parameters for the vocal fold model are given as in (c), where the parameter Q is the quantity of tension of the vocal fold. Japanese accent is controlled by the tension, and in the figure we see that the accent is on the vowel /a/.

In continuous speech synthesis, the vocal tract parameters or area functions of phonemes are required, but it is inconvenient to determine the area function from x-ray data. Then if a suitable method for area estimation from acoustic speech data is given for this synthesizer, the estimated data can be used for the synthesizer. In the next section we propose a method for area estimation.

13-6 VOCAL TRACT AREA ESTIMATION
   FROM ARMA PARAMETERS

If the vocal tract transfer function is given, the area function of the vocal tract can be obtained easily. However, it is very difficult to extract these characteristics from only the spectrum envelope of the observed speech because the envelope changes with different pitches even in the same vocal tract shape. Since an MIS (model identification systems) or WMIS (weighted MIS) algorithm [2] can estimate ARMA spectra of speech from the input estimation, the influence of the pitch is practically eliminated from the estimated spectra, and one can use the algorithm for the estimation of these characteristics. Particularly in the vocal tract shape estimation, it is shown that the vocal tract area function can be obtained without the influence of the lip impedance by using the WMIS and the proposed inverse filter model of the vocal tract model.

A new method of estimating cross-sectional area of the vocal tract using an ARMA analysis is shown. To estimate the suitable cross-sectional area for speech synthesis using the dynamic vocal tract simulator, vocal tract parameters such as an articulatory point must be obtained at low computational cost because these parameters are need for each short period for synthesizing speech.

Our estimation method includes three correction processes as follows: The first process is performed to remove the influence of the reflection characteristics at the lips from the estimated vocal tract transfer characteristics (reference ARMA model), and a vocal tract inverse filter model with a boundary condition at the lips is proposed for this correction. The second process is performed to extrapolate the characteristics of the reference ARMA model in the high-frequency region for the correction of the estimated error in this region, and a new criterion function for our estimation method is introduced. The third process is performed to remove the glottal source characteristics and the transfer characteristics between the volume velocity at the lips and the sound pressure in the free acoustic field. A gross spectrum model is introduced and the model is estimated using a multivariable search method. In the estimation, since the step–down algorithm is used in the final calculation step, the computational cost is lower than the method using only the search algorithm.

13-7 VOCAL TRACT INVERSE FILTER MODEL
WITH A BOUNDARY CONDITION

The frequency characteristics of the reflection coefficient at the lips have a considerable effect on the measured sound pressure spectrum. To remove these effects for the estimation of the vocal tract area function, we propose an inverse filter model with a boundary condition model as shown in Fig. 13-14, where $U_g(z)$ and $U_l(z)$ are the z–transformation of the volume velocity at the glottis and the lips, and $\mu_i(z)$ is the ARMA model of lip reflection mentioned previously. The model is written as follows:

$$\mu_i(z) = \frac{Y(z)}{X(z)} = \frac{y_0 + y_1 z^{-1} + \cdots + y_{ny} z^{-ny}}{x_0 + x_1 z^{-1} + \cdots + x_{nx} z^{-nx}} \tag{13-20}$$

The loss factor in the vocal tract is presented by α at the lips end. In the case of $\mu_l(z) = 1$, the inverse model is equivalent to Atal's model [9]. If we consider the vocal tract to be divided into m sections, the transfer function $T(z)$, which relates signals at point f and g in Fig. 13-14, is defined as

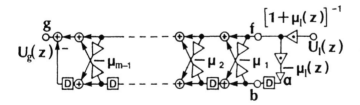

FIGURE 13-14 Vocal tract inverse filter model for area estimation.

$$T(z) = 1 + t_1 z^{-1} + \cdots + t_{m-1} z^{-(m-1)} \tag{13-21}$$

Then the transfer function between signals at points b and g is written as follows:

$$-T_*(z) = -z^{-(m-1)} T(1/z) \tag{13-22}$$

Using (13-20) to (13-22), we obtain the transfer function of the vocal tract inverse filter model as

$$S(z) = \frac{U_g(z)}{U_\ell(z)} = \beta \, \frac{X(z)T(z) + \alpha z^{-1} Y(z) T_*(z)}{X(z) + Y(z)} \tag{13-23}$$

where β is a gain normalizing factor defined as $\beta = (x_0 + y_0)/x_0$, and α is the loss factor in Fig. 13-14. If we get a suitable $T(z)$ [in other words, $T_*(z)$], we can easily calculate the reflection coefficients for the m sections and the cross-sectional areas of the vocal tract using the step-down recursive algorithm [13].

13-8 ESTIMATION CRITERION AND MODEL ESTIMATION

The reliable frequency region of the estimated ARMA model of speech sound is limited due to the bandlimitation of the pressure signal, the difficulty of accurate spectrum estimation in the high-frequency region, and the breakdown of the assumption of plane-wave modes beyond about 4 kHz. Therefore, we introduce two criterion functions to be minimized to estimate $T(z)$. One is a criterion function (CF) for the low-frequency region (LFR) where the estimated ARMA model is reliable, and the other one is a CF for the high-frequency region (HFR). Then we introduce one combined CF using these two CFs and weighting functions.

We first introduce a CF for the LFR. Let us consider that the ARMA model of the sound pressure spectrum is obtained as

$$\frac{B(z)}{A(z)} = \frac{1 + b_1 z^{-1} + \cdots + b_{nb} z^{-nb}}{1 + a_1 z^{-1} + \cdots + a_{na} z^{-na}} \tag{13-24}$$

which has minimum phase characteristics and is reliable in the LFR. This ARMA model is estimated using WMIS [2, 4]. On the other hand, we consider that the gross spectrum model is given as

$$\frac{D(z)}{C(z)} = \frac{1 + d_1 z^{-1} + \cdots + d_{nd} z^{-nd}}{1 + c_1 z^{-1} + \cdots + c_{nc} z^{-nc}} \qquad (13\text{-}25)$$

by using a multivariable search method which is mentioned in a later section. Now, using (13-23), (13-24), and (13-25), we define the CF for the LFR as

$$Q_\ell = \oint_{|z|=1} \left| \frac{B(z)}{A(z)} \frac{C(z)}{D(z)} \hat{S}(z) - 1 \right|^2 \frac{dz}{2\pi j z} \qquad (13\text{-}26)$$

Second, we introduce a CF for the HFR. We assume that the vocal tract area function is so smooth that no reflection occurs in the tract in the high-frequency region. Under this assumption T(z) becomes 1. Therefore, we define the CF for the HFR as

$$Q_h = \oint_{|z|=1} |\hat{T}(z) - 1|^2 \frac{dz}{2\pi j z} \qquad (13\text{-}27)$$

Finally, we define the combined CF Q from Q_ℓ and Q_h as

$$Q = \oint_{|z|=1} \left| \frac{B(z)}{A(z)} \frac{C(z)}{D(z)} \hat{S}(z) - 1 \right|^2 \left| \frac{V_\ell(z)}{W_\ell(z)} \right|^2 \frac{dz}{2\pi j z} + \oint_{|z|=1} |\hat{T}(z) - 1|^2 \left| \frac{V_h(z)}{W_h(z)} \right|^2 \frac{dz}{2\pi j z} \qquad (13\text{-}28)$$

based on the previous two CFs. In this CF, $W_\ell(z)$ and $W_h(z)$ are ARMA form weighting functions for the LFR and for the HFR, which are shown in Fig. 13-15. In the LFR, the first term on the right-hand side of (13-28) becomes

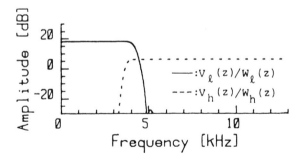

FIGURE 13-15 Weighting functions for error criterion.

dominant and the combined CF approaches the CF for the LFR. On the other
hand, in the HFR, the second term becomes dominant and the combined CF
approaches the CF for the HFR. Therefore, if we find T(z) minimizing the
combined CF, smoothed cross-sectional area functions are estimated from
the bandlimited reference ARMA model.

Now the combined CF Q is rewritten as

$$Q = [\vec{t}^T \; 1]\vec{R}\begin{bmatrix} \vec{t} \\ 1 \end{bmatrix} \tag{13-29}$$

where vector $\vec{t}$ is defined as

$$\vec{t}^T = [1 \; f_1 \; \cdots \; f_{m-1}] \tag{13-30}$$

$\vec{R}$ is a symmetric matrix, and T denotes a transpose. Then, from the following relation,

$$\frac{\partial Q}{\partial \vec{t}} = \vec{0} \tag{13-31}$$

the vector t [in other words, T(z)], which minimizes the CF, is obtained as
a solution of the next normal equation:

$$\vec{R}\begin{bmatrix} \vec{t} \\ 1 \end{bmatrix} = \vec{R}[1 \; t_1 \; t_2 \; \cdots \; t_{m-1}]^T$$
$$= [P_1 \; 0 \; 0 \; \cdots \; 0 \; P_2]^T \tag{13-32}$$

Cholesky decomposition can be used in the solving process. As T(z) is estimated, the reflection coefficients of the vocal tract are calculated using the
step-down algorithm on T(z). In addition, the minimum value of the CF is
obtained by

$$Q_{min} = P_1 + P_2 \tag{13-33}$$

Here note that Q_{min} is calculated when the gross spectrum model (13-25)
and the loss factor α are determined. In other words, Q_{min} is considered
to be a function of $c_1 \cdots c_{nc}$, $d_1 \cdots d_{nd}$, and α.

The minimum value of the combined CF is rewritten as

$$Q_{min} = F(c_1, \ldots, c_{nc}, d_1, \ldots, d_{nd}, \alpha) \tag{13-34}$$

where $F(*)$ is a multivariable nonlinear function. Therefore, we consider
the problem of estimating the gross spectrum model and the loss factor as
a problem of searching for independent variables (c_1, ..., c_{nc}, d_1, ...,
d_{nd}, and α) which minimize (13-34). Since $C(z)$, $D(z)$, and $T(z)$ must be a
minimum-phase model and $0 < \alpha < 1$ must be satisfied, we must use a multivariable constrained optimization technique. In this chapter we chose the
unconstrained Powell's algorithm for simplicity. Since this algorithm is an
unconstrained optimization technique, we increase the error-function value
by a constant factor when the constraints are violated. Jury's algorithm is
available to check whether $C(z)$, $D(z)$, and $T(z)$ have minimum-phase characteristics or not [12].

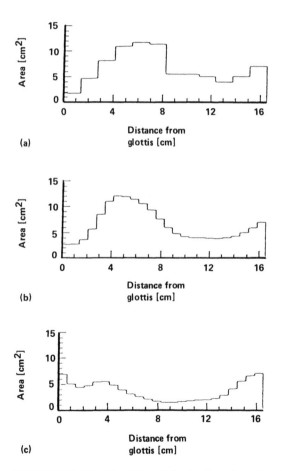

(a)

(b)

(c)

FIGURE 13-16 Experimental results of area estimation for synthesized
speech from the proposed synthesizer.

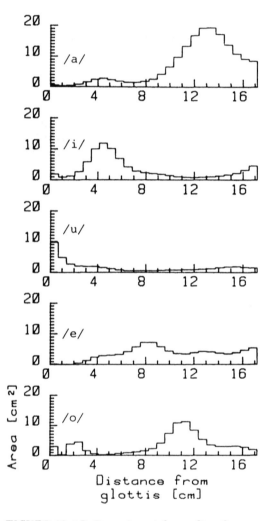

FIGURE 13-17 Experimental results of area estimation for Japanese real vowels.

Vocal tract estimation experiments were made using synthesized speech. The speech signal was synthesized using our vocal tract simulator, where the given cross-sectional area function is shown in Fig. 13-16(a), and the sampling frequency was 36 kHz. Next, the sampling frequency was reduced to 13 kHz, and the spectrum of the signal was estimated by an ARMA model using WMIS. Finally, we obtained a reference ARMA model by substituting z^{-2} for z^{-1} of the estimated ARMA model.

Note that the sampling frequency of the reference ARMA model is 26 kHz, but the model is reliable only in the low-frequency region below about 5 kHz. From the reference model we estimated an area function shown in Fig. 13-16 using the proposed method. It is seen that a smoothed cross-sectional area function of shortly divided sections can be estimated from the bandlimited reference model.

On the other hand, the bottom line of Fig. 13-16 shows a cross-sectional area function estimated under the condition where $\mu_\ell(z) = 1$. The estimated function was distorted by the effect of the inappropriate boundary condition at the lips. The proposed algorithm is applied to Japanese real vowels, and good estimation results are obtained as shown in Fig. 13-17.

REFERENCES

1. T. Nomura, N. Miki, and N. Nagai, "A digital vocal tract simulator with boundary conditions at lips and glottis," Electron. Commun. Jpn., vol. 64-A, no. 11, (1981).

2. N. Miki, S. Saga, Y. Miyanaga, and N. Nagai, "ARMA spectral estimation using weighted model identification system," J. Acoust. Soc. Jpn. (E), vol. 7, no. 1, (1986).

3. K. Motoki, N. Miki, and N. Nagai, "ARMA approximation method for radiation part in the digital model of vocal tract," Trans. Comm. Speech Res. (Acoust. Soc. Jpn.), vol. S85-06, (1985).

4. N. Miki, Y. Miyanaga, S. Saga, and N. Nagai, "Spectrum and pitch estimation of speech using a time-varying ARMA estimation algorithm," IEEE ICASSP, Tampa, Fla., vol. 3, p. 1133, 1985.

5. S. Saga, N. Miki, and N. Nagai, "A construction method of dynamic vocal tract simulator for connected speech synthesis," Trans. Inst. Elec. Commun. Eng. (Jpn.), vol. J68-A, no. 3, (1985).

6. H. Watanabe, Y. Miyanaga, N. Miki, and N. Nagai, "A note on ARMA spectral approximation methods with weighting functions," Trans. Inst. Elec. Commun. Eng. (Jpn.), vol. J67-A, no. 8, (1984).

7. H. Watanabe, Y. Miyanaga, N. Miki, and N. Nagai, "Construction of stable reduced-order ARMA model," Electron. Commun. Jpn., vol. 67, (Mar. 1984).

8. K. Ishizaka and J. L. Flanagan, "Synthesis of voiced sounds from a two-mass model of the vocal cords," Bell Syst. Tech. J., vol. 61, (1972).

9. B. S. Atal and S. L. Hanauer, "Speech analysis and synthesis by linear prediction of the speech wave," J. Acoust. Soc. Am., vol. 50, (1971).

10. T. Nomura, N. Miki, and N. Nagai, "A speech production model with digital filter," Int. Conf. Cybernetics and Society, Tokyo, 1978.

11. N. Miki, S. Saga, K. Motoki, Y. Miyanaga, and N. Nagai, "Area estimation from ARMA analysis based on a vocal-tract model," IEEE ICASSP, Tokyo, pp. 1613-1616, 1986.

12. C. T. Mullis and R. A. Roberts, "The use of second-order information in the approximation of discrete time linear systems," IEEE Trans. Acoust. Speech Signal Process., vol. ASSP-24, no. 3, pp. 226-238 (1976).

13. J. D. Markel and A. H. Gray, Linear Prediction of Speech, Springer, New York, 1976.

14. M. M. Sondhi and J. R. Resnick, "The inverse problem for the vocal tract: Numerical methods, acoustical experiments, and speech synthesis," J. Acoust. Soc. Am., vol. 73, no. 3, pp. 985-1002 (1983).

15. A. Fettweis, "Digital filter structures related to classical filter networks," AEÜ—Arch. Elektron. Übertrangungstech., vol. 25, no. 2, pp. 79-89 (1971).

Abbreviations

Abbreviations are followed by the number of chapter(s) in which they appear.

(A, b)	abbreviation of Eq. (2-1a), 2
(A, b, c)	abbreviation of Eq. (2-1), 2
{A B C D}	abbreviation of Eq. (1-1), 1
adj A	adjoint matrix of **A**, 6
a.e.	almost everywhere, 6, 11
AIC	Akaike information criterion, 8
AR	autoregressive, 8-10, 12
ARMA	autoregressive-moving average, 8-10, 12, 13
B	class of complex-valued function of bounded type, 6
BR	bounded real, 1
CF	criterion function, 13
CL	circular lattice, 8
Con	convolution, 10
Cov	covariance matrix, 8
CRC	current reflection coefficient, 12
CTC	current transfer coefficient, 12
CTL	commensurate transmission-line, 12
CUE	complex unit element, 12

DC 1	discrete Cauer's first form, 5
deg	degree, 4
det	determinant, 1, 2, 4, 6
diag	diagonal, 2, 6, 7
DPR	discrete positive real, 1
DSPR	discrete strictly positive real, 1
E	complement of the closed-unit disk in extended complex plane, 6
$E[\cdot]$	expectation, 8, 9
$E[\cdot \mid g(k-1)]$	conditional mean of $g(k-1)$, 9, 10
ELS	extended least square, 9
ES	escalator, 8
FFT	fast Fourier transform, 9
H_∞	class of analytic functions bounded in E (Hardy space), 6
HFR	high-frequency region, 13
I	identity matrix, 1-4, 6-9
iff	if and only if, 7
$Im[\cdot]$	imaginary part, 2, 3
ITI	ideal transformer interconnection, 11
j	$\sqrt{-i}$
L_∞	class of essentially bounded measurable functions of the unit circle (Lebesgue space), 6
LBR	lossless bounded real, 1
LC	inductor and capacitor, 1, 2, 5
LCR	left-coprime representation, 6
LDI	lossless digital integrator, 5
LFR	low-frequency region, 13
LICR	left-inner-coprime representation, 6
LPC	linear prediction coding, 9
LR	left (fractional) representation, 6

LSCL	least-squares circular lattice, 8
LWR	Levinson-Whittle-Wiggins-Robinson, 7, 8
MA	moving average, 8-10
MFD	matrix fraction description, 7
MIMO	multiple-input, multiple output, 7
MIS	model identification system, 9, 13
MRAS	model reference adaptive system, 9
N	Nevanlinna class of complex-valued functions, 6
N_+	Smirnov class of complex-valued functions, 6
ODF	orthogonal digital filter, 12
PARCOR	partial autocorrelation, 8
PM	polynomial matrix, 7
R^p	class of $p \times 1$ real vectors, 1
$R^{n \times p}$	class of $n \times p$ real matrices, 1
$R(\cdot, \cdot)$	autocovariance matrix, 8
rank A	rank of matrix A, 1, 4-6
RC	resistor and capacitor, 1, 2
RC	reflection coefficient, 12
RCR	right-coprime representation, 6
Re $[\cdot]$	real part of $\cdot$, 1, 2, 3, 12
RICR	right-inner-coprime representation, 6
RL	resistor and inductor, 1
RLC	resistor, inductor, and capacitor, 2
RLS	recursive least squares, 8, 9
RR	right (fractional) representation, 6
s (complex variable)	1, 2, 3, 5, 7, 11, 12
s^{-1} (an operator)	8
T	unit circle in the complex plane, 6
A^T (superscript)	transpose of matrix A, 1-13
TEM	transverse electromagnetic, 12

Symbols

Symbols are followed by the number of chapter(s) in which they appear.

$\dot{x}$ dx/dt, 2

a* (superscript)   complex conjugate of a, 1, 12

$\bar{a}$ (overbar)   complex conjugate of a, 3

f*(z) (superscript)   reciprocal polynomial [see Eq. (5-10)], 5

$A_*(z)$ (subscript)   reciprocal polynomial [see Eq. (11-2-6)], 11, 13

A* (superscript)   conjugate transpose of matrix A, 6, 12

A_* (subscript)   Hurwitz conjugate [see Eq. (7-2)], 7

a^{-1}, A^{-1} (superscripts)   inverse

$[\cdot]_{-1}$ 7 (see Lemma 7-1)

1_n $n \times n$ identity matrix, 11, 12

$A \geq 0$ matrix A is nonnegative definite, 1, 3, 6, 12

$A \geq B$ matrix A-B is nonnegative definite, 1, 3, 6, 12

$A > 0$ matrix A is positive definite, 1, 3

$\langle \ \rangle$ inner product, 7

$\dashv$ see Eq. (7-1)

$\sim$ see Eq. (7-5)

$\perp$ (superscript)   orthogonal complement, 8

$\hat{a}$ (caret)   estimated value for a, 8-10

$\hat{y}(s \mid k)$ estimated value for y(s) at time instant k, 9

$\oplus$	direct sum, 8
ϵ	belong to, 1, 3, 6, 8
λ (complex variable)	3, 11, 12
λ (forgetting factor)	8
$\equiv$	congruent with, 3
$=$	define, 2, 7
$:=$	define, 3

Author Index

Chapters in which authors are cited and corresponding references appear in brackets; pages in text follow.

Subject Index

Underscore denotes page on which a term's definition appears.

Milton Keynes UK
Ingram Content Group UK Ltd.
UKHW021850071024
449327UK00021B/1561